FOOD–CLIMATE INTERACTIONS

Food-Climate Interactions

Proceedings of an International Workshop held in Berlin (West), December 9-12, 1980

Edited by

WILFRID BACH
Center for Applied Climatology and Environmental Studies,
University of Münster, Germany

JÜRGEN PANKRATH
Federal Environmental Agency, Berlin (West)

and

STEPHEN H. SCHNEIDER
National Center for Atmospheric Research, Boulder, U.S.A.

D. Reidel Publishing Company

Dordrecht : Holland / Boston : U.S.A. / London : England

Library of Congress Cataloging in Publication Data

Main entry under title:

Food-climate interactions.

    Includes index.
    1. Food supply–Congresses. 2. Famines–Congresses.
3. Food crops–Climatic factors–Congresses. I. Bach, Wilfrid.
II. Pankrath, Jürgen. III. Schneider, Stephen Henry.
<u>HD9000.5.F5945</u>   338.1'9   81-17755
ISBN 90-277-1353-7   AACR 2
ISBN 90-277-1354-5 (pbk.)

42,287

Published by D. Reidel Publishing Company
P.O. Box 17, 3300 AA Dordrecht, Holland

Sold and distributed in the U.S.A. and Canada
by Kluwer Boston Inc.,
190 Old Derby Street, Hingham, MA 02043, U.S.A.

In all other countries, sold and distributed
by Kluwer Academic Publishers Group,
P.O. Box 322, 3300 AH Dordrecht, Holland

D. Reidel Publishing Company is a member of the Kluwer Group

Printed in The Netherlands

TABLE OF CONTENTS

FOREWORD

The present workshop is the third of a series of international conferences carried out within the framework of a research
project on behalf of the Federal Minister of the Interior. Under
this project, entitled "Impacts of Air Pollution on Climate",
the Federal Environmental Agency (Umweltbundesamt) has sponsored
so far:

- an international conference on "Man's Impact on Climate"
  at Berlin in June 1978,

- an international workshop on "Energy/Climate Interactions"
  at Münster in March 1980; and

- an international workshop on "Food/Climate Interactions"
  at Berlin in December 1980, the results of which are presented in this book.

Based on the Federal Environmental Agency's report on "Impacts of Air Pollution on Climate", on November 11, 1979, the
Federal Cabinet approved a climate research programme of the
Federal Government. Details of the programme are being worked
out by a committee on climate research under the chairmanship
of the president of the German Weather Service with the executive office placed in the Federal Environmental Agency.

This demonstrates that, by establishing a nationally co-or-
dinated climate research programme, the Federal Government recognizes the importance of climate to the national economy. Therefore the Federal Republic of Germany intents to contribute appro-
priately both to the comprehensive international programme of
the United Nations on Climate and Environment, as represented
by UNEP and WHO, and to the Climate Research Programme of the
Commission of the European Communities.

In his "Report on the Impacts of Air Pollution on the Global Climate" the Federal Minister of the Interior emphasizes
especially the present state of knowledge regarding the long-
term accumulation of carbon dioxide and other air pollutants
in the atmosphere and their impact on the world's climate.

Highlights of the report are:

"The scientific uncertainties concerning the nature and
extent of climate changes are still so great that, for
the time being, it is not advisable to take any countervailing measures. Nevertheless, there is a worldwide ne-

W. Bach, J. Pankrath, and S. H. Schneider (eds.), Food-Climate Interactions, vii-ix.

cessity to clarify, as soon as possible, the complex con-
nexions and consequences. Especially troublesome is the
variability of climate, since it may have a considerable
impact on world food production."

In view of the development of the world's population, food
production seems to be of decisive importance for securing our
future. According to recent UN estimates world population was
ca. 4.4 billion in 1979. The rate of world population increase
between 1960 and 1978 has been between 1.9 to 2 % per year. Based
on this relative constancy a world population of more than 6
billion may be expected by the year 2000.

Unlike the increasing population trend, the production of
agricultural goods is steadily decreasing. At present, worldwide some 50,000 km^2 of agricultural land are lost annually to
erosion and construction, while at the same time, the reclama-
tion of land entails steadily increasing costs. It is an illu-
sion to think that further increases in production may balance
the land losses. Agricultural production is increasingly based
on external energy for running the machines and for producing
the fertilizers and insecticides.

The problems to supply the world adequately with food are
difficult enough even without any climate changes caused by
man's activities. Such climate changes are believed to be due
to the increase in atmospheric CO_2 as a result of fossil fuel
burning thereby leading both to global and regional changes in
temperature and precipitation.

At present, about 20 % of the world's population suffer
from undernutrition and malnutrition. In the 1950s the developing countries had to import only small amounts of cereals. In
the 1960s, however, these imports reached 20 million tons, and
in the early 1970s they have increased from 50 million tons to
80 million tons - accompanied by steadily increasing prices.

We may therefore expect in the decades ahead that climate
changes, due to human activities, will lead to considerable socio-economic and political difficulties all over the world because of the increasing vulnerability of the international economy. Even under "normal" conditions and insignificant climate
fluctuations grain losses and decreasing fish catches can be
observed.

Therefore, it is very important to increase the research
efforts over several decades regarding the socio-economic effects of natural and anthropogenic climate impacts placing special emphasis on the world's food and energy needs. Research

is particularly needed in those areas which are concerned with
the development of strategies of avoiding, compensating or adjusting to climate changes. In 1979 only 4 billion $ were spent
worldwide for agricultural research. Despite this it is hoped
that the international activities of the United Nations (FAO,
WHO and others), the North South Commission and other bodies
can continue their work successfully. Within the given scope
the results of this workshop may provide valuable ideas for further action.

The Federal Environmental Agency acknowledges, with grati-
tude, the efforts of Professor Wilfrid Bach from the University of Münster in developing and organizing, and of the Aspen
Institute, Berlin, in hosting this workshop.

Dr. Heinrich von Lersner

President

Umweltbundesamt

PREFACE

    One of the most serious problems facing mankind today invol-
ves our ability to meet the increasing demand for food of an increasing world population. Providing the world with sufficient
food is a monumental task that depends on a host of interrelated
physical and societal factors, of which climate is only one. The
complex social and political constraints are most obvious in the
developing countries where the social structure often does not
allow the small farmers access to adequate financing to make use
of appropriate technology and other inputs that can minimize his
risks and thereby increase food production. The main problem is
to find social and political mechanisms that will enable millions
of undernourished people to obtain the food that is presently
beyond their reach.

    All of these problems are exacerbated by weather and climate
effects which are significant causes of food production shortfalls. Farmers must therefore learn to cope more effectively with
the adverse impacts of climatic hazards such as droughts, floods
and frost, and governments must work to maintain food security
for their people in the face of climatic variations that affect
food production. But climate is also an agricultural resource
that provides water and solar energy, shaping and conditioning
the environment in which food is produced, stored, and distributed. By using climatic resources effectively, farmers can increase
food production in years of favorable climatic conditions, and
governments can protect consumers from food shortages and price
spirals.

    The mix of natural and social scientists at the workshop,
which led to this volume, produced papers and group conclusions
that would have been difficult to arrive at without significant
cross-disciplinary interactions. For example, it is found - and
reported in the "Climate as a Hazard" working group statement -
that there is a serious mismatch between the erratic patterns
of climatic fluctuation and the often rigid annual credit cycle
that provides capital for farmers to finance their operations.
If two or more years of bad growing weather occur in succession,
farmers can be left with too little money to pay for enough fertilizer and other inputs to raise a good crop the following year.
This prolongs the period of low production and keeps consumer
prices high. It is urgent to develop ways to make agricultural
credit available on flexible terms that can be adjusted to climatic conditions. This problem, it was found, is not divisive, as

W. Bach, J. Pankrath, and S. H. Schneider (eds.), Food-Climate Interactions, xi-xiii.
Copyright © 1981 by D. Reidel Publishing Company.

it exists in both developed and developing countries. Moreover, mitigating the short-term risks of climate is also important for the long-term development of food productivity, that is, short-term climatic hazards create financial risks to producers which can reduce their investments in production inputs (e.g. fertilizers or appropriate technologies). These inputs, in turn, can raise long-term productivity of food systems. Therefore, it is important that policies be implemented both to reduce short-term climatic hazards and to increase long-term productivity by more effective use of climatic resources.

Individual papers are contained in this volume dealing in various levels of detail with most of the important food/climate interactions. Among the key issues presented in these papers are: the role of population size; nutritional levels; foreign aid, both technical, financial and food itself; systems of social and food security; access to emergency food stocks; access to necessary inputs for high yields; preserving genetic diversity of food crops; conserving genetic resources; developing appropriate new crop strains; insuring the sustainability of long-term food productivity through ecologically sound agricultural practices; exploiting and developing sustainable aquaculture and fishing; and dissemination of existing climatic information for timely use by food producers.

In addition, the workshop participants recommended that new policies be developed and implemented in the following areas:

. Providing, on a regional basis, mechanisms that maintain food security for consumers and financial security for producers despite the fluctuations of climate;

. Financing food production to match the irregular fluctuations of climate;

. Providing crop insurance;

. Establishing forecasting centers to provide early warnings of impending food shortages on a regional basis;

. Establishing research and development pilot projects on ecologically sustainable food production practices for each region of the world;

. Developing and testing farming techniques that minimize inputs of energy and chemicals.

Finally, it is well to realize that, although presently east-west tensions capture the front page of the news, it will be rather the north-south relationship which will play the decisive role in world security and world peace in the coming decades. In the interest of all it is imperative that the blatant inequities between the north and the south be removed - especially in the food sector. The principal barriers to increased food production in the developing countries are not biological, but clearly social and economic. Therefore, a significant improvement in the north-south relationship can only come from a fairer partnership which will permit the mobilization of the creative forces within the developing countries so that they can better help themselves.

Wilfrid Bach
Jürgen Pankrath
Stephen H. Schneider

INTRODUCTION TO WORKING GROUP REPORTS

The workshop recognised that problems of meeting world food needs are multifaceted, involving a complex mix of physical, biological, technological, environmental, economic, social and political factors. Climate is but one of these factors. However, in the short term climatic fluctuations often provide a destabilizing influence on the year-to-year food situation which needs to be dealt with by all nations. On longer time scales, climate information can be very helpful in designing sustainable food systems to meet world needs within local environmental, economic, population and other social constraints.

Therefore, these two aspects of climate/food interactions, (a) climate as a hazard and (b) climate as a resource, have been considered by two working groups. The "Climate as a Hazard" working group focused on shocks to the food system created by short-term climatic events, such as drought, floods, hail, frost, wind damage, hot/dry or cool/wet summers, as well as hurricanes and tornadoes. These events are part of a region's climate. When they occur they result in large property damages and great losses of lives necessitating emergency measures of food supply. The "Climate as a Resource" working group stressed long-term development by which world food needs can be more effectively met by taking advantage of climatic opportunities. In this case climate acquires an economic value derived from its variable distribution in space and time.

These groups each examined selected aspects of food/climate interactions, stressing certain issues which the participants felt were important considerations in improving our ability to meet world food needs. No attempt is made to be comprehensive or to examine the complex web of interacting factors which constitute the world food situation. Nonetheless, the issues raised and recommendations offered should materially improve our ability both to deal with climatic shocks and to develop sustainable food systems.

W. Bach, J. Pankrath, and S. H. Schneider (eds.), Food-Climate Interactions, xv-xxv.

WORKING GROUP A

MEETING WORLD FOOD NEEDS: CLIMATE AS A HAZARD

Chairmen: S.H. Schneider/W. Bach

G. Borgstrom, R.M. Ela, B. Glaeser,
C.N. Hodges, H. v.Loesch, J.D. McQuigg,
O. Moch, J. Namias, R.F. Pestel,
G. Schnatz, P. Spitz, W. Thalwitz,
P. Usher

STATEMENT OF ISSUES

Whereas climate-induced fluctuations continue to create several
percent variations in year-to-year levels of world food production
and many tens of percent variations in the levels of local food
production, it is imperative to smooth out the human and environmental shocks created by such short-term climate fluctuations.
Moreover, depending on the distribution of these climatic shocks
and their interactions with food supplies, demands and other
social and economic factors, human consequences can range from
mild price fluctuations to serious expansions of famine conditions
and associated political instabilities. Therefore, it is necessary
to identify and implement policies, from local to international
levels, which can help to mitigate the negative consequences of
such climatic fluctuations.

ISSUE AREAS

Specific areas are considered in which policies may be needed to
help mitigate the detrimental shocks to food systems resulting
from climatic events. Six such areas are discussed and some
specific recommendations are given.

1. FOOD SECURITY

Clearly, the most important human requirement is to maintain
adequate food security, even in the face of fluctuating climate.
Elements of food security systems include adequate food reserve
stocks - government as well as commercial stocks - effective
means to distribute food in localities of need, the terms of
access to food reserves and an early warning system to improve
the effectiveness of actions.

From a climate point of view, two areas of policy consideration
stand out. First, is the need to examine ways by which weather
and climate factors can be incorporated into objective indicators
of food shortages. Second, since climatic fluctuations vary considerably in space and time, it is important to determine for each
region of the world what is the liklihood and magnitude of climatic
shocks to food productivity. Such information can help to determine food security needs on a regional and local basis, from which
strategies can be developed to minimize the liklihood of regional
shortfalls in times of climatic stress.

2. TECHNICAL PROTECTION MEASURES

A number of technical measures can help to reduce the effects of
climatic stresses on food productivity, most important among these
are water and moisture management. In particular it is believed
that moisture conservation techniques (e.g. dry farming methods),
irrigation and drainage are priority areas. However, it is important to examine further how specific projects to minimize climatic
shocks to food productivity impact on regional water quality, especially in regions where little water management experience is available. We especially recommend protection measures that can serve
a variety of purposes, such as reservoirs which can be used for irrigation, flood control, fishing as well as electricty production.

Additionally, particular attention needs to be paid to the energy
and capital requirements of such projects.

3. BIOLOGICAL PROTECTION MEASURES

Management of biological factors in agriculture or aquaculture can
offer considerable protection against climatic fluctuations. At
the same time, such management can help to achieve other objectives
with regard to improving pest management, employment, soil erosion,
water quality, genetic resources and regional self-reliance. Several specific areas are offered in which much more attention is
needed to help meet these objectives.

a. Research and development for humid and arid tropical agriculture and aquaculture, including the financing of replicable
   pilot projects. Particular attention needs to be paid to
   match new techniques to traditional practices.

   Such projects should examine:

   . Multiple cropping and agroforestry systems with more resist-
     ance to climatic fluctuations, pest attacks, soil erosion
     by wind and water, etc., appropriate to regional and local
     conditions.

. Importing new varieties, collecting wild species and
  directing some of plant breeding research towards the
  task of finding high yielding varieties with wide ranging climatic adaptability (i.e. high stability) to
  minimise vulnerability to climatic disturbances.
. Comparison of performance of varieties under adverse
  climatic conditions.
. Technology transfer, including interregional experiences (i.e. testing various technologies in pilot
  projects with regard to climatic vulnerability).
. Testing various crop rotation schemes for differential
  climatic vulnerability.
. Testing organic fertilization under a variety of climatic
  conditions.

Implicit in these projects is the need to record climatic data
at each site so that the relative climatic sensitivity of various experiments can be documented.

In addition to the climatic aspects considered, such pilot
studies can also provide information to help meet other objectives, such as those which are listed below.

b. Biological pest management.

c. Crop mix including combined agriculture - animal husbandry -
   aquaculture schemes geared to nutritional requirements of a
   region to help improve food self-reliance.

d. The effects of changing levels of atmospheric carbon dioxide.

4. SOCIAL AND ECONOMIC PROTECTION MEASURES

Social, economic and legislative instruments and opportunities
for reacting to or anticipating the impact of short-term climate
variability on food systems differ strongly from country to
country. Protection strategies to deal with severe climatic
events need to be designed in adaptation to each system. In particular, in times of stress, there is need for government support
of farmers and low-income consumers. Although the following suggestions do not distinguish between different regional needs and possibilities, they were gathered with these differences in mind.

a. Maintaining continuity of productive capability.

   Just as commercial or government-held food storage can help to
   smooth out food security shocks to consumers caused by adverse
   climatic fluctuations, mechanisms which smooth out comparable
   financial risks to producers must be devised.

In order to maintain the productive capacity of food producers, inputs which help to increase productivity (e.g. fertilizer) need to be obtained for each cropping cycle. When such inputs are expensive, financing arrangements are necessary.

Climatic factors can cause food productivity fluctuations to occur irregularly over several cropping or fishing seasons. However, credit is usually available only on a regular cycle. If, for example, a few consecutive bad climatic years limit capital available to producers, then purchase of needed inputs can be restricted, thereby lowering the productivity of subsequent harvests. It is important that mechanisms be worked out to furnish credit of flexible terms that can be adjusted to account for production variations associated with short-term climatic conditions. This difficulty applies both to developed and developing countries, although detailed adjustments must be worked out to match needs. Furthermore special attention must be paid to the inequality of suffering which exists also within individual countries. Subsistance farmers, landless laborers, old people, women and children are usually among the chief victims of hazardous climatic events. Food subsidies, subsidized food sales, food stamp systems, and food-for-work programs can all help secure a minimum level of nutrition.

b.   Crop Insurance.

Another instrument to help alleviate financial risks to producers from adverse climatic events is crop insurance. If such risks can be reduced, then incentives for greater investments in production inputs cannot only help to reduce year-to-year losses, but can also work towards raising long-term levels of production.

5.   ACCESS TO WEATHER AND CLIMATE DATA AND METHODS FOR THEIR USE

An important component of minimizing climatic shocks on food systems is communication of data and methods of their use to farmers and fishermen. Farm radio programs are one example, and the use of casettes for two-way communication between people in the field and those with climate information is another. Such communication is important both for real-time situations and longer term strategies of climatic protection. The major need is to assess and then develop communications systems appropriate to local conditions, particularly in those regions where extension services are not now well developed.

One additional aspect of extension in need of emphasis is training of local extension personnel so that they can carry available climate information to local producers.

6.   RESEARCH AND ANALYSES TO ANTICIPATE IMPACTS OF CLIMATIC EVENTS

One method to anticipate climatic shocks on food systems is to
predict long-term climatic anomalies. This is still largely a
research task. However, the research itself might benefit if
researchers worked more closely than at present with those who
work with real-time data and food systems applications. Application of weather and climate data to estimate fluctuations in
granary, grazing land or fishery productivity and food security
is another method to anticipate climate shocks. The participants
acknowledged the work already going on in this area under the
sponsorship of such agencies as FAO, WMO, UNESCO (MAB) and UNEP
as well as various national and sub-national groups. In addition
to applauding such efforts, it is felt that more concerted efforts
to incorporate real time meteorological and oceanographic data into
early warning systems of food productivity fluctuations should be
considered.

A centre, or several centres - perhaps connected to existing institutions - with considerable regional outreach, to produce close-
to-real-time analyses of the developing food situation could help
to mitigate emerging problems. Participation from as wide a rep-
resentation of regions as possible would be instrumental to the
success of such a centre(s). Stressing the regional context for
real-time food/climate assessment should not exclude the need to
assess the impact of global climate anomaly coincidences on the
world food systems. In particular, attention should be focused on
the effect of production fluctuations on the mix of free and cont-
ractual trade and concessional aid, with the objective of reducing
areas of food deficits.

In order to develop methods of real-time food situation analyses,
it is essential that both weather and food data be incorporated
into the analyses as quickly as possible. Furthermore, the need
for easy access to non-real-time climatic data is high, so that
regional analyses can be made of the likelihood and magnitude of
climate impacts to food productivity on a region-by-region basis.

In view of the many alternative methods now suggested for applications of climatic data to food systems, it seems necessary that
the intercomparisons of various existing methods be undertaken to
determine their relative advantages.

Another important area is the relationship between climatic fluctuations and post-harvest storage losses. As such losses can materially affect food security, research on this topic is essential.

WORKING GROUP B

MEETING WORLD FOOD NEEDS: CLIMATE AS A RESOURCE

Chairmen: L.E. Slater/W. Treitz

J. Bardach, A. Burgers, H. Flohn,
G.Hallsworth, K. Heger, G. Hekstra,
F. Klingauf, G. Kohlmaier, C. Ludwig,
U. Nölle, D. Pimentel, C. Sakamoto,
L. Swindale, N. Wils.

STATEMENT OF ISSUES

Climate must be viewed as a resource in achieving world food
goals. It furnishes water and solar energy, shaping and
conditioning the landscape where food is grown. Agriculture was
founded and developed over the centuries by the innovative
response of farmers to climate. This long heritage, including
a wide variety of foodcrops and cultivation practises, has
almost been lost with the advent of modern, high-technology-
agriculture. Today, with intensifying demands on the land and
escalating costs for farm technology threatening the productive
potential of much of the world's arable lands, a creative
approach to utilize climate as a resource for improving and
expanding food production is proposed. The following initiatives
are recommended:

ISSUE AREAS

1. OBTAINING CLIMATICALLY SUSTAINABLE FOOD PRODUCTION

Crop shortfalls, local nutritional inadequacies and ecological
destruction often can be minimized through food system planning
and practice which is responsive to the resources offered by
climate. This applies especially in humid and semi-humid
tropical regions where solar energy and annual rainfall abundance create lush plant growth opportunities, yet where food
needs are most urgent and environmental losses stemming from
inappropriate land use most acute. Climate variations offer
the possibility of growing a wide variety of food crops, of
which only a few now furnish the main calories and protein intake
for many people. Some specific recommendations:

a. The classification of climate relevant to agricultural
    production and subsequent development and use of computer-

based and other agro-climatic information systems to
characterize, analyze and plan ecologically sound,
economically feasible and nutritionally desirable food-
production systems in localities and regions.

b. The development of programs of reforestation in the humid
and semi-arid tropics in areas marginal for food production,
offering the possibilities for a managed fuel supply,
additional food and forage through agro-forestry, the
stabilization of local microclimates and soil erosion, and
an addition to the global role of reforestation as a sink
for atmospheric carbon dioxide.

c. The evaluation of various graincrops as an alternative to
wheat as a basic food staple in regions where large-scale
wheat production may be climatically and nutritionally
inappropriate.

d. The use of climatic information in conservation and
efficient use of fertilizers, taking account of the impact
of the long-term build-up of nitrous oxide release on
climate.

e. The introduction of environmentally sustainable numbers of
ruminant livestock into regions where food production is
limited by soil and climate and as an element in environmental conservation.

f. The development of climatically adapted farming systems with
lower inputs of energy and chemicals as intercropping on
land and polyculture in fresh and brackish water as well as
integrated acquaculture-agriculture food production schemes.

2. CLIMATE OPPORTUNITIES AND GENETIC RESOURCES

If climate opportunities are to be met, the basic reciprocal
resource of the world's genetic stock must be available, in
abundanceand variety, to meet these opportunities. Appropriate
use of the existing as well as specifically improved genetic
material, taking into consideration climatic variations in the
tropical regions, would permit better use of water and solar
energy and ultimately lead to better living standards for
populations in these areas. Some specific recommendations:

a. The development of a more precise descriptive taxonomy to
characterize locally-suited sustainable crops.

b. An emphasis in genetic research on the selection and/or
genetic engineering of nitrogen-fixing foodcrop species,

both associated and non-associated.

c. The improvement, through genetic research, of salt tolerance
    in foodcrops as well as selection and development of halo-
    phytic plant species which can also be grown in arid regions
    employing saline irrigation. The overall ecological sustain-
    ability of such systems needs to be assessed.

d. The rapid expansion of the system of biosphere reserves,
    with special attention to the Vavilov Centers, (natural
    resources serving as a form of gene bank) in order to sustain
    climatically responsive diversity in food-producing systems
    as a supplement to gene banks.

3. CLIMATE RESPONSIVE PEST MANAGEMENT IN TROPICAL REGIONS

Crop pests are most abundant in tropical regions where severe
winter does not act as a control. Consistent crop losses to
pests in the humid and semi-arid tropics usually run on the order
of 20 to 30 % of yield and complete wipe-outs are not uncommon.
Further, because of the magnitude and variety of crop and livestock pests as well as their increasing resistance, effective
pesticidal control measures are becoming more intensive, costly
and environmentally damaging. Some specific recommendations:

a. The use of climatic information as an early warning in the
    forcasting of climate-related pest and disease outbreaks to
    expedite and improve the efficiency of control measures.

b. The development and introduction of climatically-responsive
    biological controls as a substitute for agri-chemicals, thus
    reducing the energy costs and ecological problems induced by
    the latter.

c. The use of genetic selection to move pest-susceptable food-
    crops into zones offering a greater likelihood of climatic
    immunity.

4. CLIMATE INFLUENCE IN DEVELOPING NEW MARINE RESOURCES

Use of marine food resources can be expanded through better
knowledge of climatic effects on physical events in the oceans.
Fish stocks respond to variations in temperature and currents.
Consequently, better predictions of weather and climate will
enable both location of the concentrations of animals for harvest
as well as less hazardous and more economical fishing operations.
Thus a specific recommendation:

a.   The preparation of climatic information coupled with
     information on ocean phenomena in a form permitting
     early prediction of major fishery upwelling shifts such as
     the "El Nino", as well as the location of migrant fish
     populations for efficient and safe harvesting.

5.   EDUCATION AND TRAINING IN AGRO-CLIMATOLOGY

Trained and skilled scientists and technicians combining back-
grounds in meteorology/climatology and agriculture are now in
short supply, especially in developing countries of the tropics
and are essential in implementing many of the recommendations
listed above. It is thus specifically recommended that measures
advocated in the report on education and training of the World
Meteorological Organisation in its planning meeting on the
World Climate Program (Food) held in Geneva, Switzerland,
10-14 November, 1980, be supported.

IMPLEMENTATION OF WORKING GROUPS' RECOMMENDATIONS

Minimizing the hazards of climatic shocks to food systems is in the short-term interests of all nations. Although the Working Groups have been separated into a short-term focus on minimizing shocks ("Climate as a Hazard") and a long-term view towards utilizing climatic resources more effectively in meeting world food needs ("Climate as a Resource") these two divisions are not independent.

Mitigating the short-term risks of climate is also important for the long-term development of food productivity, that is, short-term climatic hazards create financial risks to producers which can reduce their investments in production inputs (e.g. fertilizers or appropriate technologies). These inputs, in turn, can raise long-term productivity of food systems. Therefore, it is important from the perspective of both Working Groups that policies be implemented both to reduce short-term climatic hazards and to increase long-term productivity by more effective use of climatic resources.

Because of time constraints, the Working Groups were not able to recommend specific plans for implementation of their findings appropriate to local conditions. However, it is stressed that such implementation activities receive a high priority from nations, sub-national groups and international agencies. For example, the convening of groups of experts on the basic problem areas spelled out in the working group reports should be an important next step. Having the benefit of the reports from such expert meetings, nations could negotiate to implement policies which are urgently needed to help meet world food needs.

List of Authors and Participants

Bach, W.
Center for Applied Climatology and Environmental Studies, Dept. of Geography
University of Münster
Robert-Koch-Str. 26
4400 Münster, Germany

Bardach, J.E.
East-West Resource Systems
Institute
East-West Center
Honolulu, Hawaii 96822, USA

Burgers, A.C.J.
United Nations University
Toho Seimei Bldg., 29th Floor
15-1 Shibuya 2-Chome
Shibuya-ku, Tokyo 150, Japan

Borgstrom, G.
Department of Food Science
and Human Nutrition
Department of Geography
Michigan State University
East Lansing, MI 48823, USA

Colvin, L.B.
Environmental Research La-
boratory
University of Arizona
Tucson, AZ 85706, USA

Cooney, J.
Aspen Institute Berlin
Inselstraße 10
1000 Berlin 38
Germany

Cooper, C.F.
Department of Biology
San Diego State University
San Diego, CA 92182, USA

Cummings Jr., R.W.
The Rockefeller Foundation
Ave. of the Americas 1133
New York, NY 10036, USA

Ela, R.
Embassy of the Philippines
Argelander Str. 1
5300 Bonn 1, Germany

Eschenbach, K.-D.
Lepsiusstr. 14
1000 Berlin 41, Germany

Flohn, H.
Meteorological Institute
University of Bonn
Auf dem Hügel 20
5300 Bonn 1, Germany

Fontes, M.R.
Environmental Research
Laboratory
University of Arizona
Tucson, AZ 85706,
USA

Glaeser, B.
International Institute for
Environment and Society
Science Center Berlin
Potsdamer Str. 58
1000 Berlin 30, Germany

Glenn, E.P.
Environmental Research
Laboratory
University of Arizona
Tucson, AZ 85706, USA

Hallsworth, G.
Science Policy Research Unit
University of Sussex
Falmer, Brighton BN1 9RF,
U.K.

Heger, K.
German Weather Service
Sect. Theoretical Agricul-
tural Meteorology
Frankfurter Str. 135
6050 Offenbach/Main,
Germany

Hekstra, G.P.
Ministry of Public Health
and the Environment
Dr. Reijersstraat 12
2260 AK Leidschendam,
Holland

Hodges, C.N.
Environmental Research
Laboratory
University of Arizona
Tucson, AZ 85706, USA

Katzen, S.
Environmental Research
Laboratory
University of Arizona
Tucson, AZ 85706, USA

Klingauf, F.
Institute for Biological
Pest Control
Federal Biological Research
Center for Agriculture
and Forestry
Heinrichstr. 243
6100 Darmstadt, Germany

Kohlmaier, G.
Institute for Physical and
Theoretical Chemistry
University of Frankfurt
Robert-Mayer-Str. 11
6000 Frankfurt/M. 1
Germany

Lansford, H.
Writing-Consulting-Photo-
graphy
4430 Ludlow Street
Boulder, CO 80303, USA

Lindh, G.
Department of Water Resources
Engineering,
Lund Institute of Technology
University of Lund
Fack, 220 07 Lund 7
Sweden

Loesch von, H.
Food and Agriculture Organi-
zation of the United Nations
Via della Terme di Caracalla
I-00100 Rome, Italy

Ludwig, C.
Federal Environmental Agency
Bismarckplatz 1
1000 Berlin 33, Germany

Lückert, W.
Center for Applied Climato-
logy and Environmental Studies
Department of Geography
University of Münster
Robert-Koch-Str. 26
4400 Münster, Germany

McQuigg, J.D.
310 Tiger Lane
Columbia, MO 65201, USA

Moch, O.
Department of Meteorological
Applications & Environment
World Meteorological Organization
Case Postale No.5.
CH-1211 Geneva 20, Switzerland

Namias, J.
Scripps Institution of Oceano-
graphy
University of California
La Jolla, CA 92093, USA

Nölle, U.
German Society for Technical
Cooperation
Dag-Hammerskjöld-Weg 1
6236 Eschborn, Germany

Pankrath, J.
Federal Environmental Agency
Bismarckplatz 1
1000 Berlin 33, Germany

Pestel, R.F.
Office for Foreign Studies
Technical University of
Clausthal
Osteroder Str. 6
3392 Clausthal-Zellerfeld
Germany

Pimentel, D.
Department of Entomology
Section of Ecology & Syste-
matics
Cornell University
50-A Comstock Hall
Ithaca, NY 14853, USA

Pino, J.A.
The Rockefeller Foundation
Ave. of the Americas 1133
New York, NY 10036, USA

Roberts, W.O.
Aspen Institute for Humanistic Studies,
1229 University Avenue,
Boulder, Colorado 80302, USA

Sakamoto, C.
Center for Environmental
Assessment Services
Room 200, Federal Building
600 E Cherry
Columbia, MO 65201, USA

Santerre, R.M.
Department of Geography
University of Hawaii
Honolulu
Hawaii 96822, USA

Schnatz, G.
Battelle Institut
Am Römerhof 35
6000 Frankfurt 90
Germany

Schneider, S.H.
National Center for Atmo-
spheric Research (NCAR)
P.O. Box 3000
Boulder, CO 80307, USA

Sivakumar, M.V.K.
International Crops
Research Institute for the
Semi-Arid Tropics (ICRISAT)
Patancheru PO Box 502 324
Hyderabad, India

Slater, L.E.
Aspen Institute for Huma-
nistic Studies
1229 University Avenue
Boulder, CO 80302, USA

Spitz, P.
United Nations Research
Institute for Social
Development
Palais des Nations
CH-1211 Geneva 10
Switzerland

Stone, S.
Aspen Institute Berlin
Inselstraße 10
1000 Berlin 38, Germany

Swindale, L.D.
International Crops
Research Institute for the
Semi-Arid Tropics (ICRISAT)
Patancheru PO Box 502 324
Hyderabad, India

Thalwitz, W.
Western African Regions
World Bank
1818 H Street, NW
Washington DC 20433, USA

Tietze, W.
Geo Journal
Magdeburger Str. 17
3330 Helmstedt, Germany

Toenniessen, G.H.
The Rockefeller Foundation
Ave. of the Americas 1133
New York, NY 10036, USA

Treitz, W.
Department of Agriculture,
Forestry, Fishery and Ru-
ral Development
Ministry of Economic
Cooperation
Karl-Marx-Str. 4-6
5300 Bonn, Germany

Usher, P.
United Nations Environment
Program
P.O. Box 30552
Nairobi, Kenya

Virmani, S.M.
International Crops
Research Institute for the
Semi-Arid Tropics (ICRISAT)
Patancheru PO Box 502 324
Hyderabad, India

Wils, W.
Delft Hydraulic Laboratory
Rotterdamseweg 185
Delft, Holland

ACKNOWLEDGEMENT

The food/climate workshop was sponsored by the Ministry of the Interior of the Federal Republic of Germany through the Federal Environmental Agency and by the Aspen Institute Berlin, whose financial support is gratefully acknowledged. Co-sponsors were the German Weather Service; the Society for Technical Cooperation (Ministry of Economic Cooperation); the International Federation of Institutes for Advanced Studies, Sweden; the Commission of the European Communities, Belgium; and the United Nations University, Japan.

The Conference Rapporteur Henry Lansford deserves special thanks for preparing the discussion contributions.

Finally, we especially wish to recognize the assistance given by the staff of the Aspen Institute during the Conference and by the staff of the Center for Applied Climatology and Environmental Studies at the University of Münster in the preparation of both the conference and the conference publication.

Wilfrid Bach
Jürgen Pankrath
Stephen H. Schneider

INTERACTIONS OF FOOD AND CLIMATE: ISSUES AND POLICY CONSIDERATIONS*

Stephen H. Schneider

National Center for Atmospheric Research
Boulder, Colorado, USA

and

Wilfrid Bach

Center for Applied Climatology
and Environmental Studies
University of Münster
Münster, Germany

1. INTRODUCTION

1.1. Climate: One Factor in the Food Situation

One principal way in which climatic changes affect human affairs is by their impact on food production. But the production, distribution, storage and consumption of food depends on many interconnected factors, of which climate is but one. These include land, labor, capital, energy, level of technology, "know-how", population size, social organization, economic power, and (not least) climate. This paper concentrates on the interactions between food and climate, recognizing, however, that such interactions have restricted meaning outside of the overall context of food production, distribution, and consumption issues. Thus, anyone contemplating policy matters on food-climate issues will need to consider many of these interconnected factors.

---

*This is the executive summary of a longer report of the same title prepared for the workshop by the authors. Copies of the full report can be obtained from the Federal Environmental Agency, Berlin.

1

W. Bach, J. Pankrath, and S. H. Schneider (eds.), Food-Climate Interactions, 1–19.
Copyright © 1981 by D. Reidel Publishing Company.

1.2. The Mid-1970s Food Crises

    The example of the mid-1970s illustrates this well. In 1972
a series of adverse weather events motivated the Soviets to purchase some 20 million metric tons of grain from the U.S. and Canada. This "wheat deal" marked the beginning of a 5-year period
of rising grain prices, shrinking grain reserves and famine conditions in some countries. It was triggered, in large part, by
the combination in 1972 of Soviet and Central African droughts,
weakness in the Indian monsoon rains, floods in Pakistan, and
near-collapse of the Peruvian anchovy catch due to heavy fishing
pressure at a time of shifts in winds and ocean currents. However, the food reserves and trade policies of major exporting
nations were as important as the weather in creating the food
price spirals and shortages of the mid-1970s.

2.   CASE STUDIES OF FOOD AND CLIMATE INTERACTIONS

2.1. 1840s Subsistence Crises: The Irish and Dutch Potato Famines

    A look at a historical example of a climate-related food
disaster can show how these many factors come into play. In the
late 1840s cool damp weather contributed to the blight that reduced potato harvests in Ireland. Death and migration claimed
some 50 percent of the Irish population in this notorious famine.
The Irish were heavily dependent on this single crop, and this
dependence coupled with the size of their population pushed food
demands close to food supplies in good production years. Thus
the blight-induced production shortfall triggered catastrophe.
But at the same time as the Irish were enduring their crisis, two
other countries were coping with a similar problem, and faring
much better. Potato blight also occured in the Netherlands.
Although it led to increased deaths and decreased births, these
calamities were some 15 times less severe than in Ireland. Moreover, disaster-related migration from Holland was a trickle compared to the masses of people leaving Ireland.

    Holland's southern neighbor, Belgium, also endured harvest
failures at the same time. But it did so virtually unaccompanied
by death rate increases anywhere near those even of the Netherlands. According to economic historian Mokyr(1) the differential
vulnerability of these countries can be attributed primarily to
economic, social and political infrastructural differences among
them, in particular their relative levels of industrialization
and the diversification of their respective economies. Hence,
economic development and diversification are seen as the key to
avoiding devastating subsistence crises. Improving food processing
methods, better preservation techniques, and cheaper storage faci-

lities have meant that during the nineteenth century the cost of insuring oneself against harvest failure has declined.

But resilience against harvest failure is even more complicated than technological developments. Along with industrialization comes agricultural commercialization by which it is meant that a larger proportion of consumption is bought. This implies that fewer people would be living at subsistence levels, and also that the wherewithal would become available to allow purchase of food from others in times of local harvest failure, whether climatically caused or not. In order to trade food, infrastructure needs to be developed for trade storage and transport; and indeed such an infrastructure evolved with industrial and economic development.

2.2. Marginal Farming: Developments in NW-Europe over the past 1000 years

It has already been noted that the vulnerability of farming to climate is a complex mix of physical, biological, economic and social factors. This can be demonstrated by reconstructing agricultural settlement and abandonment. By examining aerial photographs, historical records and on-site remains in the Upland region of SE Scotland, Parry(2) found evidence that in the 12th century (i.e., the medieval climatic optimum) much of the land that is now covered by heather was farmland, and that between 1600 and 1800 (i.e., during the Little Ice Age) agriculture retreated far down the hillsides. In mapping climatically marginal areas in NW Europe, however, it is necessary to rely on highly generalized agrometeorological data, which include the number of growing months with a mean temperature above 10°C and the amount of change in average annual precipitation deficits during the months of July through September. But even in regions where climate can be shown clearly to be a strongly limiting factor in crop yields, the interpretation of the relative role of climate and other social and biological factors on the viability of farming is difficult.

2.3. Perceptions of a Disaster: The 1970s Sahelian Drought

Although it is clear that climate is only one factor in the food situation, even its supporting role can be an issue of debate. There are conceptual frameworks by which different analysts perceive food and climate problems. Their proposed solutions to such problems then naturally follow from their perception of the causes of the problem. The Sahelian drought of the early 1970s can highlight these varying perceptions, some of which are often keyed to ideological biases.

One of the most noted climatic events has been the drought

in the Sahelian region of north Africa, which decimated nomadic
tribes and their livestock, ruined the crops of sedentary farmers,
contributed to increased death rates and malnutrition, helped
create political turmoil, and led to massive international publicity and relief programs for the inhabitants of the Sahel. The
"causes" of these impacts are hotly debated.

One point of view (e.g. R.A. Bryson(3)) is that long-term
climatic stress in the Sahel is a major contributor to the human
misery that ensued, and that this climatic stress is related to
large-scale climatic trends.

Others (e.g. H. E. Landsberg(4)) have argued that droughts
in the Sahel are a regular feature of the climate, and the 1970s
case was not generically different from previous such extremes.
The problem in the Sahel, they contend, was related to lack of
foresight in anticipating and dealing with precedented bouts of
bad weather, not an adverse climatic trend.

Still others (e.g. M. H. Glantz(5)) contend that the impact
of the drought on the inhabitants of the region is more a function
of social and political factors than climate. Modern technology,
they point out, if presented in a piecemeal fashion, can actually
increase the impact of droughts.

Another group of analysts (e.g. F. M. Lappé and J. Collins(6))
see the Sahelian crisis as even more embedded in social factors.
They argue that the colonial legacy of the Sahelian countries to
produce cash crops has been continued by the elites who took over
after independence. These officials encourage annual cropping
rather than long fallow periods typical of traditional practices.
Thus in order to maintain cotton exports for the French, farmers
were forced to expand their cotton acreage by reducing the planting of millet and sorghum. Continual cultivation rapidly depletes
the soil, necessitating still further expansion of export cropping
at the expense of food crops and pasture land. This, these analysts
say, is how modern technology, infrastructure and political influence increased vulnerability of subsistence farmers and nomads
to climatic and other stresses in the Sahelian region. The Sahelian
governments over-emphasize export crops to earn foreign exchange
so that government bureaucrats and other better off urban workers
could live an imported lifestyle.

To oversimplify somewhat, the causes of the Sahelian disaster
in the latter view could be called "social structural". In an
alternative view (e.g. S. Wortman and R. Cummings(7)) of this
subsistence crisis, which could be called "technology-transfer,
marked-oriented", the implementation of more modern technology
and supporting infrastructure - adapted to spécial less developed
country (LDC) conditions of high labor intensiveness - is consi-

dered the best answer to prevent repeat disasters. The development
of market economies which provide incentives for farmers to raise
their productivity is the safest route to the elimination of subsistence crises. One objective of agricultural development, these
analysts agree, must therefore be to allow individual families
to produce a surplus for sale so that total output of a locality
exceeds total local requirements and permits sales in urban centers, other rural regions or in international markets. Imports
required for higher productivity must be purchased and markets
for products must be established. In short, traditional farmers
must be brought into the market economy.

The important issue for our purposes emerging from the Sahelian case is that different suggestions for solutions to food and
climate problems are often based on implicit conceptual beliefs.
Thus, care is needed to make as explicit as possible the separa-
tion of factual and value issues in policy analyses of food and
climate interactions.

2.4. The U.S. Plains: Making of a Food Giant

2.4.1. Increasing yields over the past century

The U.S. Plains began food production amidst controversy
as to whether this region of highly variable weather could ever
sustain high food productivity(8). After the early success during
the Homestead era of the late 1800s, drought in the 1890s created
serious outmigration of farm settlers and began a pattern of
"boom and bust" farming, where good weather years and high production were followed by drought years bringing economic ruin and
serious episodes of soil erosion. The most severe drought occurred
in the 1930s, known to history as the "dust bowl era". Average
wheat and corn yields dropped by 50-75% and much topsoil was lost.
The dust bowl deepened the great economic depression of the 1930s.

However, the lessons of the dust bowl led to the establishment of the U.S. Soil Conservation Service, whose extension agents
help farmers to protect soil reserves. Moreover, the advent of
new genetic crop strains better adapted to the climate of the
region and the increasing availability of fertilizer (and expansion
of credit institutions to allow financing of technological farming)
followed in the aftermath of the dust bowl. Grain yields have increased from two to three hundred percent since the 1930s. The
productivity of the U.S. Plains has contributed to the present
place of the U.S. as (by far) the leading grain exporting nation.

Since the 1930s the average variability of crop yields from
year-to-year in the U.S. plains has decreased relative (i.e., as
a percent of mean crop yields) to long term trends in crop yield.

(Absolute variability in crop yields, however, has increased, even though relative variability has declined(9)).

2.4.2. Climatic effects on year-to-year crop yield variability: An unresolved controversy

This decrease in relative crop yield variability has led to a major controversy. On one side some agriculturalists contend that not only are increasing trends in crop yields due to technology, but decreasing relative yield variability as well. This belief helped to motivate the U.S. to begin to liquidate government held food reserves in the early 1970s, a policy which contributed to food price spirals in the mid-1970s.

Other agriculturalists disagree, arguing that it was the unusually good weather after 1957 that led to relative crop yield variability, and that a vigorous program of food reserves is needed to hedge against a return of conditions as bad as the 1930s or mid-1950s.

Recent work by Warrick(10) seems to suggest that, if one accounts for a relative decrease in the severity of the weather, then relative crop yield variability in the U.S. Plains is roughly unchanged over the era of technological expansion. (Of course, absolute year-to-year crop yield variability still seems to have increased.)

The probability, for example, of one year of unfavorable weather (i.e., reduced yields by more than 10%) for U.S. corn is 23%, and the probability of two such unfavorable years is 11%(11).

Although this important issue of whether technology lessens or increases vulnerability to climate is still not definitively resolved, as in the 1970s the U.S. Plains have become such a major component of the world grain export markets, the impacts of fluctuations in food productivity in this granary are now felt worldwide.

2.5. The World Food Situation in the Past 20 Years

2.5.1. Recent global grain trends

With the sole exception of 1974, each year in the past fifteen saw an increase over the previous year in total world average utilization of grains(12). And, with the exception of the "bad weather" years 1972, 1974, 1977 and 1979, world total grains production and total grains yields increased annually, both at a rate of about 2-4% per year. The increase in grain utilization, also at a rate of about 3% per year, reflects both the almost 2% annual increase in world population size and a world average increase

in absolute standards of eating - particularly the indirect consumption of grains used as animal feed in Eastern Europe, Japan
and other relatively affluent places. In the developed countries
(DCs), per capita food consumption increased well ahead of population growth, whereas in the LDCs (as a whole) food production
per capita has remained about the same level over the past ten
years.

2.5.2. Per capita food production by region

It would be misleading, however, to conclude that all LDCs
merely "stood still" with regard to per capita food production(13).
Indeed, while some large regions, such as Latin America or South
Asia, maintained fairly steady per capita standards, other regions,
such as East Asia, made considerable gains whereas still other
regions, in particular Africa, saw further erosion in already low
per capita levels of food production. Furthermore, within these
large regions some nations have experienced considerably different
standards over the past ten years than the regional average; and
even within single nations vast inequities in per capita food production and consumption levels continue to exist. To refer to a
"world grain situation", then, cannot provide more than a gross
overview of a heterogeneous distribution of local food situations:
some bad, some good, some steady, some deteriorating and some improving. Thus, great care is needed before applying world average
food statistics to a regional context.

2.5.3. World grain security

Earlier it was mentioned that world food production climbed
at an average of about 3% annually. This increase is partly a
result of an increase in harvested area in the mid-1970s, but primarily it reflects an increase in grain from a spread of modern
methods worldwide. The interannual fluctuations in production,
however, (which became larger in the 1970s than over the previous
decade) are largely attributed to unfavorable weather fluctuations.
1972, for example, saw a 3% decrease in world grain production
over 1971 and 1974 saw a 4% drop over 1973(14). The 1972 shortfall
occurred as major grain exporting nations, in particular the U.S.,
were implementing policies of liquidation of government held grain
stocks and reduction of direct food aid to LDCs in favor of a new
role as grain exporter to countries willing (and able) to purchase
tens of millions of metric tons of grain (in particular, the
U.S.S.R.).

The effect of the combination of these policies and the
weather-induced shortfalls in 1972 and 1974 cut grain stocks by
about 1/3. If one measures "World Grain Security" as the yearly
grains in stock divided by the total annual utilization of grains,
then one sees that from 1970 to 1975 world grain security dropped

from more than 15% to about 10%. This drop helped create steep
price rises of two to four hundred percent in major grains and
soy beans after 1972, and contributed to an increase in famine
conditions in some LDCs.

Fortunately, increased planted area and reasonably favorable
weather conditions worldwide in major granaries in 1976-1979
allowed some rebuilding of grain stocks, and world grain security
levels in 1980 are at about the 14% level - slightly below where
they were before the food crisis of 1972 occurred. Although grain
prices have stabilized somewhat (at higher levels, of course)
since 1975, price instability - and expanded famine conditions
- are not at all unlikely in the next few years should unfavorable
weather in one or more major granary recur - unless an effective
world food reserves and distribution system is set up in the interim.

2.5.4. World grain trade

Whereas world grain production increased by about 35% (from
about 1050 mmt to 1400 mmt) since the late 1960s, world grain trade
doubled (from about 90 mmt to 190 mmt). In 1979 wheat accounted
for about 80 mmt, coarse grains for 100 mmt and rice the remaining
10 mmt of trade (mmt = millions of metric tons). Canada (about
14 mmt), Australia (about 13.5 mmt) and the U.S. (about 36 mmt)
accounted for some 3/4 of wheat exports, where the EEC (about
5 mmt), Eastern Europe (about 6 mmt), Japan (5.5 mmt), China
(7.5 mmt), U.S.S.R. (10.5 mmt) and Brazil (4.5 mmt) make up for
slightly more than half the wheat imports.

For coarse grains, the U.S. alone accounts for about
3/4 of all 1979 exports (about 70 mmt), whereas western Europe
(about 24 mmt), Eastern Europe (10.7 mmt), Japan (8.5 mmt), and
the U.S.S.R. (17.3 mmt) make up about 2/3 of the imports. These
coarse grain imports are largely used as animal feed.

Thus, in 1979 coarse grains comprised nearly half (about
730 mmt) of total grains production (1400 mmt) and more than half
of grain trade, whereas rice production (about 250 mmt - milled)
was more than 20 times larger than inter-regional rice trade. The
reason for this is that rice is a staple crop in LDCs, where most
grains are produced and consumed locally; whereas coarse grains
are the largest grain trade commodity, since they are imported
by economically advantaged countries with high meat intake in
their diets.

2.5.5. Nutritional differences between DCs and LDCs

Indeed, the principal differences in dietary standards
between DCs and LDCs can be related to animal consumption. In LDCs

total calorie intake is usually in the form of direct consumption
of grains or roots, whereas in DCs direct consumption per capita
is less than most LDCs, although total calorie intake (direct plus
indirect - i.e., grain inefficiently converted to animal calories)
per capita in DCs exceeds that of LDCs. Moreover, protein intake
per capita in DCs is typically 2 to 3 times greater than in LDCs
with animal proteins in the DC diets accounting largely for this
difference(15).

3.   CLIMATE AND ITS VARIATIONS

3.1. Basic Concepts and Definitions

Weather is the instantaneous, local state of the atmosphere
whereas climate is a time average (and other higher order statistics) of weather variables taken over a period greater than a month.
This period should be specified by the person using the term
"climate", for much confusion in the literature about climate and
climatic change occur when one author implicitly defines climate
over a different time-averaging interval than another author(16).

Climatic change is then a difference in the climatic statistics from one averaging period to the next.

The climatic system comprises the interactive elements of
the atmosphere, oceans, cryosphere (snow and ice), biota and land
surfaces which lead to the climatic state. Over short time scales
(a few months) the atmosphere, upper oceans (including sea ice)
and some surface features (e.g., snow cover) are significantly
interactive, whereas on longer time scales deep oceans, glaciers,
forests and even continental positions all can affect climate.
The composition of the atmosphere is important in regulating the
flows of radiation within the climatic system and between the
Earth and space. Both natural and human (i.e., "anthropogenic")
factors can alter atmospheric composition(17).

The causes of climate and its variations are complex and not
fully understood. These include factors external to the climatic
system, such as solar energy input, human pollutants which alter
land surfaces or atmospheric composition or volcanic eruptions
which inject dust veils into the upper atmosphere.

Internal factors, such as the redistribution of heat among
its principal reservoirs (e.g., oceans, atmosphere and glaciers),
also contribute to climate fluctuations on all time scales.

Separating climatic "signals" caused by external factors
(e.g., carbon dioxide input from fossil fuel burning) out from

internal or self-fluctuations of the climate system (so-called
"climatic noise") is a principal pursuit of climatic researchers.

Climatic models are one major means of studying both the mean
distribution of climate and the causes of its variability. Because
of the complexity of the climatic system - and the fact that some
anthropogenic perturbations are unprecedented and have no known
historical analogue from which to estimate potential effects -
elaborate mathematical models of the climatic system are developed,
tested and run on the latest computers available. A hierarchy of
these models are used, ranging in complexity from simple "back-of-
the envelope" calculations up to giant, computer-intensive coupled
ocean/atmosphere general circulation models. These "GCMs" compute
daily weather patterns from millions upon millions of computations,
and climatic statistics are then directly taken(18).

As all models are not exact replicas of the actual climatic
system, they are constantly being tested against observational
data and refined. As data collection is a global, long-term effort,
many more decades will probably be needed to provide reliable verifications of some model predictions. Unfortunately, decades is
the time frame over which present models suggest that global climatic changes from human pollutants (most notably CO_2 increases)
are likely to create a "signal" larger than natural climatic
"noise". How reliable such estimates are and what they might mean
for food productivity, water supplies, heating demands or even
global sea levels are now intense topics of research in the U.S.,
western Europe and to a lesser extent in other places. The World
Climate Program of the World Meteorological Organization is helping to organize and coordinate such needed research.

Of course, whether to act now on the basis of present information is a value judgement that depends upon whether one is more
concerned with averting the prospect of climatic impacts from
materializing or with the costs of altering or abating those human
activities (most notably fossil fuel burning) which most threaten
to alter long-term climate.

3.2. Climatic Variability

Regardless of the resolution to present controversies over
long-term climatic trends, interannual variability has (as seen
in 1972, e.g.) and is likely to continue to create year-to-year
fluctuations in food production of some several percent globally,
and up to 50% in some regions.

Predicting climatic anomalies (e.g., how one year's climate
will be different from say a 30-year mean) is now a largely empirical exercise, drawing on analogues of past anomalous years,
anomalous ocean surface temperatures, sunspots and pattern recog-

nitions of individual analysts. Consequently, despite some entre-
preneurs' claims to the contrary, such forecasts remain at marginal skill levels. Thus, the most reliable method to estimate the
likelihood of future climatic anomalies is to tabulate such anomalies of the past: so-called "actuarial forecasting"(19).

Actuarial forecasts, while not able to predict the magnitude
or timing of specific climatic anomalies, do offer an idea of the
level and frequency of climatic events which can alter crop yields.
For example, a recent study showed that over the past 50 years
the probability of unfavorable yields (i.e., yields less than 90%
of the long-term trend) simultaneously occuring in the U.S. and
the U.S.S.R. is about 8%. Such information can help food policy
analysts choose food reserves and distribution systems policies
which minimize the likelihood of a recurrence of the food
crises of the mid-1970s.

Of course, a specific forecast for next year is preferable
to an actuarial analysis, but at present low skill levels (e.g.,
60/40 for temperatures a season ahead) specific season-ahead forecasts remain of marginal use to agricultural planners.

Whether or not climatic variability is increasing is a contro-
versial question that has touched off often-bitter debate among
climatologists and agronomists. Because of a 15 year period of
good growing weather in the U.S. Plains after the 1957 drought
year, several agro-climatologists expressed (in 1973) concern that
such good "unvariable" weather could not be expected to continue.
Indeed, in 1974-1978 U.S. crop yields reversed a two decade upward
trend as poorer weather occurred.

However, as the debate progressed, many contradictory (and
often incommensurate) pieces of evidence have been cited to claim
that climatic variability has or has not been increasing recently.
Often, one analyst cites, say, five year running mean standard
deviations of surface temperature whereas another analyst supplies
interannual summer precipitation differences. These are not directly comparable.

Thus, at this time there is no clear evidence that climatic
variations (across the spectrum of time and space scales which
can define climate) are increasing or decreasing, although many
fixed-time, local-space exceptions are often cited in both direc-
tions(20).

On a worldwide crop-yield basis, however, a plot of the trend
in world total grains yields does indeed show a fairly smooth rise
from the early 1960s up to 1972, after which a several percent
interannual fluctuation in yield is evident. Thus increasing fluctuation in grain productivity has been cited by some agronomists

as evidence that climatic variations important to food productivity have been increasing through the 1970s.

4.   FOOD PRODUCTION FACTORS

     Climate, soils, nutrients, irrigation water, carefully chosen
crop strains, pest control and appropriate management practices ·
all contribute to food productivity. Intensification of farming
has led to significant advances in crop yields, but often these
advances have been accompanied by undesirable side effects such
as air pollution, susceptibility of a few high-yielding genetic
varieties to an outbreak of pests, soil erosion, water logging
or alkalinization of soil, displacement of subsistence farmers
to urban slums and, in some cases, increased vulnerability of food
systems to climatic fluctuations.

     Although considerable arable land remains undeveloped, it
is usually less economic or productive than presently cultivated
lands. Cost estimates for land and infrastructure development over
the next decade merely to maintain present per capita standards
typically range from tens to hundreds of billions of dollars(21).

     If mankind continues to increase agricultural production to
meet the rising food demands of a growing world population, great
care must be exercised to ensure that irreversible environmental
damage is avoided. Maintaining the productive potential of global
agro-ecosystems is one of the most critical environmental issues
facing humanity.

     The complexity of the interactions among physical, biological,
economic, social and political components of food production
suggests that modern agricultural development requires an integrated management approach, where all of these important factors
are viewed systematically - and in the context of the local conditions of the agricultural region being considered. Few general
statements about food-climate interactions are possible outside
of specific regional context.

5.   SOME FOOD-CLIMATE POLICY CONSIDERATIONS

5.1. Climatic Variability and Crop Yields

     It is also difficult to generalize outside of a local context
how a specific climatic change might affect world food productivity. That is because at different stages of plant development,
growth can be highly sensitive to one or more climatic factors,
a sensitivity which can alter greatly over the life cycle of that

crop. Consider a specific example: the application of one additional inch of water to spring wheat grown in the northern Great
Plains of North America. While this generally improves yields,
especially if it is added at the very beginning or the middle of
the growing cycle, it may, however, reduce yields if the water is
added near harvest time. Since climatic anomalies, such as short-
term excesses in rainfall (or drought), will, in general, occur
at different stages of growth of different crops in different places, there can be considerable compensation of negative and positive effects. That is, yield decreases in one crop in one place
from, say, less-than-normal rainfall, are likely to be compensated,
to some extent at least, by yield increases from other crops. Furthermore, droughts in one place can be compensated by excess rain
in others.

Although we might take some comfort from such compensatory
effects of climatic anomalies, large-scale (i.e., over a major
granary) negative yields can occur, as can simultaneous "bad
years" in more than one major grain region. For example, for wheat
in the United States and India over the past 65 years, the probability of a single "unfavorable" year (defined as yearly wheat
yields reduced by 10% below long-term trend line) is 17%; two such
consecutive unfavorable years occurred only 9% of the time between
1910 and 1975. In Canada, on the other hand, unfavorable wheat
years are more probable, having percentages of 33% and 17%. In
the U.S.S.R. the situation is complicated, since spring and winter
wheat regions are geographically dispersed. Nevertheless, a general characteristic is that the U.S.S.R. is more likely than the
United States or India to have a single unfavorable year, but less
likely to have a long sequence of them (again based on data from
the past 50 years). The probability, as noted earlier, of simultaneous unfavorable yields (<-10%) in the United States and the
U.S.S.R. is about 7-8% for one year, but generally less than 1%
for two or more consecutive years. Thus there is an uncomfortable
probability of both strings of unfavorable years in one region
and the possibility of simultaneous bad years in several major
granaries. It is these "infrequent" events which require policy
considerations to minimize severe impacts.

5.2. Food Reserves and Distribution Systems

Short-term climate-induced fluctuations in food production
are likely to continue at the level of several percent production
variations from year to year on a world-wide average, and many
tens of percent variations on a regional basis from year to year.
Primarily by the mechanism of sufficient regional food stocks and
distribution systems, these fluctuations in food production can
be damped out. Unfortunately, there is little world effort at
maintaining or managing food stocks as a hedge against climate-

induced regional shortfalls(22). This is because of squabbles
over who should contribute to the stocks, pay for the contributions, and who should determine where they are stored, when they
are released, and at what levels they are to be maintained.
Bickering over such non-climatic issues merely increases the probability that simultaneous negative fluctuations in food production in several grain growing areas will once again set off a
worldwide spiral of food price instability and contribute to increases in malnutrition and famine-related deaths in poorer, food-
insufficient nations.

5.3. Maintaining Diversity

Modern agriculture has tended to substitude a mono-culture
of highly productive genetic crop strains for the lesser productive, but more diverse, set of species of traditional agriculture.
The danger of such loss of diversity is potential vulnerability
to single pests or a negative climatic fluctuation. Even without
abandoning high yield varieties, agricultural policy makers can
reduce their nation's vulnerability to climatic fluctuations by
encouraging the planting of a variety of different crops in different places, planted at different times. Particularly in warmer
climates where more than one crop can be planted per year, multiple
cropping not only can raise yields, but offers some measure of
natural pest resistance and minimizes vulnerability to a short-term
climatic anomaly. The more diversity added to cropping patterns
the less likely it is that single stressful events would prove
catastrophic.

5.4. Minimizing Vulnerability by Using Existing Climatic
Information

Techniques such as the development of widely adapted crop
strains or implementation of irrigation systems can mitigate the
negative effects of climatic variations. Knowledge of the frequency and severity of climatic extremes can help in planning the
details of these strategies. Also, cultivation practices should
be consistent with local climatic conditions, in order to minimize
long-term loss of the productive potential of the land through
soil erosion or other environmental damage. For example, in places
where intense, but intermittent rainfall is common, special attention to farming practices which minimize soil erosion needs to
be considered. The application and dissemination of exsisting climatic information in a local context is an important job for extension services. Efforts to increase the effectiveness of such
services are now under way as a priority concern of the World
Climate Program(23).

5.5. Long-term Climatic Trends

One final climatic issue is the outlook for long-term climatic trends, particularly those which might arise as an inadvertent consequence of industrial and agricultural activities. There is an extensive and often contradictory literature on this subject (24). Perhaps the strongest statement that could be made which reflects the state-of-the-art knowledge (and its uncertainties) is that there is a weak (but evolving) concensus that the "greenhouse effect" from increases in atmospheric CO_2 (from the growing use of fossil fuels and expansion of deforestation and some kinds of agriculture) are likely to cause man-induced global climatic changes some time around and after the turn of this century. Such changes could be unprecedented in modern experience. Specific regional climatic scenarios from a CO_2-induced overall planetary warming are fragmentary and speculative, although preliminary investigations suggest that some regions would be adversely affected, some improved and others unaffected, depending on where one is and what activities and infrastructure are in place in the future. Sea level rises of several meters are possible, but again the probability and timing remain speculative. Although this evolving area of concern bears close watching in the food-climate context, for the next decade or so, at least, it is much more likely that minimizing our vulnerability to repeated bouts with precedented short-term climatic anomalies will be the priority food and climate problem.

5.6. Food Aid: A Question of Values and Ideologies

The volume of food aid from DCs to LDCs decreased steadily from the mid-1960s to mid-1970s, with the EEC being the only significant exception. The U.S. aid, for example, in the late 1960s averaged about 15 mmt annually, dropping to about 3.5 mmt by 1974. The EEC, on the other hand, more than doubled its annual food aid over the same period, but the total amount of aid in 1974 was only about 1.4 mmt.

Considerable controversy is generated over whether food aid is in or against the interests of either donor or recipient nations, and these controversies generally are rooted in differing ideological perspectives as to whether LDCs should opt for either independent or interdependent food relationships with DCs. Other points of controversy center around issues of population growth, with some analysts arguing that food (or other kinds of economic or medical) aid should only be given to those nations in dire need of emergency help who are, at the same time, striving to increase local food production so it has hope of coming onto balance with long-term population growth levels. The policy called for is "triage". These analysts (e.g., W. and P. Paddock(25)) would deny food (or other development) aid to any country not actively working to bring down its rate of population growth.

Other analysts (e.g., G. Hardin(26)) argue that food aid to LDCs only allows more people to live in misery over the long run and thus they advocate a no-aid policy they call "lifeboat ethics". It is severely criticized often on moral grounds.

Other analysts (e.g., those in non-governmental organizations like the Ford Foundation) argue that world food and other economic aid, coupled with transfer of high technology and market-oriented economic infrastructure, is needed to solve LDC food problems.

Yet others (e.g., P. Ehrlich(27) or E. Schumacher(28) contend that wholesale technology transfer from DCs to LDCs will create both environmental and economic catastrophes, and recommend the development of so-called "appropriate technologies" which match the local environmental, economic and social conditions of each region.

Other analysts (e.g., R. Garcia(29) or F. M. Lappé(6) contend that until social reform or revolution wrests power from "elites" and turns control of food producing capacity over to local inhabitants, no food aid or technology transfer will help the vast majority of people in LDCs. They blame the "Green Revolution" techniques for increasing the wealth differences between rich and poor in LDCs.

Finally, other analysts (e.g., S. H. Schneider(22)) suggest that aid is appropriate, but that the terms of its amount and duration should be negotiated between donors and recipients based on the principle of creating economically and environmentally sustainable levels of food (and other commodities) consumption. Such a "Global Survival Compromise" is offered as a middle course between extreme views of environmental, social or economic advocates.

While these contrasting views are clearly based on value differences of the analysts, they all depend on a critical question: What is the "Carrying Capacity" of the Earth?

5.7. Carrying Capacity of the Earth

Underlying the various views expressed in 5.6. is a belief in how many people the Earth can sustain at reasonable nutritional levels. Some environmentalists believe that soil erosion, toxic waste build-ups, and other damages to food production potential will render the long-term Earth's carrying capacity below even the present population of 4 billion, and that eventual catastrophe is inevitable if population growth continues. This belief leads to a view that any aid which allows for more people or pollution is immoral, for it only will increase the ultimate size of the inevitable collapse.

Opposed to this view are the "technological optimists", who calculate that modern methods of irrigation, fertilization, pesticide application, genetic engineering and market incentives for growers could provide a carrying capacity of up to 40 billion - ten times present! This path, to those who believe carrying capacity can be raised, is the most moral alternative. While they readily admit that the costs of such development are measured in many billions of dollars annually, they rarely admit (or even discuss) environmental or social constraints to such development.

Thus, whether the risks of development (e.g., high population growth rates and environmental degradation) are worth the benefits of improving standards of living depends both on one's values and opinion as to whether the Earth's carrying capacity will allow a given sustainable future population size. Unfortunately, there is no wholly "scientific" method to assess reliably what the future carrying capacity might be since it depends on uncertainties such as future technological breakthroughs, the long-term effects to toxic wastes, the causes of birth rate reductions, long-term climatic variations, etc.

6. CONCLUDING REMARKS

Although no more than a few global generalizations seem possible in the area of food-climate interactions, it is already clear that "actuarial analyses" of past climate-induced food production are needed to help minimize societal damage when such fluctuations inevitably recur. At the same time, developing the capacity to predict both future climatic variations and their societal impacts is needed. Minimizing the human impacts of climatic fluctuations through (1) food reserves and distribution systems, (2) diversity in cropping patterns and (3) maintaining the long-term productive potential of existing (or new) croplands through farming practices consistent with local climatic and other environmental factors, remain as urgent policy imperatives. The latter (3) policy is growing in importance, particularly in view of the need to increase food availability to presently undernourished people, let alone to provide more food for a growing world population.

REFERENCES

1. Mokyr, J.: 1980. Industrialization and Poverty in Ireland and the Netherlands. J. Interdiscipl. History, 10(3) pp. 436, 451-457.

2. Parry, M.L.: 1978. Climatic Change, Agriculture and Settlement. William Dawson & Sons, Folkestone, England.

3. Bryson, R.A. and T.J. Murray: 1977. Climates of Hunger. The University of Wisconsin Press, Madison, USA.

4. Landsberg, H.E.: 1979. In: Food, Climate and Man, ed. by Biswas, A.K. and M.R. Biswas, p. 200, Wiley & Sons, New York.

5. Glantz, M.H.: 1977. Nine Fallacies of Natural Disaster: The Case of the Sahel, Climatic Change, 1, p. 76.

6. Lappé, F.M. and J. Collins(with C. Fowler): 1978. Food First: Beyond the Myth of Scarcity. Ballantine Books, New York, USA.

7. Wortman, S. and R.W. Cummings, Jr.: 1978. To Feed this World: The Challenge and the Strategy. Johns Hopkins Univ. Press, Baltimore, U.S.A.

8. Bark, L.D.: 1978. In: North American Droughts, ed. by Rosenberg, N.J., p.14, Westview Press, Boulder, USA.

9. Newman, J.E.: 1978. op.cit. 8., p.54.

10. Warrick, R.A.: 1980. In: Climate Constraints and Human Activities, ed. by J. Ausubel and A.K. Biswas, pp.93-123, Pergamon Press, London.

11. Sakamoto, C., S. Leduc, N. Strommen and L. Steyaert: 1980. Climate and Global Grain Yield Variability, Climatic Change, 2, 349-361.

12. Foreign Agricultural Circular: 1980, US Dist. of Agriculture, FG-6-80, p.24, Washington, D.C.

13. Wortman and Cummings, Jr., op.cit. p.20

14. Foreign Agricultural Circular: 1980. op.cit. p.19

15. Gilland, B.: 1979. The Next Seventy Years: Population, Food and Resources, Abacus Press, Tunbridge Wells, England.

16. Schneider, S.H. et al.: 1981. Climatic Impacts of Energy Systems. Rept. of the Risk/Impact Panel to CONAES, Nat. Acad. Sciences, chapter 7, Washington, D.C.

17. Bach, W., J. Pankrath and W.W. Kellogg (eds.): 1979. Man's Impact on Climate, Elsevier Publ. Co., Amsterdam, Oxford, New York.

18. Schneider, S.H. and R.E. Dickinson: 1974. Climate Modeling, Revs. Geophys. Space Phys. 12, 447-493.

19. Newman, J.E.: 1978. In North American Droughts, ed. by N. Rosenberg, p.59, Westview Press, Boulder, USA.

20. Van Loon, H. and J.Williams: 1978. The Association between Mean Surface Temperature and Interannual Variability, Mo. Wea. Rev. 106(7), p.1017.

21. Ehrlich, P.R., A.H. Ehrlich and J.P. Holdren: 1977. Ecoscience: Population, Resources, Environment, W.H. Freeman, San Francisco, pp.250-251.

22. Schneider, S.H. and L.E. Mesirow: 1976. The Genesis Strategy: Climate and Global Survival, Plenum Press, New York.

23. World Climate Conference: 1979. WMO Publ. No. 537, Geneva

24. Schneider, S.H. and R.S. Chen: 1980. Carbon Dioxide Warming and Coastline Flooding: Physical Factors and Climatic Impact. Annual Rev. Energy, 5, pp. 107-140.

25. Paddock, P. and W. Paddock: 1976. Time of Famines. Little, Brown & Co., Boston, USA.

26. Hardin, G.: 1974. Living on a Life Boat, Bioscience, 24, pp. 561-568.

27. Ehrlich, P.R. and A.H. Ehrlich: 1974. The End of Affluence, Ballantine, New York, USA.

28. Schumacher, E.F.: 1973. Small is Beautiful. A Study of Economics as if People Mattered, Blond & Briggs, London.

29. Garcia, R.: 1981. Nature Pleads not Guilty, Pergamon Press, London.

DIMENSIONS OF THE WORLD FOOD AND CLIMATE PROBLEM

Lloyd E. Slater

Food and Climate Forum of the Aspen
Institute for Humanistic Studies, 1229
University Avenue, Boulder, Colorado USA

THESIS: *Understanding and dealing with the problem of climate's impact on food production and distribution is a crucial endeavor if world food needs are to be met by the year 2000 and beyond.*

1. THE EMERGING PROBLEM

If Climate were a benign force, a precise and dependable vendor of rain, wind and solar warmth, its impact on food production would be of small concern. Through intelligent use of resources man could produce his food in predictable, abundant supplies.

But history shows climate never tranquil. Indeed, there is some evidence its fluctuations may be increasing in frequency and severity in present times, although this is questioned by many climate experts. But even without any such trend, the usual variations in weather and climate are large on all geographic fronts and time scales.

Harshly variable climate leads to erratic food supplies, particularly in developing countries which lack the financial, institutional and technological means to cope with weather-induced food shortfalls. Climate thus contributes fundamentally to malnutrition and an ever-present threat of famine in much of the world. It follows that climate/food interactions must be understood and dealt with at two levels: at the global or world food system level, as well as directly in agricultural production itself.

1.1 An Unstable World Food System

Since World War II food has become an international commodity,

W. Bach, J. Pankrath, and S. H. Schneider (eds.), Food-Climate Interactions, 21–46.
Copyright © 1981 by D. Reidel Publishing Company.

behaving somewhat like petroleum, copper and other site-specific
resources. Only five of the world's 149 nations are now able to
produce enough to adequately feed their people. The remainder
increasingly require imports, mainly grain, from a few favored
agrarian giants. In essence, food production and distribution
are now an international system, complex in structure and subtle
in response to climate and climate-defensive measures. Let us
briefly consider a few of the conflicts which influence response
in this system.

1.1.1 Production vs population

A key issue in world food system behavior is that of supply and
demand. Can the system produce enough to properly supply a population which adds 200,000 new consumers each day? Demographers
project world population to reach 6 to 7 billion by the year
2000...and around 10-26 billion a century later.

At present there is no generally accepted way to reduce this future world population. True, the concept of limited family size
is now almost universal, and incentive programs for this are
operating in most of the developing countries. But the problem
lies in the age structure of many present populations. Over
half the people in Honduras, for example, are 15 years or
younger. This means a large percentage is within childbearing
age. Thus, even if all Honduras' families were to limit size to
just two children, its population would continue to increase for
the next 70 years. In a large country like India, where a similar situation prevails, its present 600 million will grow to more
than 900 million by 2050.

Expanding populations tend to settle on land where food can best
be grown and climate stress is minimal. Lester Brown estimates
that movement of people from 600,000 villages in India to urban
areas will result in the loss of 9.8 million hectares from food
production by the year 2000. He claims about 25 million hectares
of good cropland will be similarly lost worldwide during the same
period, if present trends in urban development and growth continue. While this is only about two percent of all cultivated
land, it is mainly climatically well favored and is enough
(assuming even average productivity) to nurture 84 million
people.

1.1.2 Developed vs less-developed nations

Perhaps the most sensitive and potentially explosive element
within the world food system is the increasing gap in affluence
between developed and less-developed nations. The recently issued "Global 2000 Study" (1) says this gap will almost double in
the next 20 years: now at $4,000 difference per capita, it is

expected to reach $7,900 in 2000 (omitting inflation).

The imbalance in affluence between nations shows up tragically
in how the world food system delivers to consumers. In total
world food trade, including other commodities besides grains,
the developed nations, with less than 20 percent of the world's
population, receive 68 percent of all food imports. The developing nations, with half the world's people, obtain only 23 percent. The centrally planned countries, with the remaining 30
percent of world population, import less than 10 percent of food
commodities.

Climatic impact on food supplies is greatly intensified by this
gap between rich and poor nations. When major climate-induced
crop shortfalls occur, developed or certain oil-rich nations
need not suffer, since foreign exchange is available to compete
for and obtain food imports no matter how scarce or costly. Such
was the case in the early 1970s when poor harvests led the Soviet
Union to purchase massive amounts of grain on the international
market, mainly to feed its livestock. At the same time, stimu-
lated by dwindling stocks, world grain prices rose and hundreds
of millions of people in the poor nations were pushed further
towards hunger and malnutrition.

1.1.3 Food for profit vs food for people

The problem of climate-induced distribution inequities in the
world food system is thus amplified by the actions and dynamics
of the "free" market. While many argue that food, supplied as a
minimum diet, is a "human right," the reality is that food almost
universally continues to be a product sold for a profit, and when
scarce, to the highest bidder. What the poor person eats, he or
she must be able to buy...or depend upon the questionable largesse
of food handouts. This principal would appear, in effect, to
eternally work against the possibility of low-cost, nutritious
food which can be bought by poor people in a world with decreasing surplus food supplies.

Of course, an important buffer against this inequitable aspect of
world food trade is made when a poor nation opts, as India has
done, to create its own surplus food supply and national food
"self-sufficiency." But this still does not isolate local food
prices from the world marketplace. The farmers producing food
for local consumption must be able to pay for farm inputs, which
are also tied to world prices. And they must make a profit
(living income) if they are to continue producing.

Another crucial aspect in a world with limited resources for food
production is the tendency to grow the most profitable rather
than the most economical and/or nutritious products. Many

tropical countries with large malnourished populations actually
export food and related commodities (cotton, palm oil, sugar,
etc.) grown on their best arable land, when at the same time
they import grain their poor people usually cannot afford.

1.1.4 Food for pleasure vs food for nutrition

Human nutrition should be the delivery goal of a functioning
food system, local or worldwide. A minimum amount of food nutrients -- proteins, carbohydrates, fats, minerals and vitamins --
is required for proper human' growth and development, as are a
minimum number of daily food.calories to furnish energy for life
and productivity. The balance and amounts of food needed for
physical fitness depend on such factors as sex, age, body weight
at different ages, temperatures at which people live and the
work they do. On the average the FAO estimates a reasonably
active male of 30 years requires about 2700 calories of food
energy every day...and that his protein intake (protein being
the costly limiting factor in most diets) must exceed 75 grams
per day.

The international food system falls far short of meeting this
idealized nutritional goal. Today an estimated 800 million
people obtain less calories and proteins than the suggested minimum diet. They are malnourished and suffer all its consequences. Furthermore, over 50 percent of them are children below
the age of 10, where dietary insufficiency means a lack of
proper physical and intellectual growth, if not early death.

Affluent people worldwide are not concerned with minimum nutrition. They usually eat for pleasure and often consume two to
three times the nutrients and calories required. In addition,
unlike the poor, who survive on low-cost cereal grains, the
affluent demand animal products in their menu. A typical American or European diet contains about 31 percent meat, eggs and
fish; a typical Asian diet, three percent.

Many suggest there would be no world food problem if the nutrients and calories "wasted" by the affluent were directed into a
basic food supply properly distributed to all. Most animal
products, for example, require four or more pounds of grain feed
for every pound produced. The four tons of corn required to
fatten a one-ton steer would furnish the basic nutrition required by twelve people for one year. Some 78 percent of all the
grain harvested in the United States is fed to animals. This
represents over 20 million tons of protein that could be eaten
directly by humans in a protein-starved world. Furthermore,
much of this grain is fed to household pets, rather than farm
animals. In the U.S. alone over 100 million pet dogs and cats
consume enough protein daily to sustain 4 million humans.

1.1.5 Temperate vs tropical food production

The world food system is served by two distinctly different producing subsystems, both strongly shaped by climate. There is
the modern, mechanized and usually highly productive farming
mainly practiced in the temperate zone, where climate is seasonal and weather patterns are often manageable. And there is the
primitive, largely manual and usually small-scale and subsistence farming that predominates in the tropics, where climate,
particularly rainfall variability and intensity, is usually less
predictable and more difficult to cope with.

Why does tropical agriculture in that part of the world where
food, fibre and fuel are most urgently in short supply, usually
fall so far short of its potential when compared to achievements
of the temperate zone? This complex question has many answers,
only a small share of which are truly climate-related. Superb
productivity has been demonstrated on lands subject to the intense climatic variability of the rainfed tropics and searing
heat of the arid tropics. Andrew M. Kamarck, in a recent article in Ceres, suggests that the most productive agriculture in
the world could be in the tropics, especially in its humid
areas. (2)

In the best of all worlds, with financing unlimited, tropical
agriculture could, forced by massive technology, become the
major food-generating source for our planet. But the real world
of developing tropical countries -- the world of inequitable
economics, lack of infrastructure, colonial agricultural carry-
over, long-enduring cultural constraints -- suggests the answer
must come in another direction. It most probably must come
through a new science of tropical agriculture which achieves
optimal productivity with strategies and technologies appropriate to the economics, culture and natural environment of each
developing country. Climatology and the adaptation of tropical
food-producing systems to climatic constraints will undoubtedly
be an integral part of this new science. Raaj Sah, in assessing
the value of better climatic management in the tropics, points
out that even small improvement in weather prediction services
in India, for example, could save an average of $200-300 million
in crop losses each year. He estimates this benefit would come
through a central weather prediction facility costing less than
$15 million. (3)

1.2 Food System Sensitivity to Climate

The preceding discussion hinted at how climate influences behavior of the world food system. It suggested, through brief review of conflicts within the system, that variable climate, while
impacting directly in terms of crop successes or failures,

indirectly triggers responses within the system which affect its
stability and performance. Let us now look more closely at this
problem of food system sensitivity to climate at both global and
local levels.

1.2.1 Impact of recent climate anomalies

Climatic anomalies resulting in crop failures and famine have
been well documented over the centuries. But only in this mid-
century, with emergence of the race between food supplies and
population growth and a world food trading system, has the concept of a climate-induced "world food crisis" preoccupied
mankind.

Concern with the impact of climate on food supplies became acute
in 1963, when a major Russian crop shortfall ended a long period
of abundant, low-cost surplus grain. Subsequent USSR droughts
and two monsoon failures in India increased the anxiety and led
to greatly expanded wheat production in the export countries, as
well as rapid growth in the fertilizer industry. In the late
sixties exciting progress in growing high-yielding wheat and
rice on well-irrigated and fertilized land (the "green revolution"), plus generally good weather, relieved anxieties. From
1967-'72, though malnutrition in many countries remained high,
the "crisis" essentially disappeared.

Then in 1972 a series of climate-related events appeared to turn
the situation around. Crop failures in both the USSR and India,
plus a U.S. policy reducing sown area 10 percent, resulted in a
two percent decline in world grain production. This sharply
reduced world grain reserves to 90 million metric tons (the
equivalent of only one month's supply). Continuing drought in
the Sahel in 1974, along with a poor U.S. corn crop the following year and the "El Nino" failure in anchovita protein harvest,
once again created a world "food crisis" situation. A critical
external force aggravating the crisis was the OPEC escalation
of oil prices, causing fertilizer to rise from $50-75/ton in
1972 to $300-500/ton in 1974. This tended to throttle down the
green revolution production goals in many developing countries.

In his analysis of the impact of these events of the early 1970s,
Rolando Garcia found the extent of food loss and human suffering
which followed was due more to the nature of the "system" than
to the severity or sequence of climate anomalies. (4) What happened after drought struck, he claims, was much more determined
by the structure of the food-producing and distributing system,
including its social and political components, than by the
drought itself. In other words, it is the sensitivity of the
food system to climate, as much as climate itself, which shapes
a food disaster.

1.2.2 Variable climate vs farming systems

Climate's impact on agricultural productivity depends in good measure on the farming system employed. Large-scale, non-irrigated monoculture, such as practiced on the plains of India and steppes of Russia, is especially vulnerable to climate anomalies. As recent events show, a seasonal drought or monsoon failure over those vast grain-growing areas results in crop losses large enough to destabilize the world food system. Multiple crop farms, integrated livestock/cropping systems, and the practice of intercropping, while not always as productive as monoculture, greatly reduce risk of a single crop wipeout and guarantee some food during years of climatic disasters.

When climate is unfavourable, the use of technology -- of rapid machine response, extensive pump-fed irrigation, and high fertilizer and pesticide inputs -- often is enough to guarantee reasonable yields under monoculture. However, the cost of this kind of farming, highly dependent on fossil fuels, is now too high for most developing countries.

Despite the impressive productivity of high-technology, monoculture farming, it is now recognized that good yields are also possible using less energy-intensive farming systems which are essentially "climate defensive." Over the past several years a team at Washington University, funded by the National Science Foundation, compared productivity and income in large U.S. cornbelt "organic" and equivalent-sized "conventional" farms. The organic farms employed mixed cropping systems and rotation, using no chemical fertilizers and pesticides. The "conventional" farms were typical successful midwest corn and/or soybean monoculture operations. During good weather years the monoculture farmers made more income through better over-all yields and intensive cultivation of one crop on all farm acreage. But in years with rainfall deficits, the organic farmers had better crops, and given their savings in chemical inputs, came out ahead in farm income. (5)

1.2.3 Variable climate vs agro-ecology

The environmental consequences of agriculture (agro-ecology) are strongly influenced by climate. The primary problems include massive loss of topsoil to erosion; spread of deserts; waterlogging, salinization and alkalization of formerly productive areas; flooding and silting due to deforestation. All of these problems usually come about because of ecologically improper and over-intensive farming techniques as new land is brought under cultivation in climatically difficult regions.

The "Global 2000 Report" referenced earlier has, in pages

274-298, a comprehensive analysis of the environmental consequences of ecologically and climatically unsound agriculture.
Here are some of the projections:

Erosion An average 100 to 150 tons of topsoil per hectare
are being lost to erosion each year, as sloping and hill-
side areas of the humid tropics are brought under cultivation. This loss would add up to approximately six inches
of topsoil removed from these new marginal farming areas by
the year 2000. Most would be barren by then, since they
have less than six inches of topsoil to begin with!

Deserts Losses of productive land to desert have now
reached about six million hectares a year (60,000 sq. km).
If this accelerating process continues unchecked, the possibility exists that by the year 2000 deserts will occupy
three times their present area of approximately eight million square kilometers.

Waterlogging About 225,000 hectares become wasteland each
year because of waterlogging and subsequent salinization
and alkalization. By the year 2000 an estimated 2.75 million hectares of presently irrigated land will be out of
production, an amount that could furnish food for nine million people.

Deforestation At present about 18 to 20 million hectares
of forest are removed each year as landless people in the
developing world seek fuel and subsistence farms. By the
year 2000 forests will have decreased from a present 1/5
to 1/6 of the world's land area.

1.3 Resource Constraints on Food Production

How much more food must be grown above present production levels
to meet minimal nutritional requirements of the world's population in the year 2000? The International Food Policy Research
Institute (IFPRI) has made detailed studies on this question.(6)
According to the IFPRI, the present nutritional deficit in developing countries amounts to about 100 million metric tons of
cereal grain. (Total world grain production is now about 1400
million tons.) Using the United Nations "medium" population
projection of six billion people by the year 2000, and assuming
all people would be fed the adequate nutritional equivalent offered in 1,000 pounds of cereal grain per person per year, this
means that present world grain production would have to more
than double to three billion tons in the next twenty years
(omitting any provision for livestock feed production).

What are the prospects for doubling world grain production in

two decades? Even assuming a twenty-year climate moratorium,
with weather conditions always favorable to maximum harvests,
the goal seems formidable. The answer would depend on how much
more technology could push up yields in major producing areas
such as the United States, China and Russia, as well as in large
nations such as Brazil, India and Argentina. It would also
depend on the ability to double or triple present production
levels in tropical agriculture. And, most importantly, it will
depend upon how new and expanded production will be curbed by
certain constraints, both environmental and man-made.

Environmental constraints in food production are primarily those
unrenewable, limited or stressed resources essential to agriculture: land, water, plant nutrients and energy. The man-made
constraints are those imposed by governments, landowners, farmers and consumers: they are political, economic and cultural in
nature. Stated another way, they are the policies which determine how resources are used for food production and distribution.

The discussion which follows centers on how land, water, energy
and policies place a limit on an open-ended attempt to match
food production with population growth using today's agricultural technology. They also start to suggest why creative new
approaches to the problem, such as comprehensive efforts directed towards minimizing the impact of climate on food production,
may be a major hope in extending the limits.

1.3.1 The limits of arable land

According to FAO figures, out of 13.4 billion hectares of land
in the world, only 11 percent, or 1.5 billion, are considered
arable or suitable for cultivation. The remaining land is
either too dry, too cold, too steep, or lacking suitable soil
structure for planting.

Projections from 1975 data indicate around one billion hectares
of all potentially arable land are now under cultivation, with
cereals grown on 700 million hectares. Pastures and rangeland
occupy about three billion hectares, and forests, another four
billion. According to these figures there are about 500 million
hectares of unused land that could grow crops. Most of this
potential farmland, however, is in Africa or South America,
where serious regional problems inhibit expansion. For example,
millions of hectares of sub-Saharan Africa are closed to cultivation and/or foraging because of the tsetse fly and river
blindness disease. Leaching of soil nutrients by heavy tropical
rains, and generally low soil fertility limit further agricultural development of the Amazon region.

Given the high productivity of U.S.-style farming as a model,

would land presently under cultivation be enough to produce food
for six billion people in the year 2000?   Take corn and wheat as
a food standard and assume 0.5 ton of these cereals, mixed,
would nurture one person per year.   Average U.S. yields for corn
and wheat in 1975 (a good year) were 5.4 and 2.1 metric tons per
hectare, or an average for the two of 3.75 tons/ha.   Thus, if all
one billion of the world's presently cultivated hectares were
devoted to these crops, grown with U.S. farm technology, about
3.75 billion tons could be available for six billion people.
This calculation, of course, assumes absolutely cooperative
weather worldwide for the next 20 years.   Normal climatic variations would be expected to reduce anticipated yields by as much
as 10 to 20 percent in many important food-producing regions
during "bad weather" years.

1.3.2 The availability of water

It has been observed that three-quarters of the land now under
cultivation receives insufficient rainfall at the right times
during the growing season to yield bumper crops.   About 17 percent, or around 250 million hectares, are now under irrigation.
When it is efficient, irrigation, especially in warm and sunny
regions with good land, often boosts production three to four
times over dryland farming, especially where multiple crops can
be grown.   Herbert Riehl has calculated that the small fraction
of land under irrigation accounts for one-half of the world's
harvested crops. (7)

While water covers 70 percent of the globe, excluding polar
glaciers, only one percent is the free, fresh water required for
agricultural purposes, and of that one percent, 99 percent is
underground.   Nevertheless, vast quantities of sub-surface water
flows, unutilized in irrigation, to the oceans.   At present the
average per capita consumption of fresh water runs between 2,000
to 10,000 liters per day, the amount varying according to how
industrial and urban the society, and particularly, how much land
it irrigates.   Irrigation now accounts for about 75 percent of
all fresh water uses.

A key factor in fresh water conservation, particularly in the
warm, dry regions with costly irrigation, is how efficiently it
is used by the crop.   For example, the water required to produce
sugar cane per unit of land area in Hawaii is five times that
needed for pineapple.   Large differences in transpiration exist
between species, varying on a scale of 100 for pineapple to 400
to 500 for cereals and seed legumes, to over 1,000 for some
fruits and vegetables.

The Global 2000 Study projects increasing competition for easily
available fresh water, with a 200-300 percent increase in world

water needs over the next two decades. Much of this increase in
demand will be in the tropical nations, where in many areas
fresh water for human consumption is already in short supply.

1.3.3 Emerging energy constraints

While climate, land and water furnish the basic elements in food
production, energy and its availability determine how well this
combination works. Consider first the energy requirements of
U.S.-style farming, often held as model for production goals
around the world. David Pimentel has calculated that producing
a hectare of corn in the United States requires about 8.7 million
kilocalories of fossil energy, the equivalent of about 860 liters
of fuel. Close to half this energy is used in making the needed
fertilizer. Use of machinery instead of human labor, and the
fuel to run it, takes another 20 percent; irrigation, about 15
percent. All told, if this kind of energy-intensive farming were
immediately practiced worldwide to feed four billion a U.S.-
style diet, Pimentel estimates it would gobble up all known petroleum reserves within the next 13 years. (8)

As part of the same study, Pimentel also has interesting thoughts
on the cost of fossil fuel energy in extending the benefits of
irrigation. He estimates about 12.2 million liters of water
(12,200 metric tons) are needed to produce a good yield of five
tons of corn per hectare in the subtropics. The energy required
to pump this water from a depth of 90 meters is about 20.6 million kilocalories. Thus, using corn as the standard crop, if
pump and sprinkler irrigation were applied to 1.5 billion presently dry and arid hectares to bring them into full production,
about 3,090 billion liters of fuel would be required. If all
known petroleum reserves were used for this purpose alone, they
would last only 20 years.

The Global 2000 Study offers an interesting anomaly in the oft-
repeated maxim that "U.S.-style agriculture can feed the world."
It reveals that while crop yields increased by 138 percent worldwide during the period 1945-'70, the U.S. did not once appear on
a list of 20 major world food crops and nations that achieved the
highest yields per hectare. The U.S., however, was by far the
most intensive user of energy on its farms. Clearly there appear
to be other options for achieving high productivity. The study
points out that the People's Republic of China has probably been
more successful than any other country in developing its agriculture with minimum energy requirements.

A final note on energy and its availability for agriculture. U.S.
DOE projections indicate world energy demand is expected to double by the year 2000. Oil will still be the leading energy
source, with its price increasing sharply over the next 20 years.

All known gas and oil reserves should be half-depleted by the
year 2000, with coal in a similar situation by 2100. As fossil
fuels become costly and less available in developing countries,
demand for fuelwood will rapidly increase, accelerating deforestation and causing urban families in some African cities to
spend as much as 30 percent of their income on wood.

1.3.4 The social factor

As Rolando Garcia so well emphasizes in Nature Pleads Not Guilty,
social factors, not capricious climate, usually are the crucial
forces behind success or failure in the production and delivery
of food. Government policies, often dictated by interests of a
powerful local elite, determine how land is used and where agricultural inputs are applied. More supportive policies could contribute greatly to the oft-stated but poorly-implemented goal of
"food self-sufficiency" and to the amount of food available for
local consumption. The dualistic agricultural economy in many
African nations is an instructive example of how politics influence policy. As a rule, export crops receive government credits
and supports, and benefit from irrigation schemes, water rights
and scientific extension services. On the other hand, local food
crops, often tilled with tools little advanced beyond the iron
age, receive few of these benefits.

Another example of how policy often works against maximum world
food production is the United States acreage set-aside program,
which protects farmers in surplus carry-over years from low
prices. By 1972, for example, six-and-one-half million hectares
had been taken out of production to assure better local farm
prices. Yet, this was the year of excessive drought in many
Eastern Hemisphere nations, when total world production decreased
by one percent and an international food "crisis" was reborn. It
should be noted, however, that the set-aside policy ended and
harvested acreage in the U.S. increased by 26 percent from 1972
to 1977.

2. MINIMIZING CLIMATE'S INFLUENCE

Our argument thus far has shown that unfavorable climatic variability, through direct impact on crops in major food-producing
regions, triggers responses in the world food system which am-
plify the problem of global malnutrition and hunger. The point
was made that the sensitivity of the food system, hence the degree to which climate impacts, is related to many internal conflicts, mainly social and economic in nature. Finally, the potential ability of the system to meet future world food needs
was measured against constraints imposed by limited resources
and unfavorable policies.

Given this analysis, and since the mission of this workshop is to understand and deal with climate/food interactions, what can be done to minimize climate's influence? Two strategies are obvious:

a. One is through policies which will reduce the food systems' sensitivity to climate impact and improve stability in world and local supplies.

b. A second strategy is to reduce climatic impact at the producing or farming level through applied science and technology.

A third strategy, of course, is to simply insulate food production from climate through irrigation, controlled environment farming and even non-agricultural technology. This latter approach, because it is more concerned with new possibilities in production than with climatic sensitivity per se, will be discussed separately.

2.1 Policies to Reduce Climatic Risk

Stephen H. Schneider, in his brilliant analysis of what he calls "the world food predicament," presents a compelling argument for policies which would reduce conflicts within the food system and improve its performance vis-a-vis climate (9). A few of the more obvious and perhaps easily applied are reviewed here.

2.1.1 Trade and aid policies

Most nations, unable to build up their reserves as a hedge against bad crop seasons, must rely on imported stocks, particularly in times of food emergencies. Often, because demand elsewhere has preempted surplus stocks or driven prices beyond reach, the poorer developing nation is unable to import food to meet its needs. A policy of bilateral trade, with supplies assured by agreements between nations, is an important way to overcome this problem of climatic risk.

The United States, now furnishing over half the world's surplus grain, has developed various marketing schemes which if used efficiently could adapt well to bilateral trade, particularly with developing countries. It offers the possibilities of straight commercial sales, liberal credit sales (Public law 480), and outright food aid donations for chronically food-needy poor nations. An adaptive bilateral agreement would set maximum and minimum yearly purchase levels, with the three delivery schemes flexibly used, depending on world grain prices. Thus, a poor nation with no ready cash or credit would be guaranteed import food, even when wheat or corn prices zoom out of reach.

Establishing bilateral agreements for food commodities reduces
the usual beaurocratic delays associated with other trade arrangements. Benefits to the importing nation are clear: it is
assured of a secure annual supply of grain at stable prices, no
matter how weather anomalies or trade patterns influence the
world market. The exporting nation, though occasionally for-
going windfall prices, secures a guaranteed market for continuous volume production. Further, when the importing nation does
have a bumper crop year, the arrangement forces it to build up
its own price-stabilizing food reserves. A comprehensive discussion of "Climate-Defensive Policies to Assure Food Supplies"
by Martin Abel and associates is a chapter in a book soon to be
published by Westview Press. (10)

2.1.2 Food storage and reserves

A policy of accumulated reserves is the most obvious route to
food security for nations subject to unpredictable weather and
subsequent "boom or bust" crop years. Recently, faced with impossible costs of import grain, several nations set this goal;
but it is a goal severely constrained by the amount of arable
land per capita, as well as by climate and economics. It is
interesting that India, one of the world's more densely populated
places, with only 0.2 hectare of land per person in 1980, has had
greatest success in implementing this policy. Prior to the mid-
'70s, failure of monsoons caused frequent need for massive emergency grain shipments to India. A new policy by the Janarta
Party government increased local production beyond effective demand with the goal of building grain reserves to tide the nation
through at least two years of bad weather. By 1978, India's
stocks amounted to 20 million tons, the result of three years of
bountiful harvests. This surplus almost completely sustained
India after the growing season of 1979-80, when monsoon rain
once again proved inadequate.

A critical problem in food reserve programs, particularly in
tropical nations, is lack of suitable storage facilities. FAO
has estimated that over 200 million tons of grain are lost after
harvest, most of it in the tropics, where insects and rodents
easily penetrate the poorly contained food. To understand the
human dimensions of this problem, consider the conservative figure of 25 percent post-harvest attrition in cereal and pulse
crops in semi-arid Africa. Based on 1971 production figures,
losses then amounted to about seven million tons. At $7/hun-
dredweight, this adds up to about a $1 billion loss in food in
just one poor region of the world. Since one ton of cereal offers minimum basic nutrition for three people for one year, the
food saved could supply over 20 million people.

2.1.3 Industrializing the food system

Most developing countries, particularly in the tropics, have
relatively primitive food systems. Production for internal consumption is mainly in the form of raw, unprocessed products and
facilities for storing, preserving, packaging and distributing
food, and their supporting infrastructure, are largely missing.
Hence, local food supplies come in seasonal bursts, depend on
marginal and fragile distribution paths, and are especially vulnerable to losses due to unfavorable climate.

It is suggested that appropriate national effort to industrial-
ize food would greatly improve resiliency to climatic impact in
most at-present agrarian nations. Food processing -- its preparation and packaging in preserved form -- adds value and convenience to the product and extends its life for several years
of safe storage. Furthermore, the investment in processing technology, which for efficiency should be near the raw food source,
encourages rural development and requires supporting infrastructure and distributing/marketing facilities which stabilize supplies in urban areas. Farmers, assured of income from processing
plants for their cash crops, can secure credit for agricultural
inputs, to minimize the impact of bad weather or a difficult
climate.

Many argue the capital and energy costs of an industrialized
food system negate its development in a poor country. If the
pattern of present industrial nations is followed, this might
be true. However, many new options in low-energy food preservation based on solar efficiency and on bioengineering and fermentation processes suggest a totally different kind of low-cost
food industry in the tropics, which will avoid the fancy packaging, wasted calories and merchandising excesses of the affluent nations. An added note: massively available efficiently
"pre-cooked" food in the tropics would greatly reduce the accelerating demand for firewood and its environmental implications.

2.2 Applying Science and Technology

There are numerous actions farmers and agricultural planners can
take to directly reduce the vulnerability of crops and regional
food-growing systems to climate, most requiring new scientific
insight and technology in their application. Only a few are mentioned here.

2.2.1 Modeling and information systems

Computer-based crop-climate models have been developed and mainly
employed in temperate zone agriculture. They have been successfully used to predict yields and schedule inputs, and when keyed

into world weather-based crop yield assessments, assist in commercial and national food import/export planning strategies.
There is an important need to bring this new science and technology to the tropics, where climatic variability impacts more
heavily, and where management and planning in agricultural
development are urgently needed.

Aspen Institute, through its Food and Climate Forum, has gained
some experience exploring the potential for this technology in
the tropics through a project it has had underway in Venezuela
since 1979. Led by an Aspen team, with inputs from U.S. experts, a group of Venezuelan scientists has developed and is
now testing a pilot crop/climate model for maize production
within a new farming region in the Western Llanos. A not unexpected bonus of the project has been its organization of surprisingly extensive historical, and heretofore unused, climatic
data. Proper use of this kind of information alone -- and one
must assume that similar records, as well as information-
generating points, exist in many other developing countries --
offers enough of a pay-off to justify the project. A fairly
detailed description of the Venezuela effort and its expecta-
tions has just appeared in the international trade journal,
AgriBusiness Worldwide. (11)

2.2.2 Appropriate agro-technology

A major flaw in the world food system is its tendency to grow
crops unsuited to local climate or food needs. This is especially true in tropical regions, where former colonial land use
policies still prevail. In predominantly dry Mali, West Africa,
for example, cotton and rice, which have high water requirements, are grown extensively. Yet, during the severe Sahelian
drought, lasting from 1968 to 1973, when cotton, rice and
groundnut production, all for export, were furnished precious
irrigation water and reached record highs, corn production for
local consumption fell by one-third, and many starved.

Considerable interest, mainly at the international agricultural
research centers, has developed on the question of ecologically
appropriate food crops for the various climate regimes in
Africa, Asia and South America. Crop varieties with special
water-conserving qualities, as well as food values, are being
advocated. For example, any of the millets, which are the most
drought- and heat-resistant of all the cereals, could be extensively cultivated in dry regions. The benefits of intercropping,
multiple cropping, and various tilling methods in climate-
defensive food production are also being studied. In other
words, much knowledge has been accumulated about appropriate
agro-technology, but little has been applied thus far. Politics
and privilege still cause most tropical nations to focus

research efforts and infrastructure support on export crops,
such as coffee, palm oil and rubber, rather than on locally-
consumed, climatically-appropriate foodcrops.

2.2.3 Genetics to combat climate

While the main thrust in food crop genetic research has been to
increase yields, there is much promise for seeking genetic improvements in plant resistance to the temperature and precipitation extremes of unfavorable weather and climate. A "bonus"
pay-off came in the development of high-yielding dwarf rice. In
order to carry the large grain head, a shorter, sturdier rice
stem was developed. This in turn greatly reduced losses due to
lodging caused by heavy winds and rain.

Ralph Cummings Sr., the renowned soil scientist, in a recent
workshop on climate/food interactions in the tropics, pointed
out the great possibilities for exploiting climate-defensive
characteristics found in many wild and neglected plant species
with food value. (12) Surveys by the National Academy of
Sciences on the potential of promising new tropical food and
forage crops (e.g. the fast-growing, nitrogen-fixing leucaena
tree and the high-protein winged bean) support Dr. Cummings'
optimism. (13)

2.2.4 Integrated livestock/crop systems

The livestock/feedlot monoculture extensively practiced in the
developed, industrial nations has come under heavy attack as a
"destabilizing" element in the world food system. However,
there is much evidence suggesting that proper use of livestock,
integrated with crops in food-producing systems, can actually be
an important stabilizing force. This is especially true with
ruminant livestock able to secure all their food if necessary,
foraging on land marginal or unfit for intensive food crop production.

Ruminant livestock already serve as a major hedge against climate and crop failures. Twelve percent of the world's population derives support almost entirely from the meat, milk and
by-products of ruminants, because they live in areas where food
crops cannot be grown. When rangeland feed is scarce due to
runs of dry weather, the herds are reduced and animals slaugh-
tered for food. When climate improves, the herds increase in
size.

If the itinerant livestock herding practiced in many parts of
the tropics could be integrated with production of foodcrops,
important environmental as well as production benefits should
accrue. A livestock/crop rotation contributes to soil fertility

and erosion control and helps check soil-borne diseases of both
crops and animals.

3. PRODUCING MORE FOOD DESPITE CLIMATE

Two great possibilities exist for adding significant amounts of
food to world production without the uncertainties imposed by
climate. One is to bring farming to those vast, presently uncultivated parts of the world where climate is essentially
static and predictable; namely, the warm desert regions. The
other is to develop efficient ways of producing food within
enclosed environments essentially independent of climate.

3.1 Farming Arid Lands

Hot and dry regions, where rainfall is consistently too meager
to support life, take up about five billion hectares, or one-
third of the world's land area. Climate over these lands usually offers high light intensity, high temperature, a long growing
season, and low atmospheric humidity -- all factors favoring
successful crop production. Furthermore, soils in arid regions
are generally rich in plant nutrients like potassium, calcium,
magnesium phosphate and sulfate, and if properly managed under
irrigation, they have great productive potential.

Easily available fresh water, mainly from rivers, has brought
about 200 million hectares, or four percent of the so-called
desert areas under cultivation. Irrigation has added 66 percent
to the cropped area in Asia, 19 percent in the Near East, 13
percent in Latin America, and only three percent in Africa. The
question is, how much of the remaining 96 percent of the world's
unused warm, dry lands can be made to produce food in the years
ahead?

3.1.1 New irrigation prospects

According to the Global 2000 study, irrigation will increase by
28 percent over the next 20 years, adding perhaps another one
percent of the world's desert lands to food production. Most
of the water will result from large hydraulic projects, with the
need for river flood control and electrical generation joining
irrigation in justifying the big expenditures. The study says
30 percent of the total world runoff of fresh water, or three
times the estimated 4,000 cubic kilometers now stored, will be
contained in dams and reservoirs by the year 2000.

The food-generating value of a series of proposed dams for the
Lower Mekong Basin suggests the potential of big new irrigation
projects. At present, that portion of the basin downstream
from China feeds about 33 million people on an annual rice

harvest of 12.7 million tons. New irrigation through dam construction is expected to add up to five million hectares of land
for double-cropping of rice, and furnish enough food for an
additional 50 million persons in the basin.

Not included in global study projections is the enormous scheme
by the Soviet Union to divert its northward flowing Ob River
southwards to serve crop production in the Kazakhstan and
Uzbekistan deserts. Scheduled for completion by the year 2000,
the system will irrigate an area three times the size of France.
If the Ob diversion scheme is successful, Soviet wheat could
very well become a major surplus food source in the world market.

3.1.2 More efficient irrigation

There is great hope for expanding world food production by improving the efficiency of today's irrigation. Victor Kovda,
the eminent Soviet soil scientist, estimates less than 50 percent
of all water presently deployed in irrigation is actually used
growing the crop. (14) He further notes over 60 percent of all
irrigated lands are gradually losing their fertility because of
waterlogging and salinity. Irrigation mismanagement and poor
technology are removing over 200 thousand hectares of good farmland each year (an area about the size of Luxembourg).

There are, however, many examples of highly efficient irrigation
through good water management and modest investment costs.
China, where good drainage systems and water conservation are
long-established technology, is often cited for irrigation efficiency. With about 100 million hectares under cultivation (about
one-half that of the U.S.) and about 60 percent of this irrigated, China manages to meet almost all food needs of its 900 million people. Egypt, on the other hand, irrigates all 35 million
cultivated hectares, yet requires large annual food imports for
its 42 million people. In Egypt, as a result of its High Aswan
Dam project, water is free to anyone with power to lift it to the
field, and with animal power plentiful, overwatering of crops is
widespread. Besides water waste, poor drainage and subsequent
soil salinization have steadily decreased fertility in what was
once the richest food-producing area in the world.

Techniques to improve efficiency in existing irrigation systems
are reviewed in an excellent booklet by the National Academy of
Sciences, More Water for Arid Lands. (15) At least a dozen well-
known ways to reduce water losses are covered. One of the most
promising, drip irrigation, which has been pioneered on a large
scale in Israel, almost eliminated waterlogging and saline
build-up, while also reducing the amount of energy and water
required by as much as 50 percent.

3.1.3 Saline irrigation

Promising work in Israel, California, and on the Sonora desert
in Mexico suggests an appreciable part of tomorrow's food supply
could be furnished by halophytic salt-tolerant crops, grown
under productive desert conditions and irrigated by seawater or
saline underground water. The potential for this appears enormous. There are about 20,000 linear miles of warm and sunny .
desert seacoast in the world, backed by millions of hectares
of empty land. There are also vast amounts of salty water
lying under inland arid areas that could be cultivated. Two-
thirds of the land surface of the United States lies above water
containing more than 1,000 parts per million salt; hence, unusable
in conventional agriculture.

Carl Hodges, director of U. Arizona's Environmental Research
Laboratory, describes results in the first large-scale field
tests of seawater-irrigated halophytes underway in Mexico and
Tucson. (16) The plants selected, many collected along the
coasts of the Gulf of California, include edible roots and cereal
grains. Above-ground yields of well over 1000 grams of dry
weight per square meter were registered with daily irrigation.
Preliminary trials are underway using the harvested crops as
animal feeds. Future work will include genetic selection and
improvement of the most promising species.

3.2 Climate "Invulnerable" Food Production

Food, both subsistence and commercial, has long been produced
inside structures to effectively insulate it from climate and
furnish optimal growing environments. Greenhouses, enclosed
fish "ponds," poultry coops, fermentation vessels, and even chemical factories are some of the facilities now employed to grow
or "make" food. Yet to date, with the exception of pond aquaculture, only a neglible portion of the world's food is being
produced in such essentially climate-free systems. The capital
and energy costs of isolating food production from climatic
impact are still very high, and the resulting product usually
must be limited to an affluent market.

At this point in time, however, with rising production costs and
stressed resources limiting the outlook for conventional agriculture, a serious evaluation must be made of practical possibilities for climate-invulnerable food production. Much of the
technology needed is well known, but the economies of large-scale
climate-independent production have as yet not been fully subsidized or achieved. Three of the most promising concepts are discussed here.

3.2.1 Controlled-environment production

The technology for growing crops within a controlled environment
advanced rapidly in the past decade, spurred by the worldwide
rise in the cost of food. Conventional greenhouses have become
more sophisticated, with more efficient solar collection, better
regulation of temperature and humidity, automated nutrient delivery, and hydroponic (soil-less) culture and CO_2 control.
Production inside a controlled-environment facility often is
eight to ten times more than on an equivalent growing area in
the field. However, costs for such a facility have mounted to
over $200,000 per acre of enclosed growing space, and the prod-
uct, usually fruits and vegetables, is profitable only in affluent urban markets. It should be noted, however, that big industry, notably General Electric and General Mills, is now into
unique, high-production, controlled-environment food-growing
"factories" -- a sure sign the technology will significantly
develop in efficiency over the next decade.

Progress in controlled-environment "farming" of fish, or aquaculture, has also accelerated with world food prices. Fish culture, mainly in ponds, hence subject to climate impact to some
extent, is a worldwide industry that yielded six million metric
tons in 1975. However, much greater production has been demonstrated in modern, energy-efficient facilities where fish or
crustaceans are grown rapidly and densely, with their wastes
used in satellite feed-producing systems. In such a facility at
Puerto Penasco in Mexico, Carl Hodges and his associates have
grown commercial crops of shrimp yielding over 50 tons per acre/
year. At $5/pound this food crop may be the world's most valua-
ble on a per acre basis. While a land farmer is lucky to raise
more than 100 pounds of good beef on an acre of good pasture, a
closed-system fish farmer can raise dozens and even hundreds of
tons of fish and shellfish, largely protein, on an aquatic acre
in one year.

3.2.2 Food from biosynthesis

Several years ago, when oil was much less costly, big plans were
underway to produce massive amounts of protein by growing yeast
and bacteria in huge fermentation vessels, using petroleum and
other fossil hydrocarbons as a substrate. Major ventures were
launched in England, Italy, France and the Soviet Union for factories to eventually produce as much as 100,000 tons a year of
single-cell protein, mainly for cattle feed. But OPEC economics
thwarted the future of petro-based food production.

Today, the concept of massive food through fermentation has re-
vived. But the focus is now on use of carbohydrate or cellulosic
plant materials as the carbon source needed to grow the

proteinaceous organisms. Adding incentive to this new development is the fact that alcohol is also produced in single-cell
protein fermentation. With alcohol used as a liquid fuel sub-
stituting for gasoline, the economics of biosynthesis keep
improving. Further, if more food rather than alcohol is the
process goal, the latter can be converted into glucose by further biosynthesis.

Properly organized, there could be an impressive amount of crop,
food factory and even organic urban wastes, all now relatively
unused and often causing environmental problems, to grow the
protein organisms. Eligible as fermentation substrates are the
83 million tons of bagasse and nine million tons of molasses
from tropical sugar plantations, now vastly underutilized, as
well as the 59 million tons of wheat chaff and 30 million tons
of corn cob material.

The production potential from biosynthesis is astonishing.
Properly "fed," 1,000 pounds of microorganisms will produce
100,000 pounds of protein in a single day. A one million ton
per year single-cell protein factory would occupy a few hectares of land, while equivalent production in soybean protein
would require over one million hectares of good farmland.

3.2.3 Food from chemical synthesis

During World War II, hard-pressed Germany proved that a human
food can be produced from non-agricultural materials when it
made 100 thousand metric tons of edible fat by chemical synthesis, starting from coal. All of the basic nutrients in food --
proteins, fats, carbohydrates, vitamins -- can be produced by
synthesis, and balanced, attractive, edible products fabricated
by modern food engineering. Synthetic vitamin production is
already a worldwide industry. Some of the important micronutri-
ents in protein, the amino acids, are also being produced commercially by synthesis and fermentation. Methods to build whole
proteins from these constituents have been developed, but are
not yet practical for manufacturing purposes. Techniques for
the synthesis of edible sugar, the basic food carbohydrate, have
also been developed, and its practical manufacture is predicted
by 1985. Meanwhile, a number of high-energy substitutes for
sugar have been synthesized and only await U.S. Food and Drug
Administration-type approval before widespread use.

While huge quantities of a synthetic food are not expected to
enrich the world food system during the next few decades, the
indirect impact of this emerging technology should be significant. For example, an anticipated successful large-scale production and marketing of factory-made synthetic sweeteners and
sugar could release for new local food production as much as

two million hectares of prime land in the tropics now taken up
by cane sugar cultivation. Impact on corn acreage in the United
States and Europe is not expected, since the corn now used to
produce high-fructose sweetener will undoubtedly be switched to
alcohol production. A similar intrusion on the world protein
market, however, is probable with fermentation protein and the
large-scale biosynthesis of amino acids, since their movement
into the world food market could reclaim for human food some
of the highly productive land now growing soybeans and corn for
high-quality animal feed.

4. STRATEGIES FOR THE '80s AND '90s

The aforegoing analysis suggests that a "climatological view" of
the world food problem offers some hope for its solution.
Broader understanding of climatic impact and the advocacy of
policies and technologies to deal with it could lead to substantial improvements in how abundantly food is produced and deliv-
ered. As a starting point for discussion, a few of these measures are suggested below.

4.1 National and World Policy Priorities

4.1.1 Establishing a climate-food data base

Each nation, as a first priority, should establish its own
climate-food data base or information system to integrate with a
world climate/food information system maintained by the UN/FAO.
Essential elements would include both historical and real time
climatic and crop yield data, as well as national agronomic resources and food demand information.

4.1.2 Determining climatically appropriate agro-technology

Utilizing its climate-food information system, each nation
should, after surveys and analysis, prepare a plan for an optimal system of climate-defensive food production. The national
plans, designed to meet nutritional needs and socio-economic
conditions in each specific country, would also integrate into a
coordinated effort for international climate-responsive management of the world food system.

4.1.3 Building food reserve security from climatic impact

A key element in the national food production plan, particularly
in nations subject to big climate-induced crop shortfalls,
should be a food surplus storage program. Whenever possible,
the surplus stored should be locally produced, climatically and
nutritionally appropriate food, rather than internationally marketed cereal imports.

4.1.4 Eliminating climate-induced post-harvest food losses

Concerted national effort should be directed, especially in
tropical countries, towards eliminating losses of food to predators and biodegradation, through proper storage, preservation
and distribution facilities. In essence this would establish
the food industry as a primary target in national development.

4.1.5 Reordering of national research and support priorities

Food-needy nations, particularly those in the post-colonial
tropics, should assess and change, if necessary, present policies which emphasize research and inputs to "plantation" export
agriculture, in the light of present and long-range food
security.

4.2 Research and Development Priorities

4.2.1 Climate-defensive agronomics

A science of "climate-defensive" agriculture needs to be developed, particularly for regions with wide climatic variability
and moisture stress. Tillage and planting techniques such as
intercropping, relay planting and rotation should be evaluated
and improved from the standpoint of climate resiliency, yield
and ecology.

4.2.2 Climate-defensive genetics

Present and potential foodcrops need to be systematically studied for climate-defensive genetic characteristics, and for gene
combinations which would improve their water use efficiency and
resistance to drought, freeze, wind, hail and flooding.

4.2.3 Climate-defensive food systems

Research is needed to determine the vulnerability of different
kinds of farming systems, as well as developing-country food
systems, to climatic variability, and to identify measures for
improving their climate-defensiveness.

4.2.4 Arid lands food production

The potential for halophyte food crop production and saline
water irrigation needs to be fully explored in an extensive program of research and development, perhaps through an International Center established for this purpose.

4.2.5 Climate-free food production

The technological and economic feasibility of various kinds of enclosed facility food production needs to be seriously evaluated for their present and future role in the world ·food system. If recommended, research and development programs should then be prepared for national and international subsidy to develop this new technology to the stage where it can be deployed on an economically justified and significant global scale.

REFERENCES

1 *The Global 2000 Report to the President* prepared by the Council on Environmental Quality and the Department of State. U.S. Government Printing Office. 1980.

2 Andrew Kamarck, "The Most Productive Agriculture in the World-Some Day," *Ceres*, September/October 1979, pp. 16-21.

3 Raaj Sah, *Priorities of LDCs in Weather and Climate*, a World Bank publication, June 1978.

4 Roland V. Garcia, *Nature Pleads Not Guilty*, Pergamon Press, 1981.

5 W. Lockeretz, G. Shearer, R. Klepper, and S. Sweeney, "Field Crop Production on Organic Farms in the Midwest," *Journal of Soil and Water Conservation*, May-June, 1978, pp. 13-134.

6 *Food Needs of Developing Countries*, Research Report No. 3, International Food Policy Research Institute, 1776 Massachusetts Ave., Washington, D.C. 20036, December 1977.

7 H. Riehl, *Introduction to the Atmosphere*, McGraw-Hill, 1978.

8 D. Pimentel, *World Food, Pest Losses, and the Environment*, Westview Press, 1978, pp. 1-16.

9 Stephen H. Schneider, *The Genesis Strategy: Climate and Global Survival*, Plenum Press, 1976.

10 L. E. Slater and S.K. Levin, *Climate's Impact on Food Supplies*. AAAS Symposium Series Volume, Westview Press, 1981.

11 L.E. Slater, "Modeling the Tropics: Can It Be Done?" *Agri-Business Worldwide*, August/September 1980, pp. 25-38.

12 *Climate Change and Agricultural Production in Non-Industrial Countries*. Report of a Workshop Held in Oak Ridge, June, 1980. The Climate Project, AAAS, Washington, D.C.

13 *Unexploited Tropical Plants with Promising Economic Value,*
   National Academy of Sciences, Washington, D.C., 1975.

14 Victor Kovda, "The World's Soils and Human Activities" in
   N. Polumin, ed., *The Environmental Future,* McMillan, New
   York, 1972.

15 *More Water for Arid Lands: Promising Technologies and
   Research Opportunities.* A National Academy of Sciences Publication, Washington, D.C., 1974.

16 Carl Hodges, "New Options for Climate-Defensive Food Production," in *Climate's Impact on Food Supplies,* L. E. Slater
   and S.K. Levin, eds. Westview Press, 1981 .

WORLD FOOD NEEDS AND PROSPECTS

John A. Pino, Ralph W. Cummings, Jr.,
Gary H. Toenniessen

Agricultural Sciences
The Rockefeller Foundation
New York, New York, USA

ABSTRACT. The purposes of this paper--in response to the invitation of the conference organizers--are four-fold. First, we will review the components of the food-nutrition challenge--low production levels, rapidly increasing population, and unequal income distribution. Second, we will describe the alternatives for increasing output--bringing new land into production, narrowing the "yield gap" (the difference between what yields are and what they could be, on both a per crop and per year basis),and promoting new crops. These possibilities, developed responsibly, represent an enormous food reserve on which nations now must call. Third, we will introduce some of the considerations of climate which must figure prominently in future agricultural programs. Finally, we will describe the realistic basis for hope which should lead us into action and not despair with emphasis on the response necessary to enable nations to at least alleviate, if not altogether solve, their nutrition problems.

There is little doubt that it is possible to produce the amount and kind of food which, if distributed where needed, could meet the requirements of the world's population today and even in the year 2000. Indeed, if we set our minds and resources to the task, it would be possible to feed twice or more the number of people expected on this planet by the year 2000. However, it is not a question of whether or not it is possible--but of our willingness and ability to make the necessary commitment of funds and human talent.

W. Bach, J. Pankrath, and S. H. Schneider (eds.), Food-Climate Interactions, 47–68.

PRODUCTION, POPULATION, AND DISTRIBUTION

Half of the world's population is located in the developing market economies (otherwise known as the low-income countries).[1] Yet these countries produce only one-third of the world's total food, and that is not evenly distributed among the people within the countries. This is both the problem and the challenge (1).

The Human Diet

Plants contribute over 90 percent of the edible dry matter and three-quarters of the protein in the human diet. Three cereals--wheat, rice, and maize--account for well over half of this contribution. Millets, sorghum, oats, and barley are also important food grains. Other plants supplement cereal consumption. Roots and tubers (white potatoes, cassava, sweet potatoes, and yams) and bananas are major sources of calories in some areas. Pulses (cowpeas, chickpeas, and pigeonpeas) are valuable sources of protein. Vegetables, fruits, berries, and nuts provide vitamins, minerals, and bulk. Oilseeds (soybeans, cottonseed, ground nuts, sunflowerseed, rapeseed, and flaxseed) are major sources of energy and protein throughout the developing world. Sugar cane and (to a much lesser degree) sugar beets provide energy. About one-third of world output of cereals and roots and tubers and one-half of world output of pulses are produced in the developing market economies.

Some food crops are consumed by animals which produce meat, milk, eggs, and other products. Livestock also eat grasses, shrubs, and plants which are not used directly by man. Over half of the world's livestock (three-quarters of the buffaloes and goats, over half of the cattle, 40 percent of sheep, one-third of the chickens, and less than one-fifth of the pigs) are grown in the developing market economies.

A third source of food--fish, crustaceans, and molluscs--are important sources of protein in some locations. Approximately one-third of the world's catch is accounted for by the developing market economies.

Production Increases

For the world as a whole, total food production is almost double that of twenty-five years ago; food production has risen faster than population, and per capita availability is a third higher.

Contrary to what many people believe, over the past 25 years the percentage increase in total food production in the low-income countries, although not without ups and downs, actually has averaged a bit higher than that in the developed countries (Figure 1).

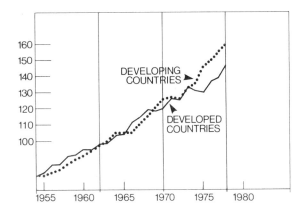

Figure 1   Index of Total Food Production
(1961 - 1965 = 100)

Source: USDA, World Agricultural Situation.
United States Department of Agriculture,
Washington, D. C., December 1980 (WAS-24)
and previous issues.

Among the commodities, the greatest production increases
have occurred in wheat, rice, sorghum, potatoes, and beans (starting
from rather low yields). The largest contributions have come from
higher yields, although in particular regions and for particular
commodities, new area under cultivation has also been important.
Generally more extensive production patterns characterize African
and Latin American agriculture compared to Asia.

Population Increase

However, increases in population in the low-income world have
also been great--rising now at more than twice the percentage rate
of the developed countries.

Per Capita Food Change

As a result, per capita food production in the low-income
world has risen slowly (0.4 percent per year) during the past two
decades. Average per capita production has increased or stayed
even for six consecutive years, 1973 to 1978.[2]   But the
progress has still been below needs in each region. Africa has
possibly lost ground, partly because of prolonged drought.
South Asia -- where the new cereal varieties have had greatest
impact -- has increased production but has been subject to
significant fluctuations in large part induced by the variable
monsoon. Latin America has depended upon increasing the area of
production as well as improving yields.

Limitations of Trade in Meeting Food Needs

Trade can and does make up for some of the difference between production and needs of the food-deficit countries. Over 10 percent (162 million tons in the 1975-77 average) of total cereal production is imported/exported. However, much of this cereals trade is between developed economies. For example, well over half of the value of U.S. agricultural exports (primarily cereals) is accounted for in purchases by Western Europe, Eastern Europe, the USSR, and Japan. Much of the imported grain, especially in the developed economies, is used for animal feed.

The low-income countries, many of whom were food exporters prior to World War II, have become increasingly dependent upon a few food surplus exporting countries for necessary cereals. But there are limits to which international trade can correct food consumption imbalances:

- Foreign exchange positions of most of the food-deficit countries are low (notable exceptions are the oil exporting countries).

- Food aid accounts for only 10 million tons of cereal annually (or about one-fourth of imports of the food deficit, low-income countries). There is little indication (especially with the current high world prices) that this quantity will increase significantly.

- The ability and willingess of the food exporting countries to expand production is limited; production costs are rising and environmental stress is being increasingly experienced as marginal lands are brought into cultivation and yields are raised. Higher domestic food costs resulting from strong export demands for food grains, are becoming an increasing cause for concern among consumers in exporting countries.

Ultimately, the major contribution to improving nutrition in low-income countries, therefore, must be from their own resources.

Net Food Situation

In some respects, progress in improving the world food situation has been encouraging. Death rates among children aged 1-4 (considered by many experts to be the best single measure of the state of nutrition) have fallen significantly. Life expectancy at birth for the less developed countries now averages 56 years.

Yet, when one looks below the surface, one can become less satisfied about the world food situation--the sense of accomplishment may be very limited. The food picture which emerges upon closer examination is one of:

o <u>Very uneven distribution of food among countries</u>. FAO estimates
that in 1975, approximately 450 million people in the world were
suffering from chronic insufficient energy and/or protein supply.
Calculating differently, the World Bank put the figure at above a
billion. Some observers feel that even this appalling situation
has been achieved under conditions of generally favorable rainfall--
suggesting that the number of persons suffering from hunger and
malnutrition under <u>normal</u> weather conditions would be much greater.
Most of the undernourished are in low-income countries. They
comprise 30 percent of the population in developing regions of the
Far East and 25 percent of Africa. The developing countries
consume only two-thirds the calories, one half the protein, and
one fourth the amount of fat per capita as the developed countries.
Furthermore, the developing market countries derive a much lower
proportion of calories, protein, and fat from animal products as
compared to vegetable products than do the developed economies.
Affluence tends to increase the demand for animal products which,
in turn, tends to increase the per capita use of grains for feed.

o <u>Very uneven distribution within countries</u>. Unmet food needs have
tended to be concentrated (a) in those sectors of the population
that are too poor to participate in the market exchange of food--
i.e. the urban unemployed, casual laborers, and landless agricultural workers; (b) among pregnant and lactating women, infants
and young children (who can incur physical and mental damage
making them unable to lead full and productive lives), and other
groups with great and/or specialized nutritional needs; (c) in
areas with a poor ratio of agricultural resources relative to
population; and (d) in areas that are poorly served by transportation and marketing systems.

o <u>Very uneven distribution of food at particular times</u>. In many
low-income countries, the seasonal rhythms of production and
inadequate marketing systems (especially transportation and
storage) combine to produce periods when food is scarce and
nutrition suffers. Adverse weather, pests, diseases, and
political disruptions repeatedly intervene--often with little
advance notice--to disrupt food production and distribution
at specific periods in specific low-income countries. For example,
in 1979 over 30 different countries with combined populations of
almost one-half billion experienced shortfalls in basic food
supplies below usual consumption requirements during at least one
month of the year. The nature of food reserves has changed significantly over the past few years: the U.S. no longer assumes the
role of residual supplier for the world, and most stocks in the
U.S. are not in any formal reserve program. There has been little
progress in building stockholding capacity in developing countries;
their share of stocks has even decreased from almost one-half of
the total held in 1976 to one-third in 1979. The low-income
countries are very vulnerable when production shortfalls occur.

Projections of Future Food Requirements

Unless improvement of the present trends are realized, the world food-poverty population situation is unlikely to ease, and probably will become worse.

Although there are encouraging signs of declining birth rates, population growth in the low-income countries averages over 2 1/4 percent annually and is not expected to fall below 2 percent until the end of the century. Income trends are more difficult to predict. However, together projected population and income suggest that food production in the future must rise at 3 1/2 to 4 1/2 percent per year to satisfy demand--a rate of increase greater than that previously attained.

Based on continuation of past trends, the International Food Policy Research Institute has projected that production of staple food crops in the developing market economies could fall short of the projected demand in 1990 by 120 to 145 million tons, as much as three times the imports of these countries in 1975 (2). The food deficit of the low-income, food short countries where the major problem exists and who lack foreign exchange to purchase imports, is projected to increase by six to seven times the present levels by 1990 (Table 1).

The low-income countries will have to make an extra ordinary effort just to maintain their current posture.

POTENTIAL FOR PRODUCTION INCREASE

The projected shortfalls need not and should not become a reality for when one looks at the potential for increasing food production in the developing world, the prospects are encouraging (3). Not all of the potentially arable land is being cultivated or otherwise utilized effectively. Yields on most farms are much below those which might be obtained. On a worldwide basis raw materials for chemical fertilizers are not in short supply. The potential of many "new" crops is yet to be exploited.

Exploiting New Areas

Arable Land. Calculations of the amount of potentially arable land in the world are highly speculative. The usual procedure is to identify those areas in which the physical, chemical, and biological properties of soils, the temperature patterns, and the amount and distribution of rainfall would support crop production with existing technology. Soils already under cultivation are included, as are those which could be put into production, some after clearing or irrigation and drainage.

Table 1   Projected Deficits of Selected Important
               Developing Countries - 1990

| | Actual 1975 | | Projected 1990 | |
	(million metric tons)	(percent of consumption)	(million metric tons)	(percent of consumption)
India	1.4	1	17.6-21.9	10-12
Nigeria	0.4	2	17.1-20.5	35-39
Bangladesh	1.0	7	6.4- 8.0	30-35
Indonesia	2.1	8	6.0- 7.7	14-17
Egypt	3.7	35	4.9	32
Sahel Group	0.4	9	3.2- 3.5	44-46
Ethiopia	0.1	2	2.1- 2.3	26-28
Burma	(0.4)*	(7)*	1.9- 2.4	21-25
Philippines	0.3	4	1.4- 1.7	11-13
Afghanistan	. . .	. . .	1.3- 1.5	19-22
Bolivia & Haiti	0.3	24	0.7- 0.8	35-38

*Surplus

International Food Policy Research Institute, Food Needs of
Developing Countries: Projections of Production and Consumption
to 1990. Research Report No. 3. Washington, D. C.: International
Food Policy Research Institute, December 1977.

    In one comprehensive exercise using this methodology, the
U.S. President's Science Advisory Committee (PSAC) in 1967
estimated that there are over 13,000 million hectares of ice-free
land in the world, of which 3,200 million hectares, or 24 percent,
were considered to be potentially arable (4).   Of these potentially
arable lands, 1,400 million hectares, or 44 percent, were already
under cultivation--that is, under crops, temporary fallow, temporary
meadows, market and kitchen gardens, fruit trees, vines, shrubs,
and rubber plantations. About 350 million hectares, or 11 percent
of the total potentially arable land, require irrigation to produce
any crop. On the remaining 2,800 million hectares, at least one
crop could be grown without irrigation (Table 2).

    Multiple cropping--the more intensive use of land by growing
several crops a year through the use of short-season varieties,
inter-cropping, and relay cropping--has immediate potential for

Table 2   Estimates of potential for expanding cultivated area in the developing and industrialized countries, 1974.

Developing and Industrial Countries	Actual Arable Area	1974 Calculated Potential Arable Land					1974 Arable Area as a Proportion of Potential Arable Land				
		FAO (IWP)	Club of Rome	Alan Strout	USA	Wagen ingen	FAO (IWP)	Club of Rome	Alan Strout	USA	Wagen ingen
		(million hectares)					(percent)				
Latin America	127	570	429	586	680	692	22	30	22	19	18
Sub-Saharan Africa	186	304	423	470	733	658	61	44	40	25	28
N. Africa/Middle East	80	80	86	80	n.a.	122	100	93	100	n.a.	66
Asia (Market Economy)	274	296	278	330	627	360	93	99	83	44	76
Asia (Centrally Planned)	132	n.a.	122	201	...	388	n.a.	92	66	...	34
Developing Countries	799	1250	1338	1667	2040	2220	64	60	48	39	36
North America	235	...	392	347	466	533	...	60	68	50	44
Oceania	46	...	150	70	154	275	...	31	66	30	17
Europe (Market Economy)	116	...	163	137	174	164	...	71	85	67	71
Europe (Centrally Planned)	278	...	382	280	356	491	...	73	99	78	57
Industrialized Countries	673	n.a.	1087	834	1150	1463	n.a.	62	81	59	46
World	1472	n.a.	2425	2501	3190	3683	n.a.	61	59	46	40

Source: Peter Oram, "Global Food Grain Production, Consumption, and Trade Trends in the 1980's." Paper presented at CIMMYT Long-Range Planning Conference, April 10-12, 1980, El Batan, Mexico.

extending the effective land base in densely populated areas. PSAC estimated that, over a considerable region, without additional irrigation, multiple cropping could increase the gross cropped areas (the cultivated area times the number of crops) to 4,000 million hectares annually, about three times the land presently being cultivated.

Furthermore, PSAC estimated that an additional 3,600 million hectares, 27 percent of the ice-free land of the earth, had some grazing potential, even though it was not potentially arable.

Irrigation and Drainage. Much of the land area of the world has adequate rainfall to grow one, two, or even three crops per year. In some of this area, drainage is a major problem. However, in other areas, supplementary irrigation is needed to grow even one crop or one high value crop. In almost all areas, controlled water management, by removing excess surface water or by adding to or supplementing deficient or unpredictable rainfall, can improve crop yield, increase the number of crops cultivated, and raise the level of management applied. Making more effective use of water for agriculture is becoming increasingly important as other demands compete for available supplies.

Attempts to estimate the world potential for new irrigation development, like estimates of potential arable area, are highly speculative. Methodological and research problems notwithstanding the PSAC study estimated that the gross cropped area of the world could be increased by the equivalent of 2,500 million hectares if irrigation water could be made available for two or three crops per year.

The gross cropped area of the earth could thus be extended to the equivalent of 6,600 million hectares or almost five times the land now cultivated.

Regional Land and Water Availability. Global data on land and water have little value except to establish a perspective from which to examine practical possibilities for increasing cropped area in specific locations.

In Latin America the physical resources in most countries are adequate to permit a substantial increase in food output over the next several decades.

Rugged terrain, often with inadequate transportation infrastructure connecting the fertile valleys, limits cultivation in Central America.

Great areas of potentially arable land, about which cropping characteristics are relatively little known, are yet to be brought under cultivation in South America (6). In particular, the Cerrados of Brazil, the Llanos of Colombia and Venezuela, and the forested Amazon Basin, are dominated by acid infertile soils often with concomitant toxicity due to high aluminum saturation.

Irrigation development and water management can play a very important role in growth of the agriculture of Latin America. In some areas on the western coast of South America water resources will soon be developed nearly to their potential. The same is true for the drier parts of Mexico. However, in other parts many large rivers have been only partially harnessed. From a climatic standpoint the large Amazon River flood-plain is ideal for rice production with two or more crops per year certainly possible if soil nutrient and toxicity problems can be resolved. Development of irrigation and drainage in the Guayas Basin of Ecuador could extend the introduction of a modified Asian-type rice culture to that area.

Africa also has vast underutilized lands available for agricultural development (7). There are significant areas of tropical rainforest, mainly in Zaire and Gabon. However, most of the potentially arable land in Africa is classified as semi-arid with lack of water being the most important physical limitation. There is considerable potential to expand irrigation in some of this area, but such development will be a long-term process. In most of the semi-arid countries of Africa, there is a severe shortage of national institutions and trained personnel working on agricultural development. Diseases of livestock and people are another major deterrent.

In Asia the potential for expanding cultivated land is more limited. For most countries, future food needs will have to be met by more intensive use of existing farm land. The most notable exception is the outer islands of Indonesia. Although the quality of the agricultural resource base on the outer islands is inferior to that of Java, with problem soils being a major constraint, development of sustainable permanent agricultural systems on these vast underdeveloped islands could significantly reduce Indonesia's food deficit and help to stem the current process of non-sustainable forest exploitation.

In summary, there clearly is great potential for expansion of cultivated area in some developing parts of the world. Except for a few instances in Africa where disease limits development, it is probably safe to say that nearly all of the world's prime agricultural land is now in production or has been converted to other uses. However, large expanses of not-so-prime, but potentially arable land remains highly underutilized in many developing countries. More intensive use of these resources without environ-

mental damage will require large investment in roads, inducements
to people to settle permanently, provision of irrigation or
drainage, organization of social services, much additional
research, and government policies which provide appropriate
incentives and remove nontechnical constraints.

Closing Yield Gaps

Contributions of area and yield to increases in total
agricultural output will vary from country to country depending on
the ease with which additional land can be planted to crops and
the spread of improved technology. For most of the densely
populated countries, however, there is only one option: yields
per hectare must be raised. With appropriate policies and inputs
this should be possible since yields of staple food crops in most
developing countries remain only a fraction of what they could
be.

Differences Among National Yields. Comparison of the
highest and lowest national average yields suggests a 10 to
20-fold spread between extremes (Table 3). The average yields
for the developing countries as a group are much lower than the
world average. This is revealed by an examination of the yield
levels currently being achieved by nations for some representative
crops. For example, for wheat, the highest national average
yields--over 5 tons per hectare--are obtained by The Netherlands
and Denmark. Seventeen other countries, mostly in Europe, have
yields of 3 to 5 tons. The remaining 73 wheat growing countries
for which FAO reports yields harvest less than 3 tons, of these,
37 obtain only 1 to 2 tons, 22 achieve less than 1 ton, and
four are under 0.5 ton! As another example, yields of cassava,
one of the major sources of calories for the poor in the drier
tropics, are reported by 71 countries. Average yields exceed 10
tons per hectare in 22 countries, the highest being Malaysia at
17.9 tons. Thirty-nine countries average 8 tons or less, and 16
of them less than 4 tons.

Best National Yields vs. Experimental Potentials. There also
is a wide gap between the highest national average yields and the
top yield achieved experimentally. It would be unreasonable to
expect national averages to match experimental potentials which
are usually achieved under exceptionally favorable circumstances
which cannot be duplicated over wide areas. But examination of
these data suggest the substantial scope for improvement existing
in most countries.

Improving Yield Potential. Substantial improvements
in potential productivity continue to be made through the traditional breeding of new varieties. This capability is being

Table 3   National and world average yields
(average 1974-76)

	Number of producing countries	Yield in tons per hectare		
		Highest national average	World average	Lowest national average
Cereals				
Wheat	92	5.4	1.6	0.3
Rice	111	6.1	2.4	0.4
Maize	131	8.0	2.8	0.4
Sorghum	72	4.3	1.2	0.4
Millets	65	3.9	0.7	0.3
Barley	74	4.6	1.9	0.3
Rye	39	4.2	1.7	0.2
Oats	51	4.6	1.6	0.2
Food Legumes				
Dry beans	82	2.4	0.5	0.1
Broad beans	36	3.5	1.1	0.2
Dry peas	54	3.8	1.2	0.2
Chickpeas	33	1.9	0.6	0.3
Lentils	28	1.6	0.6	0.2
Soybeans	39	2.0	1.4	0.2
Groundnuts	85	5.6	1.0	0.3
Oilseeds				
Castor beans	26	1.9	0.5	0.2
Sunflowers	39	1.9	1.1	0.3
Rapeseed	32	2.9	0.8	0.4
Sesame	53	1.3	0.3	0.1
Linseed	33	2.1	0.4	0.2
Cotton seed	81	3.2	1.2	0.2
Root and Tuber Crops				
Potatoes	102	39.8	13.4	2.3
Sweet potatoes	78	24.2	9.2	0.3
Cassava	71	17.9	8.9	1.8
Fibers				
Flax	11	1.1	0.4	0.3
Jute	23	3.0	1.5	0.7
Hemp	18	4.8	0.6	0.3
Sisal	23	1.3	0.8	0.2
Vegetables				
Beans, green	41	13.9	6.0	0.6
Cabbage	57	59.5	17.9	1.7
Carrots	46	47.0	21.0	5.4
Cauliflower	32	44.8	12.9	6.6
Cucumbers	40	206.5	14.9	2.1
Eggplant	23	32.8	12.5	2.5
Garlic	32	19.9	5.1	1.4
Melons	33	38.5	13.3	4.0
Onions, dry	77	34.7	11.3	1.8
Peas, green	41	25.0	5.9	0.7
Peppers, green	37	32.7	8.2	0.9
Pumpkin, squash	41	38.7	5.3	2.0
Tomatoes	92	164.0	20.0	0.7
Watermelons	45	40.1	12.2	3.5
Beverages				
Coffee, green	63	3.5	0.5	0.2
Cocoa beans	46	2.2	0.3	0.1
Tea	27	2.1	1.0	0.4

Source: FAO, _Production Yearbook_. Rome: The Food and Agriculture Organization
of the United Nations, Volume 30, 1976.

* National yields must be compared cautiously. To compensate for
fluctuations due to weather or other causes, three years -- 1974, 1975,
and 1976 -- have been averaged. A crop may be grown under quite different
conditions in different countries; for example, wheat in some countries
depends on scanty rainfall while in some others it is grown with irrigation.
Only countries with significant areas under production provide meaningful
measurements of country yield levels (we have chosen at least 100 hectares).
But these cautions notwithstanding, the data as a whole probably represent
a useful indication of the orders of magnitudes of the gaps.

further strengthened by exciting new biological cell culture and genetic techniques. The value has already been proven in the case of triticale--a man-made cereal grain resulting from wide crosses between wheat and rye. Triticale combines the grain quality of wheat with the stress tolerance characteristics of rye. Its commercial production continues to expand.

Much of the increase in productivity attained in past years was achieved without having knowledge of the fundamental biochemical and physiological mechanisms involved. An improved understanding of these mechanisms is now being rapidly generated and offers the promise of future increases in yield and stability based on manipulation of basic components of life systems. Basic research has already enhanced our understanding of photosynthesis to the point where it is feasible to explore methods for improving the efficiency of the fundamental but inefficient process on which essentially all food production is based. Investigation is continuing on all components of the nitrogen cycle. Fundamental knowledge of the hormonal systems of plants and animals is contributing significantly to agriculture. Knowledge of the mechanisms by which resistance to pests and pathogens is accomplished is expanding and providing opportunity for improved plant protection. Through further research, a wide variety of pest control techniques will become components of integrated pest management systems.

Supplying More Fertilizers. Substantial amounts of nutrients for plant and animal growth are supplied by nature. Nutrients are returned to soils by decomposition of vegetative and animal matter and are moved from place to place by irrigation water and runoff. These natural supplies by themselves, while valuable, are not adequate to support intensive production, let alone the higher levels of production needed in the future. Therefore, chemical fertilizers will remain essential to improvement of agricultural production.

Chemical fertilizer consumption is still quite small in most developing countries and on many small farms is not used at all. Average use in the developing countries of Africa (5 kg/ha) is only about one-twentieth of that in the U.S. (85 kg/ha) and one-fortieth of that of Western Europe (190 kg/ha); average dosages in developing Asia (20 kg/ha) and developing Latin America (30 kg/ha) also are low (Table 4). Moreover, in these developing regions, a high proportion of the fertilizer is used on export crops. There is considerable scope for increasing basic food-crop yields in the developing countries through higher fertilizer applications.

Availability of raw materials should not be expected to pose major problems in the expansion of production of nitrogen fertilizers. Although the principal feedstocks--natural gas, naptha, heavy oil, and coal--are in heavy demand for generating power, heating buildings, running vehicles, and manufacturing petro-chemicals, constantly expanding world supplies appear to be adequate for fertilizer manufacture, at least until the end of the century. Prices are, however, expected to rise.

Phosphate rock is available in large quantities in several areas--the U.S., the USSR, and North Africa. Shortages of sulfur (used in the manufacture of phosphatic fertilizers), one of the most plentiful minerals, should be overcome because of price incentives and new mining techniques now in the development stage.

Potash, the third major fertilizer, is mined directly. Abundant deposits are located in Canada.

Promoting New Crops

As noted earlier, man gets most of his nutrition from a relatively small number of commodities. Yet throughout history, as many as 3,000 plant species have been used for food. Most of these are still being grown and, in some localized areas, "new crops"--crops which are not really new but which are not now widely known as food sources--still play an important role in the food system of the people.

The apparent advantages of cereals and export crops over minor tropical plants often have been the result of the dispro-portionate research attention the former have been given. Many of the minor tropical plants appear to be efficient sources of energy, protein, vitamins, and minerals and can thrive, or at least survive, in difficult agro-climatic environments. For example, in its review of "Underexploited Tropical Plants" The U.S. National Academy of Sciences cites quinua, one of the most productive sources of plant protein (8). It grows high in the Andes where few other crops can survive. The Spanish introduced wheat and barley. Subsequent agricultural research has focused on these crops which eventually replaced quinua. Despite its intrinsic nutritive and economic value and the fact that protein deficiency is a serious problem in its native region, the agronomy of quinua has advanced little in the past four centuries.

The potential of many of these tropical crops has never been adequately explored.

Table 4   Fertilizer consumption per hectare, 1972/73 *

Over 200 kg/ha		25-50 kg/ha (cont.)	
Developed Nations		Developing Nations	
Netherlands	France	Brazil	Algeria
Belgium	Czechoslovakia	China	Mexico
Germany, West	United Kingdom	Peru	Guatemala
Switzerland	Denmark	Kenya	Honduras
Japan	Austria	Chile	Indonesia
Germany, East	Poland	Vietnam (North)	Nicaragua
Ireland	New Zealand	Uruguay	Panama
Norway			
Developing Nations			
Korea (South)			

0-25 kg/ha	

100-200 kg/ha	
Developed Nations	

Developed Nations		Canada	
Hungary	Bulgaria	Developing Nations	
Sweden	Itlay	Turkey	Burma
Finland		Bangladesh	Saudi Arabia
Developing Nations		Pakistan	Iraq
Korea (North)	Egypt	Morocco	Cameroon
Israel	El Salvador	Philippines	Afghanistan
Lebanon	Trinidad and	India	Uganda
Albania	Tobago	Venezuela	Khmer Republic
		Thailand	Tanzania
		Iran	Ghana
50-100 kg/ha		Nepal	Ethiopia
		Ecuador	Mali
Developed Nations		Syria	Haiti
Greece	Spain	Tunisia	Nigeria
Yugoslavia	Romania	Sudan	Zaire
U.S.	Portugal	Mozambique	Yemen (North)
Developing Nations		Madagascar	Upper Volta
Dominican	Costa Rica	Bolivia	Burundi
Republic	Colombia	Angola	Jordan
Cuba	Malaysia	Laos	Central African Rep.
Rhodesia	Sri Lanka	Argentina	Congo
South Africa	Jamaica	Chad	Guinea
Vietnam (South)		Benin	Lesotho
		Ivory Coast	Libya
25-50 kg/ha		Liberia	Rwanda
		Malawi	Sierra Leone
Developed Nations		Senegal	Togo
Australia		Somalia	Paraguay
U.S.S.R.		Zambia	Papua-New Guinea
		Indonesia	

Source: FAO, Annual Fertilizer Review, 1973. Food and Agriculture
         Organization, Rome, Italy.

* Fertilizer as NPK; based on total arable land; includes all countries
with over one million population in 1973. Figures for the Netherlands,
West Germany, East Germany, France, the U.K., and Denmark are somewhat
misleading because much of this fertilizer is used on pasture land and
not on crop land.

CLIMATIC FACTORS

Climate has always been the most unpredictable and hazardous component of agriculture and will continue to be so in the foreseeable future. While it may be difficult for man to control climate, as such, agricultural practices attempt to mitigate the adverse effects of climate on plant and animal growth, for example, through irrigation or changes in cropping systems.

One of the major accomplishments of modern agricultural research is the development of technologies which increase crop yields and stability and do so under a broader range of climatic conditions. For example, the new high yielding varieties of major grain crops are less susceptible to climate variations than are traditional varieties and they are grown in numerous and very different geographic regions. However, expanded use of more stress tolerant varieties may not necessarily reduce the overall risk of climate-induced food shortfalls since it has allowed the expansion of agriculture into areas and situations where there is a greater probability of adverse climate.

During the last few years, considerable concern has been expressed that the earth's overall climate may be changing in ways which will present a different environment for future food production. For example, due to the burning of fossil fuels, estimates project a doubling of the CO_2 content of the earth's atmosphere by some time in the first half of the next century. This topic will be discussed in detail during the remainder of the workshop. For present purposes, the probable climatic changes which might perhaps have a significant effect on agricultural production could be grouped into four categories:

o CO_2 enrichment of the atmosphere. The present atmosphere CO_2 concentration is a limiting factor in the photosynthetic process, hence the anticipated direct effects of atmospheric CO_2 enrichment on agriculture would be expected to be favorable, although the magnitude of the effect cannot be predicted.

o   Rise in temperature. The projected rise in temperature in the equatorial region is expected to be on the order of 1° to 2°C. On the balance, the direct effect of a temperature increase of this order of magnitude need not cause much concern in the low-income countries which are primarily located in the tropics and subtropics. On the other hand, greater changes projected for temperate countries could increase evapotranspiration rates significantly, and in the absence of offsetting increases in precipitation, could cause some intensification of moisture stress, especially in the sub-humid, seasonally dry, semi-arid, and arid regions.

o Changes in moisture regimes. An indirect effect of the
   overall rise in temperature, unevenly distributed over the
   earth's surface, could be changes in air circulation patterns
   and precipitation rates and their distribution over time and
   over the earth's surface. This could either increase or
   decrease moisture stress in various parts of the LDCs, with
   more critical changes likely in the arid, semi-arid, and
   sub-humid regions. Scenarios depicting geographical distribution of increased and decreased precipitation have been
   attempted but are considered to be speculative. Likewise,
   the future effects on production can only be guessed.

o Increase in moisture variability. There is a possibility
   that, associated with the temperature changes, moisture
   variability could increase, resulting in changes in seasonal
   distribution of precipitation. The occurrence and frequency
   of such changes and the effects on production can only be
   guessed. In times of drought, the already poor and malnourished are usually the first to suffer the effects of the
   shortage.

   There is also increasing evidence that the extent and
intensity of human activities, including agriculture, may alter
long-established relationships between the biosphere and the
atmosphere. The burning of fossil fuels, large scale industrial
fixation of nitrogen, extensive forest cutting, and increased
atmospheric dust levels have all been cited as possible anthropogenic modifiers of these relationships. Hence agriculture may
influence climate as well as being influenced by climate. This
possible feedback from agriculture to climate must be an important
consideration in planning for the future.

THE RESPONSE

   Elimination of hunger and malnutrition is a -- perhaps the
-- major challenge facing the world today.

   Achievement of the goal of adequately feeding the world is
technically feasible: many elements of successful approach are
known, it is likely that the additional problems can be solved as
they occur, a strategy for taking the offensive against hunger and
rural poverty is emerging, and the time is appropriate to embark on
a far-reaching effort. Several points reinforce this optimism:

o The physical and biological potential of the world to produce
   food can still be expanded. In some cases, technologies must
   be developed or adapted to realize this greater potential. A
   three-tiered, mutually reinforcing international system --
   comprising national agricultural research and production

programs,international agricultural research institutes, and
laboratories and research centers in developed countries --
now covers most of the major food crops and animals and extends
to most areas of the developing world. It has, and if adequately
supported, can continue to make a major contribution toward the
development of necessary new technologies.

o Experience clearly demonstrates that rural stagnation is not
   inevitable. Several poor countries with widely divergent
   ideologies -- for examples, China, Taiwan, India, Turkey,
   Colombia, Mexico, Kenya and the Philippines -- have achieved
   accelerated agricultural progress.

o Recognition and commitment by national governments is improving.
   Although there are not yet adequate data for a thorough analysis,
   World Bank and FAO figures for twenty-three developing countries
   indicate that average annual investments in agriculture increased
   in all but two between 1971-1973 and 1974-1976; seventeen of the
   twenty-three countries had increases in real terms. For a number
   of countries, the share of agricultural investment in total
   investment also increased.

o International assistance agencies have increased their commitment
   to agricultural and rural development. The external financial
   resources required to raise the agricultural production growth
   rate in the food-deficit countries to the U. N. target of 4
   percent annually are estimated by the World Food Council to be
   $8.3 billion in 1975 dollars, of which concessional flows would
   have to account for $6.5 billion. The Overseas Development
   Council recognizes encouraging signs that the concessional target
   may be reached as early as 1981 (10). Total external resource
   flows directly committed to food and agriculture in 1978 were
   estimated to have reached $5 billion in constant 1975 prices, an
   increase of one-fourth over 1977.

    It is fair to ask why, in the face of these positive factors,
so many programs fall short of declared goals --why so many people
in the world remain hungry and malnourished. Among the reasons are
the following:

o   Short-range planning
o   Unrealistic approaches,
o   Inadequate political and professional organization,
o   Burden of bureaucracy which inhibits change and timely action,
o   Cultural and institutional impediments, and
o   Some limits to specific technologies.

    With regard to the subject matter of this conference, although
theoretical studies confirm the prospect of a significant CO_2-
induced warming of the earth with subsequent influences on

moisture patterns over the next 50 to 75 years, there are large
uncertainties in the projections. The most that can be said is that
"there is enough possibility of long-term risk for a prudent person
to take reasonable precautionary action" (9). For low-income,
food-deficit countries this means acquiring and maintaining a
substantial degree of resilience and adaptability to shifts in
rainfall patterns, changes in temperature, and to the extremes of
weather that may accompany climate change. Modeling and simulation
may help to define the problem, but preparing an appropriate
response will require extensive applied research and field testing.

However, the pressures and needs of the low-income countries
have already made agricultural development a major world goal. The
possible threat of climate change, uncertain though it is, only
reinforces this need. It is clear that agriculture in the developing countries must become more resilient and adaptable to cope with
increasing stress and rising population and demand, as well as
extremes of climate.

Better understanding of climate impact on food production is a
necessity. The need to establish an international buffer supply
takes on increasing importance. Every nation, especially the environmentally and climatically sensitive ones, should establish its own
food reserve. Efficient management of water resources is essential
to the viability of farming. Irrigation and drainage should be
extended and improved wherever possible and needed, inasmuch as
this offers some security against variable precipitation. Cropping
patterns can be adjusted and agronomic practices can be improved.
Identifying, preserving, and more effectively using genetic resources
-- both to improve present crop plants and develop new species --
is an important and urgent goal. Storage should be improved; waste
should be reduced. In short, the best insurance against possible
future climate change are the kinds of improved agriculture and
agro-industrial systems that must be developed in any case.

However, if the objective of eliminating hunger and malnutrition is to be achieved, major changes must be made in how
development assistance is pursued. To be more specific, the
following steps must be taken:

o The goal of eliminating hunger and malnutrition through
   the world (or at least in as many countries as are willing to
   cooperate) must be adopted as an explicit target against
   which success or failure can be measured.

o The assistance agencies must reorganize and mobilize their
   resources -- human and financial -- for an all-out assault on
   food problems in the most critical areas.

o Each developing country must decide whether it is willing to
   make the commitment in terms of money, policies, and people
   necessary to pursue the goal.

o Those assistance agencies and those developing countries which
   truly wish to achieve the goal must start now to plan and
   implement action programs.

     The implications of these steps are clear -- priorities
must be established and pursued:

o The developing countries must mobilize -- almost on a war
   footing -- to eliminate hunger and malnutrition within their
   borders. What needs to be done is reasonably well understood.
   The problem is how to do what we know. Hard choices must be
   made in order to implement the requisites for increasing
   productivity through commodity production programs, defined-
   area campaigns, and synchronization and reorientation of
   services. The underlying social systems, including land tenure
   and organization of farm enterprises, must be evaluated
   and, if necessary, changed. In some areas, nutrition intervention schemes, including subsidies to increase purchasing
   power of the poor, might be introduced temporarily to supplement
   production efforts. Recognition must be given to the fact that
   the basic response must be made by the low-income countries
   themselves; the developed countries can assist -- but that is
   all that they can or should do.

o The assistance agencies must be placed under the leadership of
   the most knowledgeable and experienced people in the field and
   staffed with trained agriculturalists pursuing long-term
   careers. Hard choices should be made regarding what countries
   to support and what aspects (and in what ways) of the programs
   should be given support. Political or national security
   motivations must be kept separate from genuine efforts
   to improve agricultural and rural development.

     The response is simple and logical yet it has defied positive
action. The reason is also probably simple and logical. Accelerated agricultural development involves a process of economic
participation which seems to many to be even more difficult and
painful to achieve than political enfranchisement. Many countries
(and assistance agencies) have elite minorities, often in positions
of power, who perceive change as being inimical to their interest
and therefore to the national and international interests.
That need not be. Equity as well as prosperity, is in the
long-run interest of most, if not all, elements of society.
That message must get through.

In summary, to deal in global figures does not, in reality, deal with the food and nutrition problem where it occurs. To ask agriculture to sustain three or five percent annual growth in production or to double production in eighteen years ignores the complexity of the food system. The amount of food which is ultimately consumed is the culmination of a multitude of factors affecting production, harvest (and/or loss), storage, marketing (trade), and finally its accessibility to consumers. There is no formula of what to do to combat hunger and malnutrition which is universally applicable. The best way to respond will differ among countries, among regions within countries, and even among localities within regions. Careful consideration must be given to the socio-economic realities and natural resource base unique to each situation. The world has the capability to deal effectively with the myriad of complex factors which must be taken into consideration to respond to the challenge. The question before us is whether the world can mobilize the commitment to respond to the challenge.

NOTES

1. Another (almost) quarter of the world's population lives in the Asian centrally planned economies, notably China, which are also "low-income." These countries produce approximately 20 percent of the world's cereals. We generally will not focus on this very important group of countries in this narrative.

2. Even though food production declined in 1979, prospects for 1980, at this early date, appear promising.

3. Comparable estimates from other sources are shown in Table 2.

4. About 14 percent of cultivated land is irrigated presently.

5. This section (and later comments on climatic factors) borrows, in part, from the final report of the AAAS Workshop on "Climate Change and Agricultural Production in Developing Countries" to which the authors contributed (9).

References

1. Sterling Wortman and Ralph Cummings, Jr. To Feed This World: The Challenge and the Strategy. Baltimore: Johns Hopkins University Press, 1978.

2. International Food Policy Research Institute, <u>Food Needs</u>
    <u>of Developing Countries: Projections of Production and</u>
    <u>Consumption to 1990</u>. Research Report No. 3, Washington,
    D. C.: International Food Policy Research Institute,
    December 1977.

3. DSE, GTZ, BMZ, AND The Rockefeller Foundation. <u>Agricultural</u>
    <u>Production: Research and Development Strategies for the</u>
    <u>1980's</u>,

    o   Conclusions and Recommendations of the Bonn Conference,
          October 8-12, 1979.
    o   Soils by C. F. Bentley, H. Holowaychuk, L. Leskiw, and
          J. A. Toogood.
    o   Water by Gilbert Levine, Peter Oram, and Juan Zapata.
    o   Biological Resources by Haldore Hanson.
    o   Energy by Gerald Leach.

4. U.S. Presidents' Science Advisory Committee. <u>The World</u>
    <u>Food Problem</u>. Washington, D. C.: U.S. Government <u>Printing</u>
    Office, May 1967.

5. Peter Oram. "Global Food Grain Production, Consumption,
    and Trade Trends in the 1980's," paper presented at CIMMYT
    Long Range Planning Conference, El Batan, Mexico, April
    10-12, 1980.

6. Michael Nelson. <u>The Development of Tropical Lands: Policy</u>
    <u>Issues in Latin America</u>. Baltimore: Johns Hopkins
    University Press for Resources for the Future, 1973.

7. Uma J. Lele. <u>The Design of Rural Development: Lessons</u>
    <u>from Africa</u>. Baltimore: Johns Hopkins University Press,
    1975.

8   National Academy of Sciences. <u>Underexploited Tropical</u>
    <u>Plants with Promising Economic Value</u>. Washington, D. C.:
    Report of an Ad Hoc Panel of the Advisory Committee on
    Technology for International Development, Commission on
    International Relations, National Academy of Sciences,
    1975.

9. Lloyd Slater, rapporteur for Report of Workshop on "Climate
    Change and Agricultural Production in Developing Countries,"
    sponsored by American Association for Advancement of Science
    and held at Oak Ridge, Tennessee, June 19-20, 1980.

10. John W. Sewell and staff. <u>The United States and World</u>
    <u>Development: Agenda 1980</u>. New York: Praeger, for The
    Overseas Development Council.

POPULATION GROWTH, NUTRITION AND FOOD SUPPLY

Georg Borgstrom

Department of Geography and
Department of Food Science and Human Nutrition
Michigan State University, East Lansing, Mich. USA.

ABSTRACT. The world food scene is dominated by four powerful forces: the population growth, the accelerating affluence in the industrialized countries, the rapidly increasing armies of the destitute in the developing countries, and the unrestraint urban growth. The lack of historical and biological perspectives has led to two basic fallacies in the evaluation of the world food problems. Historically Europeans were able to temporarily resolve their food and population problems through emigration. In biological terms all people living now could get an adequate diet if feed crops were not substituted for food crops. It is shown that such metaphors as triage, lifeboat operations and the abuse of the commons are misleading and that they distract from the real nature and magnitude of the world food crisis.

Four mighty forces are dominating the world scene, all of which are highly upsetting, not least on the food front:

(1) the population growth which will add almost one billion to the world in the brief span of ten years, 90 per cent of which will belong to the developing countries. Asia, prior to the year 2000, will be adding hundreds of millions to the total of those that now live in the developed world (1.2 billion);

(2) the accelerating affluence in the industrialized world (a minority of one-third and a tiny upper middle class in the poorest nations) which is chiefly responsible for the growing use of fuel, metals, forest products and other natural resources at a rate of 2 to 3 times that of the increase in world population;

W. Bach, J. Pankrath, and S. H. Schneider (eds.), Food-Climate Interactions, 69–79,
Copyright © 1981 by D. Reidel Publishing Company.

(3) the multimillion and rapidly growing armies of destitute
(homeless, penniles, jobless, landless, and hungry) who are the
prime target victims when the world is hitting harsh limits in
land, water, energy and other resources. Their true requirements
are not recognized in any field, certainly not as far as food is
concerned, where our programs are primarily based on "effective
demand"(translated purchasing power). Food production needs to
treble or rather quadruple within this century to meet nutritional needs and merely keep up with population growth;

(4) the impact of these forces is further compounded by a
rampant urbanization, which, within a brief period of 20 years
will add more than 1,350 million to the cities, no less than
1,100 of them in the poor world. The greater portion will ensue
from a strident rural influx. Many major cities are growing 2,3
to 4 times as fast as the country in which they are located,
doubling in 8 to 12 years. The slums are in most cases growing
twice as fast as the cities. All evidence points to collapse.
Food and water are taken for granted and assumed cared for through
traditional market forces. No viable urban ecosystems are created,
as is evidenced by the enormous backlog in housing and sewage
systems.

Two basic fallacies in evaluating the world food issue origi-
nate in our lack of historical and biological perspectives. We
westerners tend to believe that we have shown much more restraint
than the other branches of the human family and that we are so
much smarter in caring for our needs. We forget that in the hundred years from 1850 to 1950 we grew fast in numbers and swarmed
over the entire globe to find outlets for Europe's surplus population and food and other commodities for the Europeans that
stayed. Some hundred million left Europe, half of them for North
America, and some 15 million for other parts of the western hemisphere, 25 million returned. In this grand operation western man
doubled his tilled land, trebled his pasturelands and increased
manyfold his forestlands, and thus temporarily resolved his food
and population dilemma.

In biological terms the capacity to feed the world is not 4.5
billion people, but rather 21 billion, when including the livestock and poultry, as gauged in their protein intake. Of these
9 billion (42%) belong to the "Satisfied World". The United States
has a feeding burden which in these terms amounts to a population
which is equivalent to 1.5 billion. This is 0.3 billion more
than India (1.5 billion) with three times more people than in
the US. Hence the Satisfied World(SW), due to excessive consumption, deprives the overpopulated areas not only of food but of
many other resources. What the US gets from the world household
would suffice to provide for the following number of East Indians:
food, 980 million; other resources such as metals, 10,5 billion;
energy, 15 billion.

Another way of illustrating the food gap as part of the poverty gap is by looking at our average diet. In recent years the US
has reached the world pinnacle of an average intake of 72 grams
of animal protein per person and day, which is three times more
than any nutritionist would consider needed. But people do not
eat averages, and we have on one hand detrimental excess intakes,
not only of protein but also of largely empty calories in the
form of sugar and fat, on the other hand about 30-40 million
Americans who do not earn enough to buy a satisfactory daily
diet - food stamps, school lunches, and other programs notwithstanding. In global terms there are less than 500 million people
who can afford to eat on the US level. If all the food now available on the globe was distributed equally, it could feed 970
million people at the US level, or less than one-fourth of the
present world population. If food crops were substituted for
feed crops in world agriculture, all people living now could
get an adequate diet. But this would not be true for very long
in view of the rapid population growth. The magnitude and the
dynamics of this development has to be taken into account, even
if a minimal quality of life is our goal.

We should put an end to the empty talk of the United States
feeding the world. North America now accounts for 80 to 90 per
cent of the deliveries to the world cereal market. But two-thirds
of the grain flow is to Europe, Japan, and the Soviet Union, not
to stave off hunger but to underpin their nutritional prodigality (to feed livestock rather than man). The most conspicuous
case is soybeans often touted as a major US contribution to alleviating world hunger, but again Europe and Japan are the chief
recipients. Currently less than 5 per cent of the exported soybeans serve as human food, and even ·less reach the "Hungry World"
(HW). Furthermore, what that world receives in food through trade
and aid is in most cases counterbalanced by its deliveries of
cash crops or feed products, i.e. to the well-fed nations, more
than 2 million tons of the available oilseed protein.

How often do we ponder the fact that through mechanization
we added yet another continent to the feeding base of the affluent world? More than 500 million ha became available for the
growth of crops and food producing livestock. Besides, this one
time trick, the benefits of which were reaped almost within one
single generation, could hardly be repeated. This was a potent
force in Europe and the USSR as well as in North America. No
wonder nine-tenths of all tractors are in the SW. France has
more tractors than all of Asia. A similar involute transgression
of limits - with lasting burdens on the running expense account -
is exemplified in key operations such as large scale irrigation
and the use of fertilizers (around 25 million ha). Frequently
figures are given computing the acreage gains which can be attributed to these innovations but rarely are they judged as basic

indicators of our evasive operations in the face of limitations -
but at what price?

Far more fundamental to evaluating our proximity to the limit
is the fact that the third force, the hungry millions mentioned
at the beginning, figures poorly in most analyses, as a rule
restricted to arbitrary demand modules. Besides, glib rhetoric
is evading the real issue by simply claiming that unemployment
is replacing food as the overpowering issue of today, thus failing to see the harsh realities of both.

Both the phenomena of famine and mass starvation need to be
redefined. Through centuries, whenever crop failures hit, whether
through droughts, floods, pests or diseases and in some cases
due to wars, these phenomena were clearly regional and hit almost everyone in the area. Transportation, in particular within
the long-range category, changed all this. Food scarcities nowadays only rarely emerge in this manner. This has led to the
prevalent use of embellishing terms and an evasive phraseology.
New definitions are urgently needed to restore meaning to such
basic terms as self-sufficiency, food reserves, surpluses, and
others. Loose talk along these lines befuddles the debate but,
still worse, affects aid and policies. People do not eat averages and even in well-to-do countries many are on sparse, inadequate diets. There are many in Italy who subsist or starve on
Asian diet levels, and others who eat as well as the affluent
in the US. In both Argentina and Brazil there are millions on
the North Korean or the Indonesian level of food intake. Hunger
is the lot of many Americans.

The situation is far more critical among the vast masses of
Asia, Africa and Latin America. More food does not necessarily
mean feeding more people or even filling crucial deficits. It
may merely mean feeding some - often only a small minority -
better. Distribution needs to be given far greater consideration on all levels, globally, regionally, and nationally as well
as within cities, communities and villages. Talking in global
terms, the postwar era until quite recently has been, dwelling
within the lofty confines of average statistics, rarely ever
making an effort to count the victims, the forgotten millions.
It is among these poor and undernourished that we find the suffering caused by our extravagance during the seventies. Finally
the common use of percentage figures contributes to a highly
misleading illusionary picture of reality.

It is even more startling that affluence, the second force
and the predominant one, is earmarking two-thirds of the gains
in the world household. The bitter lesson of the seventies is
that, despite all rhetoric, when the world was moving to the
brink, we in the affluent world allowed ourselves not only to

continue our feasting, the level of which had been escalating
ever since the mid-sixties (meat production in SW increased
49 %), but it culminated in the most lavish banquet the world
ever saw. Despite the energy and money crises, beef production
in the SW went up by 6 % and pork by 5 % from 1973 - 1975.

Limits can be clearly perceived in the realm of nutrition.
Measured in terms of animal protein intake the US has continued
to climb hastily up the ladder of nutritional prodigality, reaching a pinnacle in the world of no less than 72 grams per person
per day, 3.5 times more than any nutritionist would claim is
necessary. This is the most spectacular transcendence of limits.
A fully acceptable diet can well be composed exclusively on the
basis of plant protein. Limits have arbitrarily been drawn on
the basis of our affluence and, besides, are resulting in over-
consumption.

When analyzing the individual categories of domestic animals
one finds that although Asia has most of the world's cattle and
goats, the ratio between them and man is extremely unfavorable.
There is no question that in human history a constant adjustment
has been and is being made in this ratio in order to allow a
maximum of efficiency in utilizing the harvests of the soil.
On the whole, the world has already moved very far in this respect and today can be classified as being almost vegetarian.
Very few people living in the well-to-do largely industrialized
world realize that only one-tenth of the calorie intake of the
world household consists of animal products. In terms of protein,
the ratio plant to animal products is 2:1, the latter largely
being enjoyed by one-third of the human family.

One essential biological conclusion that appears to be justifiably drawn from these figures is that a complete eradication
of all livestock would immediately create the means for the feeding of an additional 14.5 billion people. Such reasoning is, however, fallacious, since it overlooks the basic fact that large
areas of the world could not be ploughed or cropped or harvested
to provide plant products for direct human consumption. Secondly
the plant kingdom basically is not organized to fill man's need.
Man with his monogastric system therefore belongs to those animal groups that have considerable difficulty utilizing plant
products directly. The plant cell is encased and not immediately accessible to him. Elaborate food processing, either through
milling, fermentation, or some other method, is required in order to make available the nutrients of the plant cells. In this
respect the ruminants constitute an invaluable beachhead to man,
which, early in history, broadened his survival base.

The protein of each of the cereals has its own characteristic
pattern of amino acid composition and in each a few essential

amino acids are below optimum dietary requirements. A percentage
increase of such unbalanced protein can be utilized effectively
by the ruminants (cattle, sheep, etc.), because of transforma-
tions during the digestive process. However, the monogastric
animals, which includes man, require that the protein ingested
contains the essential amino acids of approximately the required
balance if the major dietary needs are to be satisfied.

The protein percentage of grain and the kilogram of protein
produced per hectare can be modified both by management and breeding. Nitrogenous fertilizers regularly produce increases in yield,
measured in protein produced per hectare, but as a rule render
no improvement in the pattern of amino acid composition, but
frequently a deterioration in this regard, e.g., by an increase
in the inferior zein as a component of corn production. For a
given weight of grain the protein value is thus reduced.

Breeding offers more promise for increasing both quality and
quantity of protein. High protein strains are known for each of
the major cereals. Mostly they have the same amino acid balance
as the normal types. This route, therefore, offers little promise for the development of quality improvement.

One notable exception has been encountered in corn, where
two genes affecting endosperm texture (opaque 2 and floury 2)
long known to geneticists, have been shown to have a better amino acid.

The world food issue has sometimes been termed paradoxical.
This is generally shown by adding up the computed amount of plant
protein being produced each year. The amount of protein and for
that matter also that of calories shows that the amount available is twice that which is required.

On the other hand, when the traditional ratio of 5 is used
as to what is required to produce each unit of animal protein,
the animal production alone would demand an input of 121.5 g
per day as feed protein – that is more than the total availability of plant protein in the global plant crop.

The hunger gap also comes into clearer focus on the fat front.
Per capita the SW is consuming almost four times (3.7) more fat
than the HW, but the amount of fat originating from plant sources
differs only slightly (1.3). Contrast this with the fact that
the affluent world consumes, per person, almost 8 (7.7) times
more animal fat. Rather complex calculations going back to primary calories, some of which would be carbohydrates, would be
required to establish in a similar manner as above how many calories actually are needed as feed input to produce all this
animal fat. Nonetheless, this askew ratio towards that of animal

origin in the fat intake of the SW only renders deeper relief
to the calorie gap.

There are further great differences between optimum quantities and minimum requirements, i.e., the minimal intake indispensable to avert deficiency symptoms. There is also a considerable difference between theoretical experiments based on defined pure ingredients and regular foods containing the traditional ingredients. With regard to protein this becomes particularly critical as many proteins bind themselves heavily to fats
and carbohydrates. A diet of chiefly grain, beans, and other
pulses invariably means that the quantity of protein consumed
partly evades regular digestion. It should also be kept in mind
that physical work, still quite prevalent in the poor world, requires a greater input of nutrients even when temperature conditions are favorable. Reports from the field clearly show that
in the poor world protein deficiency diseases are not only rampant but that the situation has deteriorated in recent years
in many countries.

The most essential features of the reeducation urgently needed in the United States as well as in the entire world are that
we not only make ourselves acquainted with how much we need per
day in various categories of food as measured in calories, protein, fat, vitamins and minerals, but it is imperative that we
go beyond these figures and take a look at what this daily food
represents in water requirements, energy inputs from the farm to
the consumer's table, in forest products(packaging) and in steel
and other metals. Each American receives on average 175 grams per
day(gpd) from the faucet (his physiological need is 2 qts.), but
the food we eat demands 4,200 gpd of which animal products account for four-fifths, largely to raise the feed crops. A strict
vegetarian generally gets by with one-tenth of the US average.
The increase in US meat consumption since 1965 of one ounce per
day requires an additional 145 gpd of water. We "drink" indirectly 120 gpd water via alcoholic beverages (19.5 gallons per year),
smoking represents 20 gpd. The forest products (paper, timber)
require 350 gpd. This all projects onto the world scene, remind
us that on top of all other gaps there is also an enormous water
gap. It is not only in energy we live extravagantly.

What are the chief limitations facing the world? This can
best be brought into focus by drawing attention to some prevalent
fallacies: we should put an end to that loose talk of the US
feeding the world. It is indisputable that North America now
accounts for 80 to 90 % of the deliveries to the world cereal
market. World trade has, however, never accounted for more than
a minor percentage of world production, and deliveries outside
of the chief recipients of the SW are only marginal, accounting
for a few percent of their total consumption and chiefly pro-

viding receiving ports or very limited areas. Most amazing in
this context is the statement that "in recent years the US and
Canada have been feeding the 600 million Indians". At the very
best US sales to India have served as a supplement to possibly
50 million and perhaps truly fed 10 million. Counted over the
entire population of India this amounts to 7 to 8 kg per capita
which is about 4% of the minimum requirements.

A good sobering exercise is to relate the rice exporting potential to the global needs. It is frequently stated that the US
is top-ranking in this regard, vying with Thailand for the number one position. But world trade in this staple commodity for
about one and a half billion people nowadays plays a very minor
role, accounting for only a few percentage points of the total
production (4.5%). The "huge" rice export of the US, more than
half of the total harvest (67%, 1976-73), should in effect be
more than doubled merely to fill the normal annual increment
in world need, and then continue to double each year for the
foreseeable future in order to keep pace with the population
growth. It sounds like a paradox, but what appears big is thus
a pittance in the real world.

Few analyses go from the impressive tonnage figures to per
capita data. The Caribbean and the Middle East then become top
ranking as cereal recipients but vying for these positions with
several European countries with 120 to 400 kg per capita. This
contrasts with some 7.5 kg for each China and India. Corresponding data for wheat is 65 to 154 kg in the top group. Only per
capita figures convey a clear picture of the degree to which
an individual country is transcending its limits and relying
on external support.

Moving back to the overall picture, trade has predominantly
been discussed and analyzed on the one hand in terms of grain
transfers and on the other hand as aid deliveries. This has
greatly contributed to the false notion that the world cereal
trade is predominantly moving toward the needy world. The
truth is evidenced by the fact that feedgrains have increasingly gained in prominence, currently with a feed/food protein
ratio of 6.7. Furthermore, what goes to the HW is in most cases
counterbalanced by its deliveries of cash crops or feed products to the SW. Two-thirds of the grain flow is to Europe, Japan and the USSR and in all three instances the purpose is not
to stave off hunger but to underpin their nutritional prodigality (to feed livestock rather than man). The most conspicuous
case is soybeans as pointed out above.

When placed in a global context US export of soybeans and
soybean products in protein terms is 50% above the production
of 1.6 billion in China and India.

No aspect of the dietary debate is further removed from global and national realities than its land and water dimensions. If heeded at all, such matters are as a rule obfuscated or blurred due to the lack of awareness of their indispensable role both in crop and livestock production. This study is focused on the US scene, in particular revealing that the US managed to climb to the world pinnacle in terms of animal protein intake per capita and day. One result is that the US now also holds the world record in the amount of consumptive water required for daily food, with animal products occupying the lion's share (85%). The US has furthermore in the 1970s emerged as the global leader of this animalization drive, largely within the affluent but dwindling minority (27%). This ever growing dominance of feed as contrasted to food especially needs to be analyzed in terms of water requirements. The concept of self-sufficiency badly needs to be redefined along new lines. The role of aquatic protein then emerges in an entirely new light.

Water is unquestionably the most limiting factor in world agriculture. Yet this century might truly be called that of irrigation within which the watered acreage has increased more than four-fold to reach some 230 million ha. A further doubling is anticipated before the year 2000. The International Hydrological Decade has shown this to be wholly inadequate and established that another 270 million ha as indispensable with a price tag of some 2300 billion dollars (1980). Further tapping of groundwater will be called for, but must in the long run be put in reasonable line with replenishment. So far the exploitation of such waters has been proceeding at rates exceeding refill.

Australia is right in the midst of this crisis and is already unable to make any commitments for future grain deliveries and is on the whole not in a position to use water to increase yields further but only to set a more modest goal of securing crop levels. The competition with industry for water enters as another decisive force. Australia cannot afford to return water to the hydrological cycle via crops when it can be recycled for industrial and urban use, a dilemma many sections of the arid and semi-arid US are also facing. Industry and cities are on direct collision course with the food survival base. This is happening in several corners of the globe as, e.g., in parts of W. Europe (Denmark, W. Germany), Japan, and elsewhere.

Here we encounter the most rigorous limit to man's feeding. An expansion in the range of 270 million hectares would require massive desalination of ocean water. The energy demands are prohibitive but even more critical is the salt factor. Each American would accumulate more than 400 kg of salt per day on

such a basis. Each acre-foot of such irrigation water results
in 42 tons of salt.

Man already is in an ominous battle on the salt front in
most areas of perennial irrigation. A growing percentage of so-
called irrigation water has to be used for salt removal (one-
third of the irrigation water of the Nile).

Energy inputs into agriculture came late in history (1920-40),
but have reached such dimensions that despite impressive gains
in collecting solar energy via the crops, the subsidies via
fossil energy have surpassed them in most Western countries,
in US farm production by a factor of almost 3. Add to this storage, processing, and marketing and the US food system requires
subsidies of 11 to 12 calories for each calorie consumed. What
we may call reasonable becomes unattainable to a world in poverty. To copy our procedure would earmark 67% of the present
world energy account.

The lack of awareness of both the nature and the magnitude
of the world food crisis is most evident in the misleading metaphors used in the debate, such as triage, lifeboat operations,
the abuse to the Commons, and similar such expressions.

Triage, a term borrowed from the battle surgeon's dictionary
classifying the war victims, in food terms means that only those
countries that according to US judgment can "make it" would get
help. Apart from the cynicism of this attitude, it is political
naivité and a considerable degree of nationalism lies behind
this recommendation. The world would certainly not adhere to
our unilateral decision. The lifeboat notion is equally absurd.
We in the developed world are certainly not in lifeboats. We
are steaming full speed ahead in luxury cruisers with delivery
ships shuttling between our liners with "goodies". It is the
hungry nations that are in "lifeboats", and we the rich are
hauling in large loads of feed and food from them.

The Commons as a concept has a partial validity for our dilemma. But the debate has led to the entirely false conclusion
that private exploitation has been as destructive and careless
in despoiling and squandering resources regardless of this.
The way mankind is treating water, air, and the oceans are good
cases in point, so are the vast destroyed forestlands, coffee
and banana lands.

REFERENCES

Most data are computed by the author on the basis of data available in the FAO Production and Trade Yearbooks. For further references see:

Borgstrom, G. (1964). The Human Biosphere and its Biological and Chemical Limitations. Pp. 130-165 in Global Impacts of Applied Microbiology (Ed. M.P. Starr). Almquist & Wiksell, Stockholm, and John Wiley, New York, NY, 572 pp.

Borgstrom, G. (1971). Too many -- An Ecological Overview of Earth's Limitations. Macmillan, New York, NY, xiii + 360 pp.

Borgstrom, G. (1973). Focal Points -- A Food Strategy for the Seventies. Macmillan, New York, NY, xii + 320 pp.

Borgstrom, G. (1973). The Food-People Dilemma. Intext Publishers, North Scituate, Massachusetts, vii + 140 pp.

Borgstrom, G. (1979). Agriculture and World Feeding Alternatives, 323-380. In Growth without Ecodisasters? (Ed. N. Polunin, Macmillan, London, England, 580 pp.

EDITOR'S NOTE

For those interested in further information on the topics touched upon by Dr. Borgstrom we recommend consulting the following sources:

Ehrlich, P.R., Ehrlich, A.H. and Holdren, J.P. (1977). "Ecoscience: Population, Resources, Environment", W.R. Freeman, San Francisco.

Lappe, F.M. and Collins, J. (with Fowler, C.) (1978). "Food First: Beyond the Myth of Scarcity", Ballantine Books, New York.

Wortman, S. and Cummings, R.W., Jr. (1978). "To Feed This World: The Challenge and the Strategy", Johns Hopkins University Press, Baltimore and London.

Brown, L.R. (1978). "The Twenty-Ninth Day: Accommodating Human Needs and Numbers to the Earth's Resources", W.W. Norton, New York.

SEAWATER-BASED AGRICULTURE AS A FOOD PRODUCTION DEFENSE AGAINST CLIMATE VARIABILITY

C.N. Hodges, M.R. Fontes, E.P. Glenn, S. Katzen and
L.B. Colvin
Environmental Research Laboratory
University of Arizona, Tucson, Arizona

ABSTRACT. A limiting factor for food production in the sunny, productive regions of the world is the scarcity of fresh water. Also, many irrigated soils are becoming saline. This paper reviews the use of seawater to increase agricultural production and reduce climate vulnerability by adapting conventional crops to salt tolerance, by using seawater for environmental control, by culturing aquatic animals, and, in the work described in detail, by domesticating wild halophytes—plants which have evolved in hypersaline conditions—for livestock feed. Halophytes irrigated exclusively with seawater equalled or surpassed alfalfa in yield and protein, excess salts were removed by leaching, and initial animal trials were promising.

The elemental climate control of agriculture is irrigation. Although sunlight is the ultimate energy source for all food production, the major agricultural regions of the world have not been the sunniest; it was necessary to strike a balance between available sunlight and available rainfall, the primary source of water. By delivering water artificially to plants, however, it has been possible to develop crops in areas of maximum sunlight. Such irrigated regions have become prodigiously productive, as there is minimum cloud cover to shade the land and inhibit photosynthesis.

Despite the fact that specific agricultural lands (and frequently prime lands, in the USA) are being removed from production by urbanization and other factors, available data show a steady increase in world croplands, particularly in developed nations, with an estimated total of approximately

W. Bach, J. Pankrath, and S. H. Schneider (eds.), Food-Climate Interactions, 81–99.

1450 x 10^6 hectares (1). Of this amount, irrigated croplands are estimated at 230 x 10^6 ha (2), or approximately 16 percent.

In the arid and semi-arid regions of maximum solar radiation, and where the land is susceptible to agriculture, virtually all usable surface water has long since been appropriated for crop irrigation. There are continuing political and technical refinements of this appropriation, some of which have international implications, but such regional water resources are limited.

In many arid lands, however, crops are irrigated with pumped groundwater. As a desert, by the simplest definition, is a place which gets less rainfall than it can lose by evaporation, groundwater aquifers are recharged very slowly; they are severely finite. When such aquifers are pumped for crop irrigation, it represents a mining of an almost non-renewable resource, in terms of human life cycles. Water tables drop at alarming rates. In the Central Valley of Arizona in the USA, the annual overdraft caused by irrigation pumping, which represents 90 percent of the regional water use, has dropped the water table as much as 10 meters/year (3). This is not only hydrologically alarming, but, particularly when coupled with the rising costs of energy, results in the involuntary retirement of croplands which can no longer be economical under pumped irrigation.

Also alarming is the inevitable phenomenon associated with crop irrigation in arid and semi-arid lands: because of evapotranspiration, the fields tend to become loaded with residual salts. Most food crops are harmed by salt concentrations greater than one tenth the salinity of seawater (4). An estimated one third of the world's irrigated croplands are increasingly affected by salinity (5), with as many as 2 x 10^5 ha/year taken out of production because of this type of artificial desertification. In the USA, salinity is an increasingly severe problem in the great valleys of California (6, 7), affecting to some degree one half of the fields which produce a significant percentage of the nation's fruit and vegetable crops. While it is difficult to estimate the extent of the world's saline soils, due to a lack of uniform standards, the total may be as high as 950 x 10^6 ha (8).

In addition, many of the groundwater resources of the world are unusable for conventional agriculture because of naturally high salinity. For example, of the estimated 1.5 x 10^{17} m^3 of groundwater in the USA within 700 meters of the surface (9), at least two thirds are classified as "slightly saline" with total dissolved solids at 1,000-3,000 parts per million, while "saline" groundwater areas (in excess of 3,000 ppm) total 65 x 10^6 ha or one twelfth of the continental land mass of the USA. This probably exceeds the present fresh water irrigated area of the nation.

Further, a large quantity of available fresh water is rendered saline by industrial or agricultural processes, including cooling tower blowdown from thermoelectric power plants, and agricultural wastewater drained from the fields. And, in the ultimate analysis, the majority of the water on earth is hypersaline: the sea itself, with 35,000-40,000 ppm.

Given the extent of saline soils and saline water, and in some regions the lack of any other resources, it was inevitable that agriculturalists would try (as they have for centuries) to produce food crops under these conditions. Working with available strains of conventional crops, which had been developed for entirely different environments, they had little success, particularly with full-strength seawater.

Further, during most of the past half century, in the more developed nations, an abundance of land and fresh water, and the availability of low cost energy, provided little impetus for saline water agricultural research. But the increasing awareness of salinity problems, the increasing pressures of political environmentalism, and the tremendous economic impacts of the energy crunch of the 1970s, have changed all that. Biosalinity research has suddenly become popular, with an emphasis upon the use of seawater.

The inevitability of requiring more food production in the future, coupled with the sensibility of utilizing land and water not previously employed for this purpose, suggests a new attractiveness for the thousands of kilometers of desert seacoast in the world, backed by millions of hectares of barren land. These landscapes have remained unused because of the lack of fresh water, which is traditionally frustrating to the relatively few inhabitants of such regions, who stand on empty coastlines, viewing the immensity of seawater at hand.

One obvious use of seawater is to put it into shoreside confinements for the production of marine animals and plants; this is an old art, particularly in the Asian tropics, and is one aspect of aquaculture, or mariculture. Such pond culture already accounts for a very modest percentage of the world production of animal protein. An increasing number of scientific groups is working to improve this contribution. As an example, our Environmental Research Laboratory of The University of Arizona is developing higher technology systems for the dense culture of valuable crustacea (10).

A second approach toward the use of seawater as a defense against climate in food production is found in controlled-environment agriculture (CEA). Our laboratory has developed systems of sophisticated greenhouses, primarily for coastal desert applications, in which raw seawater is used to evaporatively cool the aerial environment, and maintain high humidities. Evapotranspiration requirements are thus greatly reduced, with fresh water being used in trickle irrigation systems essentially as a carrier for plant nutrients (11). Only one tenth to one fiftieth of the fresh water needed for open field production is required in CEA greenhouses, varying as a function of the vegetable crop produced. Such systems are obviously capital and energy intensive, but we have demonstrated their economic viability when the only alternative for many consumers is even costlier vegetable importation (12).

A third approach to the use of seawater is based on modern research toward achieving that ancient dream, the growing of conventional food plants in hypersaline environments. Selective breeding for such crops was suggested at the beginning of World War II (13), with significant genetic research beginning about a decade ago. Epstein's work in California (14) is perhaps the best known of these efforts in genetic and environmental manipulation of familiar terrestrial food crops, but there are now reported to be more than five dozen researchers active in this field (15).

A fourth and newer approach toward the use of seawater for plant production is the development of unconventional crops from those wild species of plants which have evolved naturally in hypersaline environments (16,17). Found along seacoasts, estuaries, salt marshes and inland deserts with saline soils, these plants are commonly categorized as "halophytes," and are represented by hundreds of species in scores of genera, ranging from small salt grasses to large trees. There appear to be several physiological mechanisms employed by halophytes to store or reject the high levels of salt in seawater, but these mechanisms are not clearly understood (18) and crop research is necessarily pragmatic.

Epstein et al. (14) have recently cited the existence of halophytes as evidence that other plants (food crops) may be capable of being manipulated to exhibit halophilic capabilities, tolerating higher salinities than at present. Others have noted that with few exceptions, one of which is the beet, conventional crop cultivars have been developed from plants which evolved in fresh water environments, and which are therefore not particularly salt tolerant (16).

Our laboratory elected to work directly with halophytes,

seeking those which may be irrigated solely with seawater to produce human food, livestock feeds or portable energy.

Halophytes were part of the diet of indigenous nomadic coastal desert tribes in at least one region (19), but have never played a significant role in human nutrition. In modern times, some halophytes have been described as potential vegetables and potherbs (20), and Somers (16) has speculated that, as halophyte seeds generally do not accumulate excessive salts, some species may be developed as grain crops.

Halophytes have been (and still are) grazed extensively by domestic livestock in some saline areas where other forage is unavailable (21), but they are usually considered to be of low quality, and, unless the land is far too salty for conventional forage grasses, they have customarily been eradicated in most developed nations (22). Saltbush species of the genus _Atriplex_ are, however, regarded as important forage crops in parts of Australia, and Goodin (17) considers _Atriplex_ spp. as significant high quality forage plants.

To initiate our own halophyte research, a plethora of pumped seawater was available at the site of our controlled-environment shrimp research on the Gulf of California, in Sonora, Mexico. This water, pumped via four wells from a shallow saline aquifer 50-100 meters inland from median tide level, was hypersaline (40,000 ppm) and, after discharge from the shrimp raceways, was enriched by 30-70 ppm of organic matter, composed of shrimp feces, food residues and algal debris, and 0.01-0.03 ppm of dissolved ammonia.

In sharp contrast to Somers' (16) temperate zone halophyte test site in Delaware, the Mexican coastal desert location at Puerto Peñasco can be described as harsh. Precipitation is 9 cm/yr with evaporation at approximately 250 cm/yr; the average high temperature is 43 C and relative humidity is normally low during daylight hours (23). With high temperatures, high evaporation, low rainfall and low humidity, there was little opportunity for open field halophyte plantings irrigated solely with seawater to obtain moisture from other sources.

Two soil types were represented in the preparation of test plots; beach sand, with a saturated infiltration rate of 2.5-3.0 cm/hr; and sandy desert soil (5 percent clay) with an infiltration rate of 1.3-2.0 cm/hr. Irrigation with seawater was not expected to increase salt buildup in the soil, based on Boyko's (24) data from irrigating desert soil with hypersaline water.

For the first set of field trials, halophyte seeds and

transplants were obtained from the collections of The University of Delaware, the USDA Plant Introduction Center, individual investigators in Israel, and, most importantly, from more than a dozen collecting trips by our own plant scientists along the estuaries and salt flats of the Gulf of California.

A total of 32 species representing 19 genera were tested in an on-site CEA greenhouse and in small scale preliminary test plots. Eleven candidate species were then selected for planting on a one hectare site of sandy desert soil, and evaluated for total yield under different types of irrigation with the hypersaline seawater. Yields and other data reported here are from the one hectare test farm, unless stated otherwise.

Halophytes grown exclusively on seawater may be expected to contain higher levels of salt than plants grown in less saline water, and other nutritional components may also be affected. Therefore, the present research has included studies of simple methods for removing excess salts from harvested materials, and detailed analyses of nutritional values of halophyte products, followed by conventional animal feeding trials.

Candidate species included _Atriplex barclayana_, _A. lentiformis_, _A. linearis_, _Salicornia europaea_, and _Batis maritima_. Plots of two potential grain crops were also established, _Distichlis palmeri_ and _Cressa truxillensis_, but had not produced seed in time for our laboratory analyses and animal trials. (Seeds in quantity from the latter two species were collected, however, from wild plants in a seawater estuary in Northern Mexico.)

Except for _S. europaea_, which was direct-seeded into the test plots, seedlings were established in a greenhouse on fresh water and conditioned to seawater irrigation before transplanting into the field. 30,000 seedlings were transplanted, with a survival rate exceeding 80 percent. This initial fresh water usage was held to a minimum (500 ml/plant) by using compact silviculture trays and recycling the water. We did not attempt initially to develop mechanical or chemical treatments to permit seed germination in full strength seawater; this will be pursued during a subsequent phase of the project, and, because simple and effective techniques have been developed for other desert plants with specialized germination criteria, we are hopeful of ultimate success.

Plots were flood-irrigated with seawater from the shrimp raceways to a depth of 6-10 cm at three different time intervals: 12 hours, 24 hours and 72 hours. In the less frequently flooded plots (72 hour intervals) salt accumulation increased in the soil surface, and moisture content decreased. Plant yields were reduced in the same plots. As had been anticipated from the literature, however, there was no significant salt buildup in the more frequently irrigated plots (Table I). Of equal interest, the addition of inorganic fertilizers did not increase plant yields. As the levels of soluble nitrogen and phosphorus in the seawater were low, plant nutrients were probably made available by the breakdown of the shrimp metabolites and other debris in the wastewater from the raceways.

TABLE I

MOISTURE AND SALT CONTENT (mg/g SOIL)

Moisture and salt content (mg/g soil) in desert plots watered every 12, 24 and 72 hours with 40,000 ppm seawater.

Watering Frequency (hrs)	Moisture (mg/g)		Salt (mg/g)	
	Surface	15 cm	Surface	15 cm
12	73.6	68.2	4.0	2.6
24	77.6	72.0	5.3	2.9
72	18.4	54.9	10.3	2.3

Measurements were made on soil samples from the surface and 15 cm depth, immediately before a scheduled irrigation. Each entry is the average of 4 determinations.

Annual yields of above-ground dry matter ranged from 895 g/m^2 (A. lentiformis) to 1365 g/m^2 (S. europaea) for the best producers among the candidate species (Table II). New harvest data from the Mexico test farm are reported to include an even higher yield for S. europaea of 2,270 g/m^2 (25). These yields exceed those of the most popular conventional fresh water irrigated forage crop, alfalfa, which in the USA had a 1977 average annual yield of 648 g/m^2 (26).

TABLE II

ABOVE-GROUND YIELD OF HALOPHYTES
(g dry-weight/m^2)

	Above-ground Yield (g/m^2)		
Watering Frequency (hrs):	12	24	72
Salicornia europaea	1365	**	**
Batis maritima	1137	**	**
Atriplex linearis	927	1134	245
Atriplex barclayana	901	733	431
Atriplex lentiformis	895	620	230

** Not determined

Halophytes were planted on 0.61m centers irrigated exclusively every 12, 24 and 72 hours with 40,000 ppm seawater. Atriplex spp. were transplanted from greenhouses to 18 desert plots (12.3 x 18.5m) on 23 April 1979. Plants were harvested at t=217 days. Batis maritima was planted in beach plots on 31 July 1979 and harvested at t=355 days. Salicornia europaea was sown on 16 February 1979 and harvested at t=265 days.

Harvested plant material was dried to constant weight at 75 C and ground in a Wiley Mill prior to analyses. Crude protein, fiber, fat and ash were determined by standard analytical procedures (27) in our own laboratories, and by a commercial analytical laboratory (EFCO, Inc., Tucson, Arizona). Minerals were determined by atomic-absorbtion flame-spectrometry, and oxalate (soluble plus insoluble) was isolated by the method of Moir (28) and quantified by the determination of Ca in Ca-oxalate.

Leaching experiments to remove excess salts were conducted with 5 g dry weight samples suspended in 50 ml dH_2O and gently mixed (50 rpm) at room temperature. Plant material was recovered by squeezing the suspension through three layers of cheesecloth. In addition, larger amounts of dried and ground plant material were batch-leached to produce sufficient quantities for animal trials, which were conducted by the Department of Animal Sciences, The University of Arizona, in Tucson, and by a large commercial feedlot, Mesquital del Oro. in Sonora, Mexico.

It will be seen from Table III that the composition of Atriplex and Salicornia spp. varied by the developmental stage of plants at harvest, with A. lentiformis material lower in ash and fiber, and higher in protein, in immature plants, vis-a-vis mature harvested material. A. barclayana showed similar results, except that ash contents were equal in both stages. It will also be noted in the table that immature Atriplex material can provide forage with levels of protein and soluble carbohydrates comparable to alfalfa and other high protein forage crops, as Goodin (17) had concluded, with S. europaea also high in nutritional value. As expected, however, seawater irrigation had increased the ash content of the products, which would limit the use of untreated material in animal diets.

Table III.

NUTRITIONAL COMPOSITION OF SEAWATER-IRRIGATED HALOPHYTES

	Crude Protein (%)	Crude Fat (%)	Crude Fiber (%)	Ash (%)	Total Phosphorus (%)
A. lentiformis					
Immature	18.7	1.2	16.9	22.4	0.27
Mature	14.9	1.3	35.1	33.9	0.33
A. barclayana					
Immature	12.4	1.3	15.4	33.5	0.20
Mature	9.8	1.3	36.3	33.2	0.40
S. europaea					
Immature	14.4	--	--	--	--
Mature	5.4	0.4	19.9	42.4	0.18

Immature plants were harvested 3-4 months after planting and before flowering. Mature plants were harvested after 7 months, after flower and seed set (seeds not included.)

The ash and cation contents of these plant materials were significantly reduced, indeed brought down to levels not much higher than conventional forage crops, by soaking ground product in 10 volumes of fresh water for 16 hours (Table IV). Subsequent time-course experiments with A. lentiformis showed that salts were removed more readily than protein, resulting in enhanced protein values at optimal soaking times; and that

salts were removed more effectively from finely ground
material, rather than from coarsely ground or unground
products.

Table IV.

ASH, P AND CATION CONTENT
OF SOAKED AND UNSOAKED HALOPHYTES

	A. lentiformis		A. barclayana		S. europaea	
	U	S	U	S	U	S
Ash (%)	20.2	6.7	33.5	15.4	42.4	6.5
Na (%)	10.5	1.0	11.1	5.0	13.0	3.5
K (%)	8.0	0.3	7.5	0.3	1.05	0.05
Ca (%)	.415	.55	1.095	1.07	.495	.61
Mg (%)	0.4	0.3	0.9	1.3	0.7	.25
Cu (ppm)	5	5	10	5	10	7.5
Zn (ppm)	30	25	27	40	33	40
Fe (ppm)	60	35	105	110	85	65
P (%)	.27	.06	.20	.06	.18	.06

Dry material was ground in a Wiley Mill (10-mesh screen)
and soaked for 16 hours in 10 volumes dH_2O.

That the simple desalting procedure was effective may
result from two phenomena associated with the genera
represented in Table IV. Atriplex spp. deposit salts on
outside leaf surfaces, from which the salts may be removed
without rupturing cells (29). Salicornia spp. are believed to
store excess salts in cell vacuoles from which they may leave
the cells along an osmotic gradient during soaking (30).

The use of fresh water for this desalting is relatively
minimal and represents only a minor fraction of the water
needed to grow a conventional irrigated forage crop. In many
regions of the world, where at least limited fresh water or
brackish water is available, such usage would not physically or
economically inhibit the development of seawater irrigation.
However, in other regions where fresh water is scarce or simply
unavailable, it would be desirable, and should be possible, to
soak the halophilic materials in raw seawater to reduce the ash
content. It can be calculated that products soaked in
seawater, and subsequently pressed to reduce the moisture
content to approximately 40 percent, would have an ash content
only two or three percentage points higher than plant material
soaked in fresh water.

It was also seen that the soaking procedures resulted in some losses of organic matter. The data are not shown here, but protein losses from the three species ranged from 15-56 percent. There are indications, however, that protein loss may be reduced by adjusting the pH of the soaking solution (31) and it is also possible that lost protein could be recovered using procedures developed for the preparation of leaf protein concentrates from other types of plants (32).

Protein quality indices showed high levels of essential amino acids in A. lentiformis and A. barclayana for chick and rabbit requirements (Table V), but whatever the results of laboratory analyses, actual animal trials, in which other factors may appear, and in which product acceptance and palatability play important if subjective roles, are indispensable.

Table V.

PROTEIN QUALITY OF A. LENTIFORMIS AND A. BARCLAYANA.

Plant

	A. lentiformis	A. barclayana
% Whole Egg Standard		
Essential Amino Acid	49.7	54.5
M+C	20.2	25.1
% Chicken Requirement		
Essential Amino Acid	72.7	80.5
M+C	38.1	47.5
% Rabbit Requirement		
Essential Amino Acid	76.9	83.3
M+C	41.1	51.2

Indices were determined by comparing amino acid
spectra of samples with whole egg protein, and
known amino acid requirements of chickens and rabbits.
M+C=methionine plus cysteine.

Initial animal testing was conducted with untreated (unleached) halophilic plant material containing high ash levels. In experiments with both mice (Table VI) and broiler chicks (Table VII), it was seen that A. barclayana ground leaves and stems, incorporated at levels up to 15 percent of diet, had no adverse affect on growth. The lower weight gains at higher levels of incorporation may be attributed to the lower energy value of the plant material due to high salt and fiber content.

Table VI.

PERFORMANCE OF MICE FED A. LENTIFORMIS AND A. BARCLAYANA

Mouse Diet	Body Weight Gain (g)	F.C.R.
Basal	17.4	5.07
5% A. barclayana	17.0	5.21
10% A. barclayana	17.2	4.90
15% A. barclayana	15.8	4.83
5% A. lentiformis	15.0	5.05
10% A. lentiformis	12.2	5.24
15% A. lentiformis	9.8	6.30

Animals fed at 0, 5, 10 and 15 percent of diet. Each group consisted of 20 mice fed a basal diet plus halophyte material for 4 weeks. Diets were balanced for protein, Ca and P.

Table VII.

PERFORMANCE OF CHICKS FED A. LENTIFORMIS AND A. BARCLAYANA

Chick Diet	Body Weight Gain (g)	F.C.R.	% Mortality
Basal	574	1.72	0
5% A. barclayana	608	1.77	0
10% A. barclayana	555	1.80	5.6
15% A. barclayana	467	1.93	5.6
5% A. lentiformis	333	2.10	0
10% A. lentiformis	98	3.00	72.2
15% A. lentiformis	56	4.78	94.4

Each group contained 20 birds grown for four weeks on the indicated diets. Halophyte material replaced milo in the diets, which were balanced for protein, Ca and P.

A second series of chick trials recently completed compared unleached and leached A. barclayana to a control diet. All formulations containing leached material or small percentages (5-10 percent) of unleached product, are said to have performed as well as the control (33), although increasing the untreated A. barclayana to 15 percent of diet depressed growth and resulted in inferior effective feed conversion ratios.

Returning to Tables VI and VII, it will be noted that unleached A. lentiformis tissue inhibited growth of both mice and chicks, and, when incorporated at 15 percent of diet, caused 94 percent mortality of the chicks. Those conducting the mice trials concluded that the poor performance of A. lentiformis was not due to increased salt uptake by the animals, because there was no significant increase in water consumption. They noted that the animals ate well for two or three days, and then rejected the feed, resulting in high mortalities (34). This implied the presence of some unknown toxic material in the unleached plant material.

A second series of broiler chick trials utilizing A. lentiformis, leached and unleached (treated plants had been soaked in a mildly acidic aqueous solution), at 10 percent of diet, reconfirmed the presence of this toxic factor and, almost inadvertently, appeared to suggest a method of eliminating or neutralizing it (Table VIII). After four weeks of the six week trial period, birds receiving the A. lentiformis diets were 62 percent smaller than the control group. In addition, those receiving the unleached halophyte had experienced mortality in excess of 23 percent; clearly the unknown inhibiting factor was at work.

At the beginning of the fifth week, the decimated group on the diet with untreated A. lentiformis was switched to a 10 percent diet containing plant material which had been leached in alkaline, rather than acidic, water. As seen in the same table (VIII) the results were remarkable. Mortalities in the group receiving alkaline-leached plant material dropped to zero during the final two weeks of the test, and these chicks gained 119 percent in body weight, while the control group gained 57 percent in weight.

Table VIII.

PERFORMANCE OF CHICKS FED   A. LENTIFORMIS
LEACHED AND UNLEACHED MATERIAL
(INCLUDING A CaOH WASH)
AT 10 PERCENT OF DIET

	Treatment Period	
	Wks 1-4	Wks 4-6
Group A (control)		
Weight Gain (g)	669	409
Mortality (%)	1.7	0.0
F.C.R.		2.68
Group B (HCl-soaked)		
Weight Gain (g)	223	148
Mortality (%)	8.4	10.4
F.C.R.		6.53
Group C	Untreated	CaOH-soaked
Weight Gain (g)	219	326
Mortality (%)	23.3	0.0
F.C.R.		2.82

HCl-soaked material was leached for 24 hours
in 3% HCl, washed, and re-dried. CaOH-soaked
material was leached for 24 hours in a saturated
solution of CaOH, washed, and re-dried. Controls
contained sorghum flour in place of halophytes.
All diets were balanced for protein, Ca and P.
At the end of 4 weeks, Group C was switched to
CaOH-soaked material and evaluated for an additional
additional 2 weeks. Each group contained 55 birds.

In recently completed goat studies, animals on diets
incorporating 25 percent of untreated A. lentiformis showed no
significant growth during the experimental period, although
digestibility of organic matter was equivalent to that of
alfalfa. This lack of growth was interpreted as a palatability
problem; i.e., the animals ate only enough for maintenance
(35).   In concurrent rabbit studies (a useful test animal in
its own right and with a digestive system similar to that of
the horse), all rabbits fed a high level of unleached A.
lentiformis (50 percent) died within two weeks, and lesions
found in the gut of such animals suggested the presence of a
saponin as the toxic factor (36).

Saponins are gylcosides found in some plants (including

alfalfa, to some degree) which can yield steroid compounds in the bloodstreams of animals. Saponins have been found in some species of Atriplex (37), although A. lentiformis is not known to have been examined in this regard. Interestingly, it has also been reported that leaching Kochia spp. seeds (a drought resistant potential turkey feed in Australia) in a one percent alkaline solution "greatly reduced saponin" (38). Thus, while no direct data are yet available on the cause of the toxic manifestations of untreated A. lentiformis, circumstantial evidence begins to suggest the presence of a saponin.

We are concerned with the forage potential of A. lentiformis for obvious reasons. It will be seen from the earlier tables on crop yield and proximate analysis, that while dry-matter yield per unit of land area is approximately the same for both immature A. lentiformis and A. barclayana, the former is higher in crude protein and lower in ash. From the data in the tables, it can be calculated that even on an ash-free basis, A. lentiformis can produce 2.09 MT/ha/yr of total protein, compared to 1.46 MT/ha/yr for A. barclayana; an improvement of 70 percent.

The Atriplex spp. are not the only interesting forage candidates. While the crude protein content of S. europaea is somewhat lower than that of A. lentiformis, and the unwashed material contains more salts, its very high yield as shown initially in Table II, and the prodigious 60 percent improvement reported subsequently, appear to give this species a remarkable potential as a (leached) forage crop. The yield of S. europaea when irrigated exclusively with seawater may be 3.5 times better than the average USA yield of alfalfa irrigated with fresh water. In addition, S. europaea is the only halophyte tested to date which was successfully direct-seeded into the field with full strength seawater, not requiring initial greenhouse germination in fresh water. It remains unknown, however, whether it is non-toxic to livestock, or whether they will readily accept it. We have not yet begun animal trials with S. europaea, simply because it was not available in sufficient quantity when the first animal experiments were established.

Research needs for the identification and domestication of wild halophyte crops are obvious. While working with the more promising species first collected in the coastal deserts of northwest Mexico, we have been collecting seeds and whole plants of hundreds of additional species from coastal and inland deserts in South America, South Africa and Australia. These are now being screened in our controlled-environment greenhouses in Tucson for germination characteristics and salt tolerance, and a halophyte germ-plasm bank is being

established.   Additional   collecting   expeditions   are   in
planning.

Our first phase animal trials must be followed by larger
scale testing of beef animals and other ruminants, and new
series of first phase trials must be initiated with different
plant species, as they are grown and harvested. The first
pragmatic leaching experiments must be followed by carefully
structured trials to determine more effective techniques of
reducing salt content while minimizing protein loss, or of
recovering protein from the leaching liquids.   Methods of
seawater leaching must be devised and tested.

Basic research is required to better understand the
singular physiological mechanisms of these plants, particularly
of those species encountered which may be described as "salt
thriving" rather than merely "salt tolerant."   A   wider
knowledge of these mechanisms will be essential to those
researchers seeking to genetically manipulate conventional food
crops to improve salt tolerance, as well as to those of us
domesticating wild halophytes.

A cardinal precursor to all of these needs, and to the
total concept of developing seawater agriculture as a defense
against climate variability in arid lands, is the establishment
and maintenance of suitable halophyte test-farms. Large plots
are needed to grow out plant material, not only to develop
yield data and to continue to evaluate various saline water
irrigation systems and techniques, but also to simply provide
plant materials in quantity for laboratory work, and, in terms
of the greatest bulk required, for the expansion of leaching
and livestock trials.   In our work at The University of
Arizona, we are planning now for additional farm sites in
Mexico, in cooperation with several Mexican federal agencies,
and in Arizona and other states in the USA southwest desert,
where saline water and/or saline soil is unused and, until now,
has been considered unusable.

REFERENCES

1. Simon, J. L. 1980. Resources, population, environment: An oversupply of false bad news. Science 208, pp. 1431-1437.

2. Wittwer, S. H. 1979. BioScience. 29, pp.603.

3. Mann, D. E. 1963. "The Politics of Water in Arizona." Univ. of Arizona Press, Tucson. p.44.

4. Richards, L. 1954. "Diagnosis and Improvement of Saline and Alkali Soils," USDA Handbook 60.

5. Maas, E. V. and G. J. Hoffman. 1977. "Managing Saline Water for Irrigation," H. E. Dregne, Ed. Texas Tech Press, Lubbock. p.326.

6. San Joaquin Valley Interagency Program. 1979. "Agricultural Drainage and Salt Management in the San Joaquin Valley," California Department of Water Resources, Fresno.

7. Robinson, F. E., J. N. Luthin, R. J. Schnagl, W. Padgett, K. K. Tanji, W. F. Lehman and K. S. Mayberry. 1976. "Adaptation to Increasing Salinity of the Colorado River," California Water Resources Center, Davis.

8. Massoud, F. I. 1974. "Salinity and Alkalinity as Soil Degradation Hazards," FAO/UNEP Expert Consultation on Soil Degradation, Food and Agriculture Organization, Rome.

9. Glaubinger, R. S. 1980. Groundwater regulations: trouble from the deep? Chemical Engineering. July 28, pp. 27-31.

10. Salser, B., L. Mahler, D. Lightner, J. Ure, D. Danald, C. Brand, N. Stamp, D. Moore and B. Colvin. 1979. Controlled environment aquaculture of penaeids. "Food and Drugs from the Sea — Myth or Reality," University of Oklahoma Press, Norman. pp. 345-355.

11. Fontes, M. 1973. Controlled-environment horticulture in the Arabian desert at Abu Dhabi. HortScience 8(1), pp. 13-16.

12. Jensen, M. 1977. Energy alternatives and conservation for greenhouses. HortScience 12(1), pp. 14-24.

13. Lyon, C. B. 1941. Botanical Gazette, Chicago. 103, pp. 107.

14. Epstein, E., J. D. Norlyn, D. W. Rush, R. W. Kingsbury, D. B. Kelley, G. A. Cunningham and A. F. Wrona. 1980. Saline culture of crops: a genetic approach. Science 210, pp. 399-404.

15. Epstein, E. 1980. Reported as results of a survey made for a report being prepared by the USA Congressional Office of Technology Assessment, in press.

16. Somers, G. 1979. "The Biosaline Concept: An Approach to the Utilization of Unexploited Resources," A. Hollaender, Ed. Plenum, New York. pp. 101-115.

17. Goodin, J. R. 1979. Atriplex as a forage crop for arid lands. "New Agricultural Crops," G. A. Ritchie, Ed. AAAS Selected Symposium 38. pp. 133-148.

18. Flowers, T. J., P. F. Troke and A. R. Yeo. 1977. The mechanism of salt tolerance in halophytes. Ann. Review Plant Physiol. 28, pp. 89-121.

19. Felger, R. S. 1979. Ancient crops for the 21st century. "New Agricultural Crops," G. A. Ritchie, Ed. AAAS Selected Symposium 38. pp. 5-20.

20. Mudie, P. J. 1974. The potential economic uses of halophytes. "Ecology of Halophytes," R. J. Reimold and W. H. Queen, Eds. Academic Press, New York. pp. 565-597.

21. Chapman, V. J. 1960. "Salt marshes and salt deserts of the world." Interscience, New York. 392 pp.

22. Ludwig, J. R. and J. McGinnes. 1978. Revegetation trials on a salt grass meadow. J. Range Management 31, pp. 308-311.

23. Weather data for Puerto Peñasco, Sonora, Mexico, compiled by Alison Dunn, The University of Arizona, from data for 1952-1967 recorded by The Environmental Research Laboratory.

24. Boyko, H., Ed. 1966. "Salinity and Aridity: A New Approach to Old Problems," W. Junk, The Hague.

25. Glenn, E. P. 1980. Unpublished data (University of Arizona).

26. USDA. 1978. "Agricultural Statistics." Washington, D.C.

27. Association of Analytical Chemists. 1975. "Official Methods of Analysis," 12th edition. Washington, D.C.

28. Moir, K. W. 1953. The determination of oxalic acid in plants. Queensland J. Ag. Sci. 10, pp. 1-3.

29. Waisel, Y. 1972. "Biology of Halophytes," Academic Press, New York. 395 pp.

30. Black, C. C. 1971. Ecological implications of dividing plants into groups with distinct photsynthetic capacities. Adv. Ecol. Res. 7, pp. 87-114.

31. Katzen, S. 1980. Personal communication on the development of experimental animal feeds at feedlot of Mesquital del Oro, Sonora, Mexico.

32. Edwards, R. H., D. de Fremery, and G. O. Kohler. 1976. Use of recycled dilute alfalfa solubles to increase the yield of leaf protein concentrate from alfalfa. J. Ag. Fd. Chem. 26, pp. 738-741.

33. Reid, B. L. and L. B. Colvin. 1980. Unpublished data (University of Arizona).

34. Weber, C. and L. B. Colvin. 1980. Personal communication (University of Arizona).

35. Swingle, S. R., S. Wiley and L. B. Colvin. 1980. Unpublished data (University of Arizona).

36. Schurg, W. and L. B. Colvin. 1980. Unpublished data (University of Arizona).

37. Askham, L. R. and D. R. Cornelius. 1971. Influence of desert saltbush saponin on germination. J. Range Mgt. 24(6), pp. 439-442.

38. Coxworth, E. C. M. and R. E. Salmon. 1972. Kochia seed as a component of the diet of turkey poults: effects of different methods of saponin removal or inactivation. Canadian J. Animal Sci. 52(4), pp. 721-729.

AGRICULTURAL DEVELOPMENT PROSPECTS UNTIL THE YEAR 2000

Heinrich von Loesch

Food and Agricultural Organization of the United Nations,
Rome, Italy

ABSTRACT. In order to provide governments with a long-term perspective of food and agriculture, FAO circulated a study titled
"Agriculture: Toward 2000" in 1979. It is a detailed assessment
of the implications for agriculture of a faster rate of economic
growth in each of 90 developing countries as well as a continuation of trend rates in developed countries to 1990-2000. Between
now and the end of the century food production in the developing
world must approximately double if significant nutritional improvements are to be secured without further deterioration in
food self-sufficiency. The scope for production strategies,
including problems of investment and assistance requirements, is
scrutinized along with other aspects such as opportunities for
expanding international trade. The study finds, however, that
malnutrition cannot be reduced significantly unless accelerated
production is complemented by improved distribution.

INTRODUCTION

When the World Food Conference of 1974 demanded the eradication of world hunger within the brief time span of ten years,
experts and laymen alike were surprised. The target seemed too
ambitious, and it can be safely assumed that it will not be
achieved. A study has been submitted by the Food and Agriculture
Organization of the United Nations (FAO) to its 20th Plenary Conference session in 1979 which - without proposing a new target
date - attempts to prove the feasibility of greatly reducing
world hunger before the advent of the 21st century. A revised
and final version of the study will be completed by the end of
1980 and submitted to FAO's 21st Conference in 1981.

W. Bach, J. Pankrath, and S. H. Schneider (eds.), Food-Climate Interactions, 101–116.
Copyright © 1981 by D. Reidel Publishing Company.

The declaration of the World Food Conference of 1974 was a
spontaneous and noble gesture but it did not offer a well-designed,
comprehensive and realistic strategy for attaining this target.
FAO's report "Agriculture: Toward 2000" (1) embodies quite the
opposite approach to the global problem. Innovative in its metho-
dology but cautious in its judgement, FAO has assessed what progress in the field of food, agriculture and trade could reasonably be achieved by the year 2000 if an effort is undertaken to
speed up development. The study arrives at the conclusion that
an annual growth-rate of gross agricultural production in the poor
countries close to 4.0 percent - seen against the background of a
2.6 - 2.8 percent growth rate during the period from the early
sixties to the present - could be achieved and it would greatly
contribute towards getting closer to the objective as described in
the forword of the FAO report:" The world could free itself from
the scourge of hunger largely by substantial advances in food production in developing countries; yet, it cannot all be done within the next five or ten years. What is needed is a sustained
effort to the end of the century; there is no new major technology that can be relied upon; there are no easy shortcuts."(2)

ALTERNATIVE SCENARIOS

The study measures the forest by examining the trees. Global
conclusions were reached combining new projections of future demand and supply of major agricultural commodities, as well as international trade and food aid, for each of 90 developing countries
and the main region of the developed world. For the purpose of
illustrating the differing impact of a complacent laissez-faire
attitude and a strong concerted effort, this study analyses two
alternative development paths.

A "trend" scenario projects what could happen over the next
two decades if development continues to follow patterns of the
recent past.

A second, "normative", scenario attempts to trace a more
desirable path of faster growth for which governments of the world
would have to implement a package of strong policies to improve
agricultural performance and to restructure their economies for
a new international economic order.

THE TREND SCENARIO

The trend secenario does certainly not describe the worst
possible outcome of events, yet assuming only modest improvements
in diets, it projects rapidly declining self-sufficiency in cereals
of the 90 developing countries studied. Cereal deficits of the net
cereal importer countries (83 out of 90) to be met by imports or
food aid would increase from around 50 million tons in the mid-

seventies to 115 million tons by 1990 and a staggering 175 million
tons by 2000. The deficit in meat production would increase ten-
fold.

The total agricultural trade deficit of the 90 developing
countries would soar from a surplus of around $12,000 million to
a net deficit of close to $30,000 million by 2000 (in 1975 dollars).
If past trends continue to prevail, as assumed by this scenario,
the report concludes that "all in all, the results paint a depressing picture. Agriculture would fail, in these circumstances, to
make any worthwhile contribution to widely accepted national and
international growth and welfare objectives. The major cause is
the inadequacy of food and agricultural production." In view of
the huge cereal deficits which this scenario projects, even the
moderate growth of food demand per capita to be expected from
higher incomes might not be fully satisfied because of scarcity
and rising prices with the result that the corresponding improvement in nutrition would also not materialize. A major reason for
this disappointing course of development is, of course, the expected increase in population, which FAO throughout its calculations
has assumed to follow the official "medium variant" of the UN
population, although slightly modified in the light of latest data.

THE NORMATIVE SCENARIO

Seen against this sombre background, the normative scenario
paints a somewhat brighter picture.

This scenario embodies patterns of economic growth which
would help to significantly improve the living standards of most
people in poor countries and to arrest their increasing dependence
upon food imports from developed countries. The scenario, therefore, examines the feasibility of, and requirements for, a distinct
improvement in the performance of agriculture in developing countries. Balanced emphasis is put on both increasing domestic food
supplies and on developing agricultural exports of the poor
countries.

NEW DEVELOPMENT STRATEGY FOR THE THIRD DEVELOPMENT DECADE

The scenario represents, in other words, an assumption of
accelerated overall economic growth and a drive towards improving
the degree of food self-sufficiency or at least checking any
tendency of its reduction, but with an allowance for sizeable export
development, which in turn would require modifications in the
agricultural import policies of developed countries. This set
of assumptions and targets broadly corresponds to the philosophy
and objectives of a New International Economic Order and falls within
the likely range of growth targets in the new international development strategy of the UN for the Third Development Decade,the 1980's.

The report "Agriculture: Toward 2000" represents in this respect
FAO's input to the formulation of the new strategy. Most of the
250 pages of "Agriculture Toward 2000" are devoted to discussing
the targets of the normative scenario and the conditions and policies related to their attainment. Major aspects of employment,
nutrition, world trade, food aid and external assistance are also
discussed. The study will also prove useful to national economic
and agricultural planners, since it provides a globally consistent
perspective of world agriculture.

DEMAND

     The normative or main scenario of the study puts strong
emphasis on the need for stimulating agricultural production in
developing countries. Increasing demand for agricultural produce,
particularly the food demand, together with renumerative producer
prices for the farmers are essential components in the drive towards
accelerated agricultural development. "The poor and underprivileged
must be able to increase their incomes, or must be otherwise enabled
to translate their basic food requirements into effective demand,"
says the Director-General of FAO in his introduction to the report:
"The productivity of the rural poor themselves, small farmers and
landless labourers, must be raised and they must contribute a large
part of the increased input."

     The growth of effective demand partly presupposes a correspond-
ing growth rate of national income. This also implies that assumptions have to be made about the growth of the non-agricultural
sector which - under the conditions of a sector study like "Agriculture: Toward 2000" - has been dealt with as an external variable
oriented toward development targets which are likely to be adopted
by the international community in the context of the new International Development Strategy in the 1980's.

     Another factor determining demand is government action to
raise the incomes of underprivileged and undernourished sections
of the population through income re-distribution. Government
policies to hold down consumer prices, particularly for food,
should, however, not interfere with the need to grant price incentives to farmers and to improve income levels of an expanding
agricultural population, the report states.

PRODUCTION

     Given the required growth of demand and the necessary incen-
tives, the question arises as to where and how the additional
production required could actually be obtained. Expanding the
land area under cultivation offers part of the potential required,
mainly in some developing regions such as Latin America and Africa
whereas in others, such as the Near and Far East, the prospects

for area expansion are very limited. Increasing yields and step-
ping up cropping intensity must thus provide most of the additional production.

Expanding irrigation together with better drainage and improving existing irrigation facilities constitutes one of the most
promising possibilities for more intensive land use. Water management, improved seeds, more fertilizer application, plant protection
and power inputs are other essential elements of the strategy for
improving agriculuture. The ratio of input to output value in the
agriculture of the poor countries would have to increase considerably in order to step up production growth. This change requires
a price incentive which makes intensification of agriculture worthwhile for the farmer. Considerable increases in the working capital
of farmers would be needed to finance the growing use of inputs.
But this represents only a fraction of the additional investment
necessary to expand agriculture along the envisaged growth path.

INVESTMENT

The study contains a detailed assessment of investment requirements under a broader-than-usual definition. Instead of dealing with investments for crop and livestock production only, FAO's
definition also includes depreciation charges for existing capital
plus major investments for the support of agriculture, such as
transport, storage, marketing, credit and the first stage of processing. Under this wider definition, total gross annual investment requirements in the agriculture of developing countries are
estimated to be $52,000 million in 1980 and expected to rise to
$107,000 million by 2000. The primary agriculture sector component
of these estimates is $35,000 and $70,000 million, respectively.

Regionally, the highest levels of cumulative investment
needed per hectare of land or per head of farming population are
expected to be required in Latin America; the lowest figures are
indidated for Africa. The Near East shows both investment ratios
in the medium range, whereas high cumulative investments per unit
of land but a low ratio per caput in the Far East reflects land
scarcity and the need for intensification.

Most of the funds would, of course, have to come from private
capital formation of the farmers themselves, the remainder from
government sources. A considerable part of the investments would,
as in the past, require foreign exchange expenditure which would
heavily tax the balances of payments of most of these countries.
In fact, already by the mid-seventies, the net export earnings of
agriculture in developing countries were not enough to cover the
combined gross foreign exchange requirements of agricultural investment and inputs. "This situation would prevail in the future even
in the face of greatly improved net trade earnings," the study says.

The gap between these requirements and the net agricultural trade
earnings would, however, grow only modestly from about $6,000
million in the mid-seventies to around $10,000 million by 2000 due
to a favourable development of the net trade balance in the scenario. This relatively favourable picture is due to two factors: investment requirements considered in these calculations exclude the
components for transport and processing, and secondly, it is
assumed that the estimated foreign exchange share of inputs would
decline from around 33 percent in 1975 to 20 percent by 2000.

In practical terms, however, the investment problem seems to
be more thorny than the overall figures suggest. In many regions,
the report says, the recent emphasis on agricultural investment
"has barely matched the rate of physical depreciation of irrigation works, soil erosion and related capital investment. Much
more is needed in the 1980s and 1990s to catch up on past neglect
of both domestic and external investment and to bring about higher
growth in this sector for balanced development."

EXTERNAL AID

The report therefore calls for a sharp increase in aid
commitments, particularly through the early 1980s, to overcome
past neglect and to provide the basis for future rapid development,
particularly in the poorest countries. The study confirms an
estimate of the World Food Council that annual investments in agriculture of $8,300 million would be required between 1975 and 1980,
and projects requirements rising to about $13,000 million a year
by 1990 and $17,000 million by the year 2000. Moreover, assistance
requirements might rise even more strongly if the developing countries themselves turn out to be unable to finance a growing share
of their foreign exchange requirements for agriculture. The terms
of foreign aid would also need to become more generous. The share
of concessional lending would have to rise more quickly than total
aid, namely from 67 percent in 1976 to 85 percent in 2000. This
would reflect the growing emphasis on the poorest countries as
the share of assistance in funding investment gradually declines
when growth in the advanced developing countries becomes more self-
supporting. More predictable, continuous and assured aid commitments, the report says, would help to increase the impact and
"cost-effectiveness" of assistance.

TRADE

Lack of ability on the part of developing countries to pro-
cure the necessary foreign exchange could be caused by an unsatis-
factory development of agricultural trade. "Total export availabilities", the measure of the agricultural trade potential of the
90 developing countries analyzed in the study have increased
sluggishly in the past at an average rate of only 1.7 percent

per year. The normative scenario, however, envisages a potential
for a 4.7 percent annual growth.

The growth of import requirements of agricultural commodities
could, on the contrary, decrease from a past rate of 4.6 percent
per annum to 3.6 percent per year during 1980-2000. Thus, the developing countries' net agriculturual trade balance with the rest of
the world could increase from $12,000 million in 1975 to $19,000
million by 1990 and $29,000 million by the year 2000, assuming that
availabilities and requirements would indeed materialize as actual
trade. The projected development of supplies and self-sufficiency
ratios in individual commodities is shown in Table 1 (3).

RESTRUCTURING THE WORLD ECONOMY

That, of course, would depend upon deliberate governmental
efforts toward a restructuring of the world economy for a New International Economic Order. Trade liberalization leading to better
market access would be vital if the developing countries' share
in world agricultural exports is to rise from 38 percent in 1980 to
48 percent in 2000. This would require more radical changes in agricultural trade policies than governments have been willing to negotiate up to now. "Many of the major adjustments", the study argues,
"are needed on the part of developed countries." Developing countries, too, would have to adjust, for instance by reducing some of
the heavy export taxes levied on commodities and by liberalizing
their own agricultural import policies, thus permitting a freer
flow of agricultural products among themselves for improving their
collective self-reliance.

The restructuring which is urgently required would presuppose
that governments find and introduce "ways of protecting domestic
interest groups without resorting to protection which
hampers the development efforts of developing countries,"
the report says.

FOOD AID

Even if all demand and production targets of the normative
scenario were reached and the foreign exchange situation of the
poor countries develops as envisaged, there would still remain a
gap between food supplies, food demand and food requirements
which would call for food aid.

Despite the acceleration of food production, 330 million
people would nonetheless remain undernourished by 1990. Assuming
that the international community would aim at eliminating the
calorie gap of at least one-fifth of the undernourished by 1990,
a minimum additional 6 million tons of cereals would be needed,
thus bringing the total food aid requirements , including an
estimate of 1 million tons for developing countries additional

Table 1.   Historical and Projected Supply Utilization Accounts (in Thousand Metric Tons) for 90 Developing Countries(3)

Commodities	Year	Food	Feed	Other Domestic Use (1)	Total Domestic Disappearance	Exports	Imports	Net Trade	Production	Self-Sufficiency Ratio (%)
1. Wheat	1963*	51414	1519	10700	63633	3972	18311	-14338	49294	77
	1975*	87873	3969	22822	114664	2567	34362	-31794	82869	72
	1990	142660	7675	23687	174022	5477	48254	-42775	131246	75
	2000	186523	13707	28874	229104	8036	73451	-65414	163689	71
2. Rice, Milled	1963*	84381	793	11238	96412	4877	4803	74	96486	100
	1975*	113346	1463	16285	131094	3324	5563	-2239	128855	98
	1990	177500	3848	24061	205409	5556	5279	278	205687	100
	2000	220478	7393	31091	258962	11141	9053	2088	261051	101
3. Maize	1963*	27423	13218	8685	49326	4654	1381	3272	52599	107
	1975*	37067	23518	11939	72524	8208	5917	2291	74815	103
	1990	57802	53212	21045	132059	14071	14175	-103	131956	100
	2000	74921	93765	31640	200326	26917	29093	-2175	198150	99
4. Barley	1963*	6938	5399	2794	15131	878	952	-73	15057	100
	1975*	9512	7719	3051	20282	345	2375	-2028	18253	90
	1990	12654	18064	5920	36638	165	9734	-9568	27069	74
	2000	15165	34791	9490	59446	190	21571	-21379	38065	64
5. Other Cereals	1963*	32842	4270	6058	43170	1215	493	722	43892	102
	1975*	38218	10494	5585	54297	3739	2464	1275	55572	102
	1990	52924	21231	9484	83639	7611	7336	275	83914	100
	2000	64310	37021	13451	114782	15003	15815	-811	113969	99

(1) This item includes industrial use, seed and waste. 1963* and 1975* include also stock changes.

Table 1. (continued)

Commodities	Year	Food	Feed	Other Domestic Use (1)	Total Domestic Disappearance	Exports	Imports	Net Trade	Production	Self-Sufficiency Ratio (%)
6. Roots and Tubers	1963*	83495	15027	28661	127183	2767	703	2064	129247	102
	1975*	112385	18607	36085	167077	7822	958	6864	173941	104
	1990	163090	39094	54732	256916	14707	953	13754	270670	105
	2000	201328	62733	71721	335782	23526	1257	22269	358051	107
7. Sugar, Raw	1963*	25469	403	656	26528	11033	3376	7657	34184	129
	1975*	38470	598	2116	41184	15083	4611	10473	51657	125
	1990	68197	951	751	69899	24145	5183	18961	88860	127
	2000	95122	1453	758	97333	35981	7908	28073	125406	129
8. Pulses	1963*	17367	1642	2770	21779	784	421	363	22143	102
	1975*	18874	1591	3251	23716	884	607	277	23993	101
	1990	29591	2730	5020	37341	1310	481	829	38170	102
	2000	39743	4394	6972	51109	2453	883	1570	52679	103
9. Vegetables	1963*	57301	330	7382	65013	1134	593	541	65554	101
	1975*	84463	607	11165	96235	1841	788	1053	97289	101
	1990	148788	1009	19320	169117	3580	275	3305	172422	102
	2000	217237	1619	27932	246788	7193	458	6736	253523	103
10. Bananas and Plantains	1963*	22149	1007	8699	31855	3522	311	3212	35067	110
	1975*	31905	1312	12456	45673	5798	449	5349	51021	112
	1990	52224	2655	18949	73828	6929	685	6245	80072	108
	2000	70760	4479	25134	100373	9272	1067	8205	108578	108

(1) This item includes industrial use, seed and waste. 1963* and 1975* include also stock changes.

Table 1. (continued)

Commodities	Year	Food	Feed	Other Domestic Use (1)	Total Domestic Disappearance	Exports	Imports	Net Trade	Production	Self-Sufficiency Ratio (%)
11. Citrus Fruit	1963*	8143	0	1197	9340	1191	127	1064	10404	111
	1975*	16246	0	2493	18739	2140	572	1568	20308	108
	1990	26981	0	4365	31346	7006	867	6139	37484	120
	2000	38377	0	6480	44857	13853	1511	12342	57199	128
12. Other Fruit	1963*	34073	1062	11272	46407	2208	831	1376	47783	103
	1975*	47273	1445	13427	62145	3632	1352	2280	64425	104
	1990	85928	2440	19003	107371	7155	596	6559	113930	106
	2000	128072	4000	25509	157581	14779	985	13794	171375	109
13. Vegetable Oil	1963*	6884	221	2162	9267	3814	1110	2704	11972	129
	1975*	10671	155	3762	14588	5556	2416	3140	17727	122
	1990	21159	269	5442	26870	11489	4311	7179	34048	127
	2000	33486	424	7856	41766	19555	7896	11659	53426	128
14. Cocoa Beans	1963*	116	0	59	175	1042	36	1006	1181	676
	1975*	146	0	139	285	1175	40	1135	1421	498
	1990	294	0	189	483	1653	79	1574	2057	426
	2000	467	0	196	663	2363	131	2232	2895	437
15. Coffee	1963*	1052	0	697	1749	2763	138	2625	4374	250
	1975*	1221	0	-238	983	3437	174	3264	4247	432
	1990	2200	0	160	2360	4243	313	3931	6291	267
	2000	3240	0	220	3460	5509	507	5001	8462	245

(1) This item includes industrial use, seed and waste. 1963* and 1975* include also stock changes.

Table 1. (continued)

Commodities	Year	Food	Feed	Other Domestic Use (1)	Total Domestic Disappearance	Exports	Imports	Net Trade	Production	Self-Sufficiency Ratio (%)
16. Tea	1963*	650	0	16	666	601	201	400	1066	160
	1975*	902	0	11	913	692	259	433	1346	147
	1990	1523	0	21	1544	950	512	438	1982	128
	2000	2211	0	27	2238	1267	811	455	2694	120
17. Tobacco	1963*	0	0	1290	1290	454	95	359	1648	128
	1975*	0	0	1643	1643	683	160	523	2166	132
	1990	0	0	2716	2716	979	173	806	3522	130
	2000	0	0	3597	3597	1604	252	1351	4948	138
18. Seed Cotton	1963*	0	0	8136	8136	6231	1276	4954	13090	161
	1975*	0	0	11791	11791	5655	1995	3659	15450	131
	1990	0	0	19323	19323	8848	3557	5291	24614	127
	2000	0	0	25118	25118	16175	5817	10358	35476	141
19. Jute and Other Fibres	1963*	0	0	3080	3080	1839	191	1647	4727	153
	1975*	0	0	3266	3266	1185	282	902	4168	128
	1990	0	0	4330	4330	915	495	420	4750	110
	2000	0	0	5588	5588	1022	792	230	5818	104
20. Rubber	1963*	0	0	217	217	2173	214	1959	2176	1002
	1975*	0	0	678	678	3100	366	2734	3412	503
	1990	0	0	1099	1099	5092	521	4571	5670	516
	2000	0	0	1558	1558	7649	756	6893	8451	543

(1) This item includes industrial use, seed and waste. 1963* and 1975* include also stock changes.

Table 1. (continued)

Commodities	Year	Food	Feed	Other Domestic Use (1)	Total Domestic Disappearance	Exports	Imports	Net Trade	Production	Self-Sufficiency Ratio (%)
21. Beef and Veal	1963*	8072	0	-61	8011	1156	367	789	8800	110
	1975*	10682	0	41	10723	1054	470	584	11307	105
	1990	18600	0	36	18636	1348	934	414	19050	102
	2000	27761	0	41	27802	2538	1672	866	28668	103
22. Mutton	1963*	2413	0	4	2417	86	70	16	2433	101
	1975*	2935	0	6	2941	104	182	-77	2863	97
	1990	5000	0	13	5013	76	783	-706	4306	86
	2000	7383	0	16	7399	152	1338	-1185	6214	84
23. Pigmeat	1963*	2446	0	38	2484	22	43	-20	2463	99
	1975*	3575	0	50	3625	30	27	3	3627	100
	1990	6896	0	56	6952	37	6	32	6983	100
	2000	11048	0	59	11107	57	10	48	11155	100
24. Poultry Meat	1963*	1523	0	-3	1520	10	40	-29	1490	98
	1975*	3641	0	-2	3639	41	183	-142	3496	96
	1990	9288	0	0	9288	104	6	99	9386	101
	2000	18823	0	0	18823	98	22	76	18899	100
25. Milk	1963*	55942	7531	5616	69089	113	4973	-4859	64230	93
	1975*	77658	10601	6496	94755	462	8770	-8307	86447	91
	1990	136885	17835	10551	165271	244	19382	-19137	146134	88
	2000	203086	26048	15328	244462	191	32800	-32607	211854	87

(1) This item includes industrial use, seed and waste. 1963* and 1975* include also stock changes.

Table 1. (continued)

Commodities	Year	Food	Feed	Other Domestic Use (1)	Total Domestic Disappearance	Exports	Imports	Net Trade	Production	Self-Sufficiency Ratio (%)
26. Eggs	1963*	1779	0	315	2094	22	25	-3	2091	100
	1975*	3369	0	581	3950	17	64	-47	3903	99
	1990	7500	0	1414	8914	39	0	39	8953	100
	2000	13641	0	2770	16411	29	0	29	16440	100
27. Cereals	1963*	202998	25199	39474	267671	15596	25941	-10344	257326	96
	1975*	286016	47163	59682	392861	18183	50681	-32497	360363	92
	1990	443540	104029	84200	631769	32880	84778	-51897	579871	92
	2000	561397	186677	114548	862622	61287	148982	-87694	774927	90
28. Meat	1963*	14454	0	-22	14432	1274	520	754	15186	105
	1975*	20834	0	93	20927	1229	863	366	21293	102
	1990	39787	0	101	39888	1566	1728	-162	39726	100
	2000	65017	0	115	65132	2845	3042	-195	64936	100

(1) This item includes industrial use, seed and waste. 1963* and 1975* include also stock changes.

to those studied individually, to 20 million tons by 1990, the
study warns. It furthermore expects that food aid requirements
of the same size would persist in the year 2000, though a precise
quantitative assessment is considered very difficult. This would
be required to close the calorie gap of the 240 million people
remaining undernourished by 2000. To help these people food aid
in project form would be particularly required. The complete
picture of envisaged changes in per capita food consumption as
measured by calorie intakes is given in Table 2 (3).

SELF-SUFFICIENCY IN CEREALS

Under the normative scenario the declining tendency of the
cereals self-sufficiency ratio of the developing countries could
be arrested: it could be maintained in the area of 90 percent.
However, the absolute volume of imports of the developing deficit
countries would still increase considerably by 2000, though partly balanced by increasing exports (to the tune of 45 million tons
by 2000) which could be forthcoming from the cereals surplus of
developing countries.

Moreover, a good deal of the overall cereals deficit should
be concentrated in the better-off developing countries, including
the major petroleum and other mineral exporters. By contrast,
such traditional deficit regions as the Indian sub-continent could
become fully self-sufficient and even turn into a modest net exporter. Such developments reflect in part also the commodity composition of demand; increases in wheat demand are largely translated
into import requirements because of production constraints in many
tropical developing countries. By contrast, rice production could
respond fully to the quest for complete self-sufficiency of the
developing countries as a group. Another factor which would be of
increasing importance in cereals gap developments is the emergence
of some of the higher income developing countries as significant
importers of coarse grains for animal feeding purposes, driven by
fast growing demand for livestock products.

The above described developments refer to country groups. An
examination of the study's results for individual countries, however, reveals that quite a few of them, particularly mineral-rich
and other threshold countries, could be facing mushrooming cereal
deficits and a rapidly declining self-sufficiency.

LIVESTOCK AND FISHERIES

Apart from the analysis of the food problem as a whole,
"Agriculture: Toward 2000" deals in considerable detail with specific sectors of food production, such as livestock and fisheries.
Whereas demand for fish is expected to outstrip supplies and price
increases seem inevitable, livestock production could rise steeply
- by nearly 5 percent per year - under the normative scenario,

Table 2. Calorie Intakes (3)

	Year	Developing Countries	Africa	Far East	Latin America	Near East	Petroleum Exporting Countries	Other Developing Countries Low Income	Other Developing Countries Medium Income	Other Developing Countries High Income
1. Calorie intake (per caput/per day)	1963	2141	2115	2035	2453	2336	2115	2041	2175	2526
	1975	2207	2197	2054	2543	2614	2357	2001	2324	2656
	1990	2452	2308	2322	2856	2755	2524	2247	2527	3014
	2000	2645	2520	2504	3080	2890	2703	2455	2698	3183
2. Calorie intake as percent of requirements	1963	93	90	91	102	94	92	90	94	104
	1975	96	94	92	106	106	103	89	100	109
	1990	107	98	104	120	112	110	99	109	124
	2000	115	107	112	129	117	117	109	116	131
3. Calorie intake from cereals (per caput, per day)	1963	1264	1004	1371	975	1520	1136	1306	1413	1069
	1975	1301	1038	1402	998	1633	1312	1293	1462	1101
	1990	1382	1081	1534	1077	1570	1298	1431	1493	1166
	2000	1395	1154	1555	1096	1496	1254	1495	1479	1121
4. Calories from cereals as per-cent of total calorie intake	1963	59	47	67	40	65	54	64	65	42
	1975	59	47	68	39	62	56	65	63	41
	1990	56	47	66	38	57	51	64	59	39
	2000	53	46	62	36	52	46	61	55	35

particularly of meat other than beef and veal.

Breeding fish in enclosed water areas, or aquaculture, could increasingly help to satisfy demand for fish in view of the limited growth potential of maritime fish catches. Utilization of hitherto unexploited resources of conventional fish or of unconventional species, better utilization of the catches and improved management of resources, particularly in regions that have become the Exclusive Zones of developing countries, could also help to maintain a slow but sustained growth of production.

Not only in fisheries but also in agriculture and forestry, badly managed production growth could collide with the need for protecting our resources and the environment, our most valuable heritage.

ENVIRONMENT

"Agriculture: Toward 2000" pays attention to environmental issues. The report particularly stresses the need for country-level long-term plans for land and water use. Such plans can be more effectively implemented by promoting understanding and acceptance by the population at large. Another vital prerequisite of environmental protection is a remunerative farm-gate price policy which helps the farmer and forester to cover all long-term costs of using natural resources.

PERSPECTIVE STUDIES

Every modelling exercise, be it predictive or normative, is of course an effort to simulate utopia. "Agriculture: Toward 2000" is no exception. As a successor to the 1969 "Provisional Indicative World Plan for Agricultural Development" (IWB(4)) and to the FAO contribution to the documentation of the 1974 World Food Conference, the present study is built on FAO's considerable experience in perspective studies. In this respect, the foreword to "Agriculture: Toward 2000" says: "All major regions include a number of countries whose record in improving agriculture demonstrates clearly that sustained high growth is possible. Slow agricultural growth in other developing countries is generally neither inherent, nor is it inescapable in the future; imposed by past circumstances, it can be reversed - though not easily - by new policies, attitudes and incentives"(5).

REFERENCES

1) Agriculture: Toward 2000, FAO, C 79124, Rome, July 1979
2) ibid. p. VIII.
3) ibid. Statistical Appendix, Tables 5 and 9.
4) FAO, Rome, 1969.
5) FAO, 1979, op.cit.,p VIII

DISCUSSION

It was suggested that Dr. Pino's view may be too narrowly technical and not concerned enough with capital. Raw material for fertilizer is abundant but expensive. Irrigation can be used to make 2.5 billion ha of arid land arable, but that too will be expensive. Where will the capital come from? Who can afford these things - Tanzania, India, even the People's Republic of China? Dr. Pino agreed that capital is the big question in places where food is a problem. He said that it is technically feasible to increase the food supply, but he does not know whether it is financially feasible.

A participant said that the problem is that fertilizer is being put on the land in Europe and America instead of in the developing countries, and the food is shipped to where it is needed with additional costs for transportation and distribution. He suggested that the international community should take the funds that are available and put them into improving agricultural production in developing countries instead of sending them food aid. Another participant said that fertilizer is simply too expensive for farmers in poor countries. The low agricultural prices and terms of trade that prevail in developing countries mean that the farmers cannot afford fertilizer, but farmers in developed countries can because agricultural prices are high and terms are favorable. It is important to see the problem as economic rather than technical. It was pointed out that most subsistence farmers are not able to pay for inputs such as fertilizer. Unless the farmer sells some of his product, he will not buy fertilizer to increase his productivity.

Dr. Borgstrom was asked how he calculates the nutrient requirements of the animal population - on the basis of grains or nutrients? Dr. Borgstrom said that he uses both kinds of nutrients. A participant pointed out that much land can be used only for growing grass, thus feed and fodder should be viewed separately in considering animal nutrients. He suggested that it is misuse of good agricultural land to use it for production of animal feed instead of human food, but that feeding animals on grassland that is not suitable for crops is not misuse. Dr. Borgstrom noted that building a city on either pasture land or cropland reduces food production.

A participant said that there seems to be something wrong with the balance between feeding humans directly with crops and using crops to feed animals that are then fed to humans. What kind

117

W. Bach, J. Pankrath, and S. H. Schneider (eds.), Food-Climate Interactions, 117–120.
Copyright © 1981 by D. Reidel Publishing Company.

of signals, he asked, are we giving to agricultural producers that
encourage them to misuse resources in this way? It was pointed
out that many farmers have no idea how much land they are depending on. A farmer who buys soybeans or hay to feed his cattle does
not consider how much land it took to raise them. The feedlot ope-
rator is the ultimate example - he does not have the slightest
idea how much land he is using. We seldom look at the real costs
of the things that we do. Recycling manure is often more efficient
than using chemical fertilizer because it takes so much energy
to manufacture fertilizer. But most farmers know very little about
that. It was suggested that two types of education are needed.
The first is a relatively long-term approach that will make ecological facts clear to decision makers and farmers (or to the farmer's children and grandchildren). The second should promote incentives to speed up the transformation from ecological irresponsibi-
lity to ecological soundness in agriculture.

Dr. Hodges was asked if greenhouses have problems with a CO_2
deficiency. He replied that if you have a totally closed system
to minimize water usage, such as greenhouses that "rain" the equivalent of 180 inches/year, you have to bring in CO_2. It is a by-
product which one may obtain from a power plant or chemical process plant. If you are not maximizing water economy, you can venti-
late and bring in outside air. With this system you can enrich
above natural CO_2 levels. With some crops you can go as high as
2400 ppm and still get increases in photosynthesis. He also suggested that the use of greenhouses might contribute to political
stability, citing the fact that the Israelis were glad to hear
that Jordan was developing a greenhouse industry because they felt
that the Jordanians might be less likely to act recklessly.

Dr. Hodges was asked if soil quality and the need for nutrients are problems in maintaining the stability of this sort of
artificial ecosystem. He replied that soil is one of the present
limitations. The soil must be porous so that you can irrigate
generously and still flush through. He said that at Puerto Penasco
they do not have to add fertilizer because they get the waste products from the aquaculture system. If it was necessary to add fertilizer, the crops would be less likely to represent the full value of the investment. Trickle irrigation with sea water can be
used without plugging up the pipes. It was suggested that a total
energy analysis of one of these greenhouse systems would be useful.
They may be more energy intensive than other methods, especially
if you count the plastic and every piece of the structure, etc.
Dr. Hodges said that he does not claim that these systems are energy efficient compared to outdoor production, but that they are
not really very far off. In reply to the enquiry of other research
on salt-tolerant plants he replied that similar efforts are in
progress at the University of Maryland and the University of California. However, Epstein at the University of California is trying
to make traditional crops tolerant of salt water by genetic selec-

tion, whereas the group at the University of Arizona is taking
plants that have naturally evolved from sea water and trying to
develop agronomic systems for them. Similar work is also being
done at Elat, Israel. To the question of whether greenhouses have
pest problems such as aphids and fungus Dr. Hodges replied that
they often have problems with white flies, aphids and fungus and
that they are hard to control. Such pest problems are potentially
serious, but integrated pest-control systems are possible.

Dr. von Loesch was asked how the scenario used in the FAO
report Agriculture: Toward 2000 compares with the one proposed
in the U.S. Global 2000 report. He replied that there are basic
differences. The Global 2000 scenario envisages rising relative
agricultural prices, whereas the FAO study assumes constant relative agricultural prices. The FAO group feels that their assumption is more realistic. He said that the Global 2000 report was
strongly influenced by the views of environmentalists. For example, the Global 2000 assumptions about forestry and climate were
very rigid and reflected more concern about these problems than
that of the FAO group. He said that projections prepared by the
U.S. Department of Agriculture do not differ drastically from
those of the FAO study, whereas the views of the environmentalists
differ from both.

A participant pointed out that one of the FAO scenarios projects an increase in fertilizer use by a factor of five by the
year 2000. He said that this would add a lot to the greenhouse
effect of CO_2 because N_2O is one of the final products of fertilizer. We do not know very much about the global aspects of nitrogen, he said, but we do know that if the N_2O in the atmosphere
increases by such a large factor, the greenhouse effect of CO_2
will be accelerated by something like 30 to 40%. This would be
a very serious problem - not in the next 20 years, but thereafter. Dr. von Loesch said that the fivefold increase in fertilizer
use was for the 90 countries included in the study, not for the
whole world. Actual fertilizer use in developing countries is so
low, he said, that even a strong increase would not bring it anywhere close to the prevailing levels in Western Europe, which run
to more than 200 kg/ha. The levels envisioned for the developing
countries by 2000 are something like 100 kg/ha. Fertilizer consumption in the rich countries will probably increase as well,
so the problem, if there is one, concerns the potential fertilizer increase in the developed countries.

Dr. von Loesch was asked how the projected increase in modern inputs by farmers in developing countries will occur - through
market forces or aid programs? Will the 90 poorest countries in
the world be able to achieve the projected increase in domestic
production in response to market forces or will they have to rely
on aid for a long time? In his reply Dr. von Loesch pointed out
that the idea of the normative FAO scenario is to reduce depen-

dence in the long term. The study favors reducing input subsidies because they benefit the rich, whereas increases in farm-gate prices are scale neutral. Governments should give higher priority to agriculture, he said, impoving agriculture's balance of trade with other sectors of the economy. In the past, the overwhelming majority of the countries that were studied actually provided disincentives for agriculture. However, he feels that governments will have to rely partly on outside aid to finance programs to improve the status of agriculture. Market forces may work very well inside a country, he said, but they do not work very well at the international level. Asked how the FAO study projected energy prices, he replied that he was not sure what the assumptions were about energy prices, but that this was one of the weak points of the study. They are now working on an additional chapter, and the new study will come out in favor of reducing the impact of rising energy prices on agriculture.

CLIMATE VARIABILITY AND CROP YIELD IN HIGH AND LOW
TEMPERATE REGIONS

James D. McQuigg

Consultant
310 Tiger Lane, Columbia, Missouri, USA

ABSTRACT   There are three major components of grain yield variability over a period of years. Technological change is the most
important in the long run. Climate variability is the most important contribution to year-to-year yield variance. Several
examples of climate variability will be presented for the USA,
Canada and the USSR for wheat and coarse grains. With rising
costs for technological inputs and continued upward pressure on
grain demand, the agricultural production systems of the world
will be increasingly sensitive to climate variability. It will
be more important than ever to monitor climate.

INTRODUCTION

The word "variability" is used quite frequently in combination with the word "climate" in recent years. The opening
sentence of the WMO Secretary General's foreword to the published
summaries of papers presented at the recent World Climate Conference is, "During the past decade large variations of climate
have occurred in many parts of the world and have had very serious, in some cases, disastrous consequences for the people living
in the areas affected.....". The words "climate variability"
appear in nine of the twenty six papers presented at this conference.

All of the agricultural regions of the world are subject to
significant temporal and spatial climate variability. The impact
of this variability on yield and production of food grain is important in each case. My assignment is to discuss climatic variability and crop yields in the high and low temperate regions. I

W. Bach, J. Pankrath, and S. H. Schneider (eds.), Food-Climate Interactions, 121–137.

will do this in terms of aggregate yields for large areas rather
than in terms of the impact of weather on yields of grain from
an individual field or a particular farm.

The fact that I have chosen to use examples of yield and
climate variability from these large regions should not be interpreted as an indication that I do not believe that the impact
of climate variability on the operation of individual farms is
important. If I were making a presentation on the subject of
climate variability to an audience of individual farmers, I would
devote most of my time to a discussion of the alternatives they
have to adjust their management to this variability. I would
also point out the fact that their individual enterprises are
impacted by climate events in other parts of the country within
which they are operating and by climate events in other parts of
the world.

YIELD VARIABILITY

The three major components of yield variability over a period
of years are technological change, climate variability and random
"noise".

Technological change is the most important component of
yield variability over a long series of years. It is easy to
name the factors that have contributed to a changing agricultural
technology. These include such things as improvements in seed,
increased use of chemical fertilizer, insecticides, herbicides,
improved management practices, better machinery, increased use
of irrigation systems, etc..

The impact of technological change is apparent in the yield
data series of most crops in most of the agricultural production
regions of the high and low latitude temperate climate zones.

Year-to-year variability of climate is the most important
contributor to short term variability of yield. For example,
US corn yield for 1979 was 68.6 quintals/hectare. In 1980 the
yield will be close to 56.0 quintals/hectare, a drop of a little
over 18 per cent. As another example, the yield of winter wheat
in the USSR in 1978 was 29.8 quintals/hectare. In 1979 it was
19.7, a drop of almost 34 per cent.

In the case of the US corn yield change from 1979 to 1980,
the major contributing factor was a prolonged drought, with very
high temperatures during the critical flowering-pollination
period in July. In the case of the USSR winter wheat yield variation, the major contributing factor was a cool, wet beginning
of the winter wheat region growth.

The interaction between agricultural technology, climate and yield of grain is a very complex one. In many instances, the detailed precise data series one would need to "sort out" the impact of each of the technological and climatic factors is not available. There are also many problems inherent in the available yield data series. Enough sample data are available to allow a reasonable good level of understanding of the yield-technology-climate interactions. The component of yield variability that cannot be specified is usually included in the "error" term in an analysis of variance table.

GENERAL DISCUSSION

Tables 1 and 2 contain aggregate yield data for wheat and for coarse grains for the USA, the USSR and for Canada for the period of 1970 through 1980. These data provide further examples of the year-to-year variability of grain yield for very large aggregated regions.

Some of the more notable weather-related low yield years for wheat in the past decade are: 1974, 1980 for the USA; 1972, 1973, 1979 and 1980 for Canada and 1972, 1974, 1975, 1977, 1979 and 1980 for the USSR. For coarse grains the low yield years are: 1970, 1974, 1980 for the USA; 1974 and 1980 for Canada; 1971, 1972, 1975, 1977 and 1979 for the USSR and 1970, 1976 for western Europe.

Table 1. Wheat Yields (a) (b)

Year	USA	Canada	USSR
1970	20.8	18.0	15.7
1971	22.8	18.2	15.4
1972	22.0	16.9	14.0
1973	21.2	16.9	17.4
1974	18.3	14.9	14.1
1975	20.6	18.0	10.7
1976	20.4	21.0	16.3
1977	20.6	19.6	14.9
1978	21.3	20.0	19.2
1979	23.0	16.9	15.7
1980	21.8*	14.5*	17.0*

a) USDA Agricultural Statistics
b) Quintals per hectare
*) Preliminary values

Table 2.   Coarse Grain Yield (a) (b)

Year	USA	Canada	USSR	Western Europe
1970	35.9	22.9	18.2	29.3
1971	43.4	23.7	16.7	33.4
1972	47.3	22.6	14.2	33.2
1873	44.9	22.3	18.3	33.9
1974	37.1	19.5	16.8	34.5
1975	43.5	23.2	11.3	32.3
1976	44.8	25.2	18.9	30.0
1977	46.4	26.4	15.3	35.2
1978	50.2	25.7	16.9	37.0
1979	56.7	26.3	13.1	37.5
1980	48.3*	23.9*	16.0*	--

a)   USDA Agricultural Statistics
b)   Quintals per hectare
*)   Preliminary values

Most of these low yield years can be associated with one or more periods of unfavorable weather during critical periods of the growing season.   One exception is the year 1970 in the USA corn-growing region.   This was a failure of technology, resulting from widespread use of male-sterile genotypes in the production of hybrid seed corn.   This development was an attempt to reduce the amount of labor required in the detasselling process.   The technology succeeded in that respect, but the resulting genotypes were more susceptible to southern leaf corn blight.   The yield depression in USA corn that year was mostly the result of an outbreak of this disease.

SPECIFIC EXAMPLES

USA Corn

The yield data series for the 1950-1980 for US corn is shown in Table 3.   As mentioned above, the year 1970 yield was greatly influenced by an outbreak of southern corn leaf blight.

With the exception of that year, the variability of yield from year to year from 1950 to 1973 was very small, leading some persons to conclude that weather really was not an important factor at all.   The trend on corn yield from 1970 through 1976 was comparatively flat, leading a number of investigators (including myself) to suspect that the technologically induced trend in yields that had been the main feature of the yield series may have peaked.

Table 3.   US Corn Yield and Average Amount of Nitrogen Applied
           to Corn (a)

Year	US Corn Yield(b)	Nitrogen Applied to Corn (c)
1950	38.5	7
1951	36.9	10
1952	41.8	15
1953	40.7	18
1954	39.4	19
1955	42.0	19
1956	47.4	19
1957	48.3	20
1958	52.8	21
1959	53.1	22
1960	54.7	23
1961	62.4	25
1962	64.7	40
1963	67.9	50
1964	62.9	58
1965	74.1	75
1966	73.1	86
1967	80.1	93
1968	79.5	104
1969	85.9	110
1970	72.4	112
1971	88.1	107
1972	97.1	115
1973	91.2	113
1974	71.4	103
1975	86.2	105
1976	87.4	127
1977	90.7	128
1978	100.8	126
1979	109.4	132
1980	86.6*	132* (preliminary value)

a)   USDA Agricultural Statistics
b)   Bushels per acre
c)   Pounds of nitrogen
*)   McQuigg estimate

     The droughts of 1974 and 1980 have removed all doubt that
climate is an important factor in USA corn yields. The rapid
runup in yields from 1976 through 1979 was evidence that an upper
limit in USA corn yields had not yet been reached.

There are many factors that have contributed to the increase
in yields that we lump together under the term "technology". One
major technological factor in corn yield is the amount of nitrogen that is applied to the crop. In Table 3 is shown the data
series on the US average number of pounds of nitrogen applied to
corn each year during the period 1950 through 1980. (The value
for 1980 is a preliminary estimate.) The years following World
War II saw a rapid increase in the use of hybrid seed corn. Much
of the increase in yields that is apparent through the decade of
the 1950's is the result of better seed. The development of
better seed has continued to be an important factor and it will
be so in future years.

The substantial increase in the amount of nitrogen applied
to corn from 1962 through 1970 was a major factor in increased
yields. The apparent leveling off of nitrogen application from
1970 through 1976 is consistent with the flattening out of the
yield trend during those years.

Regression analysis of the yield and nitrogen applied to
corn data series for the period 1950-1979 (with the year 1970
deleted) shows that the nitrogen variable "explains" approximately 90% of the total variance in the yield series.

The impact of weather on USA corn yields is most evident
during the month of July. This is the period when the flowering-
pollination process is underway and the formation of the grain
begins. Weather is important during the other portions of the
years, too, but not as important as it is in July. The weather
data shown in Table 4 are weighted averages (weighted according
to the corn harvested area) of temperatures for the principal
corn-growing states of Iowa, Illinois, Indiana, Ohio and Missouri.

When the weighted July precipitation and July temperature
were also included as variables in a regression analysis, the
"explained" variance was 94%. This is similar to the results of
other regression-based weather-yield models in that the trend
(technology) term "explains" about 70-90 per cent of the total
variance. Weather "explains" from 5 to as much as 25 per cent
of the total variability.

Some of the unspecified variance is the result of the fact
that the published yield data for most agricultural production
regions are not based on precise measurements, but are produced
as a series of estimates. The weather data used in weather-yield
modelling is rarely a precise estimate, either, especially when
the modelling is for a relatively large agricultural region.
Even where the weather-yield investigation is for an individual
field, or an experimental plot, it is usually the case that the

investigator has available the weather data only for some nearby location. In most instances the weather data available and
the historical series of yield estimates include a reasonable
portrayal of the year-to-year variability of both weather and
yield.

The statistic of most interest, as far as food is concerned,
is production. We can write

$$P = (Y) \cdot (A)$$

where P is production, Y is yield and A is harvested area. Usually, the impact of weather is on yield, with only minimal impact
on harvested area. But there are years (and particularly vulnerable climatic-soils zones) where the impact of weather is felt
on the ratio between planted and harvested area.

In Table 4 are shown a series of values of the per cent of
planted corn area in the USA that was actually harvested for
grain. The value shown for 1980 is a preliminary estimate, prepared at the time this manuscript was being written in mid-
September. Also shown is a series of weighted July temperature
data for the five major US corn states. The impact of the
droughts of 1954, 1974 and 1980 is apparent in this graphical
presentation of the data.

Regression analysis of this series of harvested/planted
ratio and the weighted temperature and precipitation data for
the five major US corn-producing states shows that July temperature is a useful predictor.

The production estimate for USA corn we prepared early
in August 1980 was based on an estimate of yield, (using
weather-yield models) and an estimate of harvested area, based
on our regression model. From this analysis, we expect both
yield and precent harvested area to be lower than the "normal
weather" levels.

Another problem that arises in trying to develop and use
weather-yield models for a major crop (such as corn in the USA)
is that the "final" estimate of yield published by the Department
of Agriculture is expressed in terms of yield per harvested acre.
It might be interesting to use a series of yield estimates that
were published in terms of yield per planted acre. (Data on
planted area is available early in the crop season.)

A further problem in relating climate variability to the
variability of production for a crop such as USA corn is that
there are a few years in the data series where the weather events
are extreme, and where these events have a significant impact on

yield and production, but where the weather events are comparatively rare. In such an instance, it is difficult to develop
a quantified weather-yield model that adequately takes these rare
events into account.

A good example is the prolonged heat wave that occurred
in the western and southern portion of the USA corn growing
region in July of 1980. A series of days with maximum temperature in the daytime in excess of 100 degrees F (37.8C) occurred
just at the time the corn crop should have been in the flowering-
pollination process. The extreme heat caused the flowering to
proceed much too early. When the tassels started throwing pollen,

Table 4. Planted/harvested Ratio, US Corn, and Weighted
          July Temperature for the Five Major Corn States

Year	Ratio (a)	July Temperature (b)
1950	.874	71.2
1951	.855	73.5
1852	.868	76.4
1953	.867	75.7
1954	.836	77.7
1955	.797	79.6
1956	.833	73.5
1957	.862	76.7
1958	.866	72.3
1959	.871	73.7
1960	.877	72.7
1961	.874	73.6
1962	.857	72.5
1963	.861	74.4
1964	.841	75.8
1965	.850	73.1
1966	.859	77.4
1967	.853	71.6
1968	.860	73.9
1969	.849	75.5
1970	.858	74.5
1971	.865	71.7
1972	.860	73.2
1973	.861	74.8
1974	.839	76.7
1975	.859	74.3
1976	.845	74.7
1977	.857	77.0
1978	.878	74.3
1979	.887	72.9
1980	.830*	77.9*

a) Source of ratio data USDA
b) Temperature in F^o
*) Preliminary estimate

the silks had not yet formed, so there are many areas where there
are no ears on the corn stalks at all. In other instances, the
extreme heat dried the pollen and/or the silks and the pollination process was a failure. In Figure 1 are shown the number of
days with temperature 100 degrees F (37.7C) or higher during the
period July 1-23, 1980. The boundary between the area which received minimal pollination damage and the area that had significant problems with pollination of corn is roughly the isoline for
five days on this map. The most substantial damage during the
pollination process occurred with more than ten days with temperatures over 100 degrees F.

Figure 1   Number of Days with Maximum Temperature
                Greater than 100°F during July 1-23, 1980

These extreme temperatures are included in the monthly mean
value which we used as input for the weather-yield models we have
in an operational mode, so some of the pollination effect was included in our initial estimates of yield, but the quantified models do not adequately include the total weather impact in this
instance.

Another rare event which is difficult to include in a weather-yield relationship is a severe freeze early in the fall,
occurring just as the corn crop is beginning to mature, and before the grain in the field is completely dried down to safe
storage levels. Such was the case in 1974 in the US corn belt.

US Winter Wheat

In Figure 2 is shown a series of US winter wheat yield values
for the period 1950-1980. (The 1980 value is a preliminary estimate.) When viewed in graphical form, the trend in yields that
is the result of a changing agricultural technology is apparent.
Again, this trend is not a steady, almost linear change of yield
from year to year. There have been new varieties of seed introduced during this period that take a few years to be completely
accepted. There have been changes in the amount of commercial
fertilizer applied from year to year. The apparent flattening of
the trend in yields that was evident in the corn yield series
during the first part of the 1970s decade is apparent in the wheat
series, together with the recent runup of yields. Superimposed
on the trend in yields that is related to agricultural technology
is the year-to-year fluctuation of yield because of climate variability.

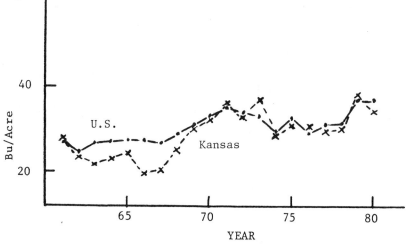

Figure 2   U.S. and Kansas Winter Wheat Yields
                    Source - USDA Agricultural Statistics

Kansas is one of the major wheat producing states. Shown in Figures 3 and 4 are March temperature and June precipitation values for a sample of years.

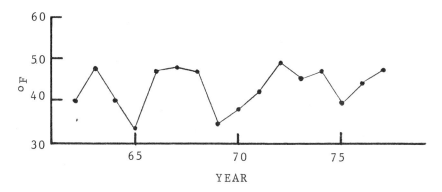

Figure 3   March Temperature for Kansas

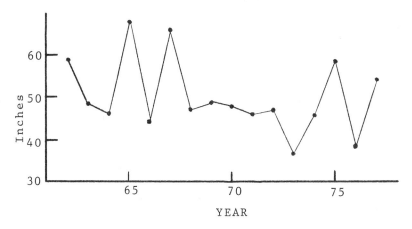

Figure 4   June Precipitation for Kansas

Canadian Wheat and Coarse Grain

In Figures 5 and 6 are shown a series of yield values for Canadian wheat and Canadian coarse grain. The general trend toward higher yields is apparent in both these series, on which is superimposed the fluctuation in yield induced by weather. Saskatchewan is one of the major wheat-producing provinces. In the case of wheat in this province, one of the most important weather variables is the amount of precipitation received in June. June

precipitation values for crop district 1, in southeast Saskatchewan are shown in Figure 7. The comparatively low wheat yield
values in 1961, 1967, 1968, 1974, 1979 and 1980 are coincidental
with comparatively low precipitation in June for those same years.

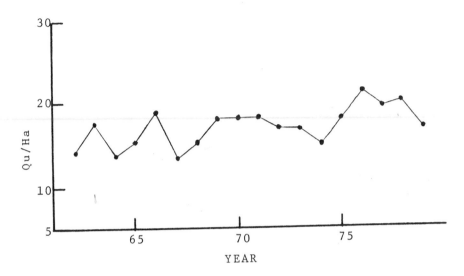

Figure 5   Canada Wheat Yields
                  Source - USDA Agricultural Statistics

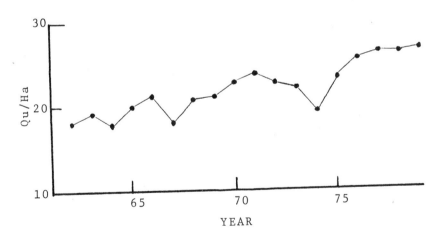

Figure 6   Canada Coarse Grain Yields
                  Source - USDA Agricultural Statistics

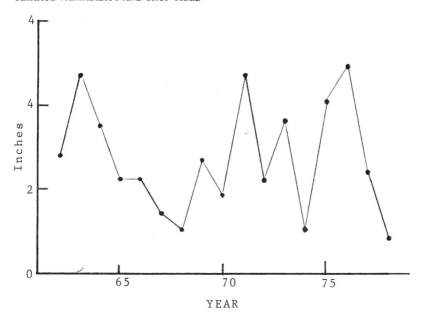

Figure 7   Saskatchewan Southeast District
                           Precipitation
                Source - Canadian Meteorological Service

    The relationship between Canadian wheat yields and weather
involves a much more complex set of weather data than just rainfall for one month in one corner of Saskatchewan. The temperature and precipitation during the fallow months, and during the
initial growth period, as well as temperature and precipitation
during the formation-filling of the grain, and during the harvest
season is important. A graph of time series of any or all of
these other weather variables would also illustrate the large
degree of year-to-year weather fluctuations that are an essential
feature of the climate of the Canadian wheat producing region.

    The graph of Canadian coarse grain yield, shown in Figure
6, illustrates the trend toward higher yields, upon which is
superimposed the impact of year-to-year fluctuations of weather.

USSR Wheat

    The USSR produces wheat over an immense geographical area,
with wide variations in topography, soil type and climate. Within a given year there are large differences of climate and yields
of grains within the USSR. Some years these differences tend to
balance out, but there are many years when they do not. These
would be the years when there were large anomalies in the general

circulation that would result in prolonged, wide-ranging anamolies
of precipitation and/or temperature that would have a major impact
on the total production-yield situation.

In Figure 8 is shown a series of yields of USSR spring and
winter wheat. To illustrate the kind of climate variability that
is a feature of the USSR winter wheat growing region, the April
temperature data series for the USSR crop district 13 (Central
District) are plotted in Figure 9 and June precipitation data
are plotted for crop district 26 (Western Siberia) in Figure 10.

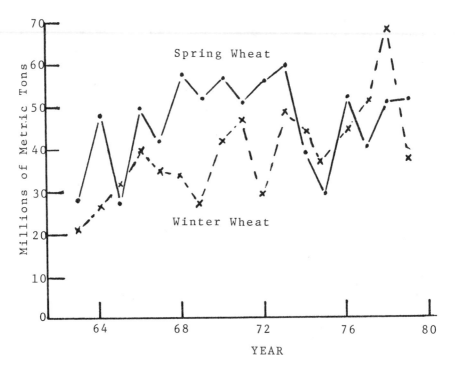

Figure 8   USSR Wheat Production
                Source - USDA Agricultural Statistics

The samples of temperature and precipitation data shown for
the several major grain-producing regions are intended to serve
as illustrations of the year-to-year variability of weather that
is the primary contributor to the year-to-year variability of
crop yield and production that is observed. Of course, there are
other months and other weather elements that contribute to grain
yield and production variability. Due to limitations of space,
these other weather factors are not illustrated here. (They are
included in most weather-yield models, however.)

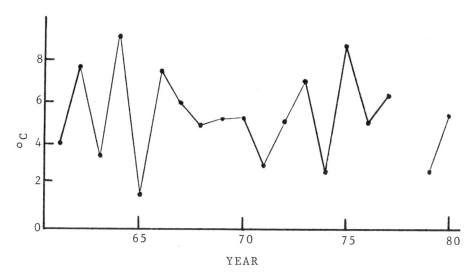

Figure 9   April Temperature Central District USSR

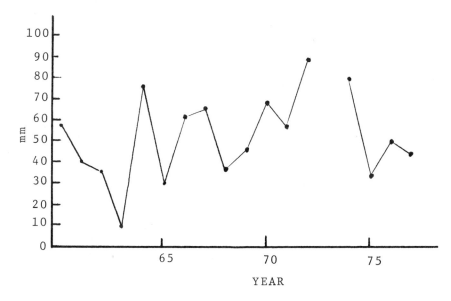

Figure 10   June Precipitation for Western Siberia
USSR

FUTURE CLIMATE AND GRAIN PRODUCTION VARIABILITY

There have been increases in the amount of land devoted to crops in several major production regions of the world in recent years, and this kind of thing will continue, as the pressure on world food resources continues to increase. (The data shown in Table 5 were included in my presentation to the World Climate Conference in Geneva in February of 1979.) As shown in Table 5, the increases in wheat production that have occurred during the last one or two decades has come mostly from increased yield. This is likely to be true during the coming one or two decades. The chief source of variability of grain yield and production will continue to be year variability of temperature and precipitation.

Table 5. Wheat yield, harvested area & production indexes (a)

(Canada, United States, USSR, China, France, Australia, Argentina, West Germany, United Kingdom, Spain)

Average of 1960-64 period = 1.00

Year	Yield	Production	Area
1960	1.023	.983	.973
1961	.998	.899	.980
1962	.906	1.025	.999
1963	.965	.945	.993
1964	1.106	1.148	1.055
1965	1.006	1.042	1.054
1966	1.256	1.316	1.061
1967	1.123	1.188	1.076
1968	1.247	1.321	1.078
1969	1.206	1.209	1.021
1970	1.298	1.234	.967
1971	1.405	1.342	.967
1972	1.381	1.279	.941
1973	1.489	1.471	1.001
1974	1.356	1.347	1.011
1975	1.264	1.304	1.050
1976	1.539	1.625	1.075
1977	1.423	1.487	1.063

a) U.S. Department of Agriculture, Foreign Agriculture Circular FG-9-76, May 1976, and Foreign Agriculture Circular FG-2-78, February 16, 1978.

With rising costs for fuel, machinery and other inputs, the

agricultural production systems of most of the world will be increasingly impacted by climate variability. The need to continue
development of systems to forecast future climate and to assess
the impact of climate variability on food production will be more
urgent in the coming decade.

When the shortfall in grain production in 1972 was recognized, some writers believed that this signalled the beginning
of a continuing catastrophe. Since that time, other writers
have been equally mistaken, when they concluded that the kind of
thing that happened in 1972 was just a rare event. This sort of
conclusion began to look reasonable, following one or two high
production crop seasons.

My conclusion, after several good crop years and some very
poor ones since 1972, is that neither the doomsayers nor the
overly optimistic interpreters of the climate-food scene are entirely right.

Climate will continue to be variable, and this will continue
to cause significant year-to-year fluctuations. These fluctuations will continue to cause large changes in the price and
availability of food grain.

SOURCES OF DATA

1. Yield and production data are found in:

    a. Agricultural Statistics, US Department of Agriculture
(a yearly series)
    b. Corp Production, US Department of Agriculture (a monthly
series)
    c. Foreign Agriculture Circular, Grains, US Department of
Agriculture (a monthly series)
    d. Foreign Agriculture Circular, Oilseeds, US Department of
Agriculture (a monthly series)

SUGGESTED READING

1. Baier, W., Crop-Weather Models and Their Use in Yield Assess-
    ments, 1977, World Meteorological Organization Technical Note
    No. 151, Geneva, Switzerland. (contains an excellent biblio-
    graphy).

CLIMATIC VARIABILITY AND CROP YIELDS IN THE SEMI-ARID TROPICS

L.D. Swindale, S.M. Virmani and M.V.K. Sivakumar

International Crops Research Institute for the
Semi-Arid Tropics, Hyderabad, India

ABSTRACT. The climates of the semi-arid tropics are characterized
by high incidence of solar radiation, high temperatures and very
variable rainfall. Droughts and floods are both common occurrences. Deficient rainfall years may be followed by similar
years or years with excess rainfall in no predictable pattern.

   Rainfed farming is risky in such conditions and farmers are
reluctant to invest in crop production. Traditional agriculture
means low but stable yields, low inputs, mixed cropping, large
families, low incomes and living standards and outmigration of
family members, both seasonal and permanent. Only about 4% of
the arable land of the semi-arid tropics is irrigated.

   Farmers try to ameliorate the effects of drought through
crop management. If the first crop fails they may plant a second
or a third. When all crops fail farmers and their families must
decrease their food intake, sell movable assets or their land.
Without government intervention many people may migrate or even
starve.

   New technologies now exist that reduce the risks of farming
in the rainfed semi-arid tropics. The technologies include
improved soil and water management on watershed-based land units,
the use of fertilizers and improved seeds, improved cropping
systems and supplementary irrigation. The productivity of traditional agriculture is declining and the threat of serious food
shortages is rising but the potential for the semi-arid tropics
to feed itself in all but the driest years does exist. Institutional and infrastructural improvements will be necessary for
this potential to be reached.

W. Bach, J. Pankrath, and S. H. Schneider (eds.), Food-Climate Interactions, 139–166.
Copyright © 1981 by D. Reidel Publishing Company.

INTRODUCTION

The semi-arid tropics covers much of the African continent
(Fig.1). It stretches in a broad band from west to east below the
Sahara desert, and includes most of eastern and south-central
Africa. In Asia it includes most of India, eastern Java and
north-eastern Burma and Thailand. It includes most of the northern
quarter of Australia, nearly all of Central America and portions
of north-eastern Brazil, Paraguay and Bolivia. More than 600
million people are estimated to live in the region, with 56% of
them in India.

Rainfall in the semi-arid tropics is highly variable. In
low-rainfall years there may be droughts; in high-rainfall years
or even for short periods in low-rainfall years there may be
floods. Deficient rainfall years may be followed by similar years
or by years with excess rainfall.

Sorghum and pearl millet are the major rainfed cereals of
the region. Rice is grown in the river deltas and wheat is grown
mainly in the winter season where irrigation is available. The
main grain legumes are pigeonpea, chickpea, cowpea and mung bean.
Groundnut, safflower, sesame and mustard provide the main cooking
oils.

Traditional agriculture in the semi-arid tropics is designed
to reduce the risk of losses in dry years, because they can be
very severe. The benefits that could accrue in good years are

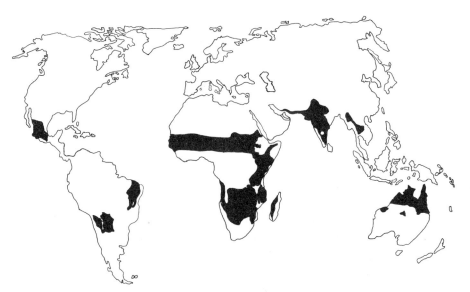

Figure 1. The semi-arid tropics, shown in black.

usually lost. Much of the effort to create new or improved
technology for the region is designed to remove the risk of loss
in dry years so that farmers will invest more in anticipation of
good years.

Food production in semi-arid India is increasing, although
not fast enough to improve standards of living very much. In
semi-arid Africa food production is declining. A recent report
from the International Food Policy Research Institute (1) predicts
substantial deficits in cereals in the semi-arid tropics by 1990.
India may have deficits of 20 million tons. Per capita deficits
in the African semi-arid tropics may be ten times as great as
those in India. A large and urgent effort in research and
development will be required to avert the human misery that such
deficits foretell. The effort will only succeed if the resources
of soil and water are used wisely and well, and if the inherent
variability in the climate of the semi-arid tropics is offset.
Although irrigation will help significantly, particularly to
increase production of rice, the potential for large-scale irrigation is quite limited. Most of the arable land must continue
to be cropped in rainfed conditions. In consequence this review
of present conditions and future prospects deals only with
rainfed agriculture.

CHARACTERISTICS OF SEMI-ARID TROPICAL CLIMATES

The semi-arid tropics is a region of high water demand and
scarce water resources. Temperatures are high, usually exceeding a mean of $18^{\circ}C$ in all months of the year. Total annual rainfall varies considerably from about 400 to 1200 mm. (Table 1).

Rainy seasons are short with most of the rainfall concentrated
in two to five months; from April to October in the northern
hemisphere and October to April in the southern hemisphere.
Seasonal rainfall is often 90% of total annual rainfall.

Of the climatic elements important for crop production -
rainfall, temperature and solar radiation - rainfall is most
variable. The coefficient of variation for annual rainfall is
20-30%, for annual temperature and solar radiation of the order
of 5%. Detailed climatological studies have shown that the
distribution of rainfall within the year is also highly variable,
and cannot be predicted with any reasonable degree of confidence.
Smaller yearly or seasonal totals are associated with larger
year-to-year variability.

In tropical latitudes there are large amounts of energy
available for evaporation of water. Since solar radiation is
the most important component affecting evaporation, and since it

Table 1. Average monthly rainfall (mm) for selected locations in the semi-arid tropics.

Location	Lat ° '	Long ° '	Jan	Feb	Mar	Apr	May	Jun	Jul	Aug	Sep	Oct	Nov	Dec	Annual*
1. Hyderabad (India)	17 27N	78 28E	2	10	13	23	30	107	165	147	163	71	25	5	761 (93)
2. Dakar (Senegal)	14 44N	17 30W	0	2	0	0	1	15	88	249	163	49	5	6	578 (98)
3. Bamako (Mali)	12 38N	08 02W	1	0	3	15	60	145	251	334	220	58	12	0	1099 (99)
4. Maradi (Niger)	13 28N	07 05E	0	0	0	4	32	60	164	260	110	12	0	0	642 (100)
5. Sokoto (Nigeria)	13 01N	05 15E	0	0	0	13	53	89	165	252	147	15	0	0	734 (100)
6. Ouagadougou (Upper Volta)	12 21N	01 31W	0	3	8	19	84	118	193	265	153	37	2	0	882 (99)

*Percentage of seasonal rainfall (April to October) to annual total is shown in parentheses.

varies little from year to year, potential evapotranspiration is
more or less constant from year to year. When rainfall exceeds
evapotranspiration, soil moisture reserves are recharged. When
rainfall is less than evaporation, soil moisture reserves are
utilized. With uneven seasonal distribution of precipitation and
with great inter-annual variability, small negative deviations in
precipitation are all that are required to initiate drought. In
semi-arid India moderate or worse droughts are likely to occur
one year in every four (Fig.2).

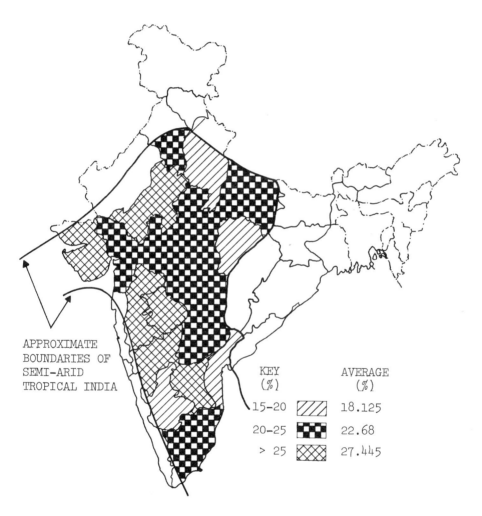

Figure 2.   Percentage occurrences of droughts of class moderate
            and worse in the rainy season in the Indian semi-arid
            tropics (2).

In most years the rainy season in the semi-arid tropics is long enough for crops to grow. Indeed there is usually excess water in the rainy season, some of which can be stored in the soil, but most of which runs off causing soil erosion. Management of land, crops and livestock are intimately associated with the inflow and outflow of water. In determining the agricultural potentials of any semi-arid area, quantification of the timing, amount and duration of rainfall, the intake, storage and release of soil moisture and the evapotranspiration is essential. For example, Virmani et al (3) using a water balance model incorporating these factors have shown that the length of the growing season at ICRISAT Center on shallow sandy red soils (Alfisols) fluctuates from 12 to 21 weeks and on deep clayey black soils (Vertisols) from 20 to 31 weeks (Table 2). The soil type clearly plays an important part in defining length of growing seasons in a given climatic situation.

FARMING THE SEMI-ARID TROPICS

There are 48 less developed countries with semi-arid tropical climates. There are about 3 ha of geographic land available per person in these countries to provide food, clothing, shelter and the wherewithal to invest in the future. In semi-arid India there is only 0.6 ha of land per person. (4)

Table 2. Length of the growing season (in weeks)[a] for three soil conditions.[b]

Rainfall probability	Available water-storage capacity		
	Low (50 mm)	Medium (150 mm)	High (300 mm)
Mean	18	21	26
75%	15	19	23
25%	20	24	30

[a] From seed-germinating rains (25 June) to end of season [time when profile moisture reduces EA/PE (Ratio of actual evapotranspiration to potential evapotranspiration) to 0.5].

[b] Low: shallow Alfisol; medium: shallow to medium-deep Vertisols; high: deep Vertisol.

India

In India virtually all farmers in the semi-arid tropics are small farmers. The resource base and endowments of farmers in the rainfed areas have been described in considerable detail by Jodha et al (5) and Jodha (6). More than 90% of the holdings are less than 12 ha in size and 70% are less than 8 ha. The land is usually held in 4 to 5 separate fields. Size of holding tends to be inversely related to rainfall or access to irrigation.

Investments are low. Fertilizers, agricultural chemicals and improved seeds are used only on land that can be irrigated. There is usually one draft bullock per farm and a total invest- ment of less than $150 on farm implements and machinery. Although only 4% of the arable land is irrigated, 50% of the farmers have access to a source of water for minor or supplemental irrigation.

Sorghum and pearl millet are the major rainfed cereal crops. All arable land is cropped at least once per year, i.e. the crop- ping intensity is at least 100%. Where there is access to water the cropping intensity is greater, but nowhere seems to exceed 120% (6, 7). Nearly 18 million ha, or 24% of the net sown area of semi-arid India, are fallowed during the rainy season and cropped in the post-rainy winter season.

Intercropping, i.e. growing two or more crops simultaneously is a common feature in the region. It is most common amongst small farmers and where farmers have least access to irrigation. It gives higher returns per hectare and may spread labor require- ments more evenly. It is also a hedge against disaster. Rao and Willey (8) examined the stability of 94 experiments involving sole pigeonpea, sole sorghum and sorghum/pigeonpea intercrops. Using a gross return of Rs.1,000 per hectare as a 'disaster' level, they found that sole pigeonpea failed one year in five, sole sorghum one year in eight, but intercropped sorghum/pigeonpea only one year in thirty six.

As many as 84 different crops are used in traditional inter- cropping systems in semi-arid India, but seldom more than three at one time. The systems used fall into four main categories (6), viz: mixtures of differing maturities, mixtures of drought- resistant and drought-sensitive crops, cash crop - food crop mixtures, and legume-non-legume mixtures.

The farmers of semi-arid India live in caste-dominated villages in nuclear families that usually include dependent parents, brothers and sisters. Standards of living are low but improving. Per capita annual income and literacy rates are generally below the national averages which in 1976 were $150 and 36% respectively (9). Malnutrition and poor health are widespread, but life expectancy has risen substantially in recent years. The infrastructure for development exists and is being improved.

West Africa

Newman et al (10) have comprehensively reviewed the literature on rainfed farming in semi-arid West Africa. Subsistence farming appears to be more common than in India. Much of the land is owned communally, but worked by family units which have usufructuary rights to as many as 16 separate fields on soils of varying quality around the village. The range in size of holdings is similar to that in India, and the main cereal crops are, again, sorghum and pearl millet.

Fertilizers and improved seeds are hardly known and seldom used, but animal manure is used on the fields nearest the village. The use of animal traction is not common. Donkeys are the usual source of animal power when it is used, except in Mali where oxen are fairly common. Because the power available is low, soils are seldom ploughed; and all agricultural operations are time-consuming.

Fallowing and burning are used to return some fertility to the soil. Cropping intensities on commonly used land seldom exceed 75% except for the land closest to the village which is used to grow the most important staple cereals. Land far from the village may be cropped only one year in four. Intercropping is common, usually of cereals with cereals or cereals with cowpea. Cash crops such as groundnut, bambara nut and sorrel are grown on small fields as sole crops.

Traditional villages in semi-arid West Africa have a strong sense of community and often an hierarchical system of control (10). Family units are large and often contain several adult males and 10 to 20 people, but nuclear family units comprising a single adult male and his dependents are increasing. Standards of living are low and food production is declining. Because population growth is high the pressures on land are increasing.

ADJUSTMENTS TO DROUGHT

The traditional farmer in the rainfed semi-arid tropics lives with the possibility of drought. He has learned by experience how to adjust to the inherent variability of climate. Each year he must decide what to plant, where and when. If he is a subsistence farmer he will consider mainly what will survive until harvest to meet the needs of his family. If he is also a commercial farmer - as virtually all farmers are in India - he

will also consider what price he will get for his marketable
produce. He must choose amongst several crops and within each
crop amongst several types and varieties. He has several fields
with differing fertilities and water-holding capacities at
differing distances from home, a limited time period in which to
plant and insufficient labor to do everything at once. He starts
with a bewildering array of choices that are rapidly reduced in
number by their primary determinants; the timing, intensities
and frequencies of the early rains.

When a farmer first perceives that his crops might fail he
will try to improve weed control or thin the stand. Both actions
conserve soil moisture. In soils that crack, creating a dust mulch
on the surface has the same effect. If any supplementary water is
available it will be used on the most important crops. If the
first crop does fail he may replant it or plant a second or a third.
Near Hyderabad, India, early planted sorghum may be replaced by
finger millet in mid-season and even by cowpea or horse gram in
the last month of the monsoon. At Jodhpur, in a much drier
climate where pearl millet is the only suitable cereal, the
farmer has no high-calory alternative if that crop should fail.

Jodha (11, 12) describes the adjustment mechanisms used by
farmers in northwest India when crops fail due to drought. First
the farmer reduces his social consumption, i.e. he will delay a
family ceremony (a marriage) or make lesser inputs into annual
festivals. He may also decide to sell off some nonproductive
assets. He will come to no permanent harm, nor will his family,
and his productive resources will remain intact and available for
use when conditions improve.

When the drought is more severe he must adjust in more
drastic ways. His first recourse is to reduce the consumption of
food. Table 3 shows that the greatest reduction tends to occur
in the more expensive, protective foods such as milk, sugar, meat,
and vegetables. Cereals and pulse crops are the last to be given
up. Obviously a serious decline in the consumption of protective
foods will seriously affect the farmer and his family. This is
particularly likely to be true where the cultural tradition is
to deprive the children first (13). Faulkingham and Thorbahn (14)
have estimated for an area of eastern Niger that 25% of the children under age 5 died in the last Sahelian drought because of food
deprivation.

When the farmer has reduced consumption as low as he dares,
he disposes of his assets - first his animals, as he cannot work
with them and he cannot feed them. Then he disposes of his farm
implements and machinery. Assets disposed of may take several
years to recover. In spite of the substantial governmental assistance provided to farmers in Maharashtra during the 1971-72

Table 3. Indexes of Consumption Expenditure by Households in Drought years Relative to Post-Drought Years in Different Areas (Post-drought year situation = 100), from Jodha (12).

Items of Expenditure	Jodhpur (Rajasthan)			Barmer (Rajasthan)			Banas Kantha (Gujarat)		
	Small Farms	Large Farms	All Households	Small Farms	Large Farms	All Households	Small Farms	Large Farms	All Households
Total Food Items	95	99	98	93	110	103	90	97	94
Protective Foods	52	85	64	43	74	58	87	98	72
Clothing, Fuel, etc.	79	85	84	84	83	85	72	76	69
Socio-religious ceremonies	23	56	36	69	50	52	63	18	66

drought, farmers in some of the drier regions, such as Sholapur, had not fully replaced their lost animals in 1978.

The penultimate step is forced outmigration of families. In Gujarat and Rajasthan during the 1972 drought, 40% or more families migrated from 50 to 243 kms from home for periods from 100 to over 200 days (12). Seasonal migration for work or in search of feed is normal in transhumant societies, but in severe drought the migrating families may never return, as was true for the American Great Plains in the twenties and for the Sahel in the seventies.

The final act of adjustment and the most destructive is to sell one's land. No redress is possible and the seller is condemned to a landless life either in the rural labor force or in shanty towns on the fringes of cities. Clearly it is wise for governments to bring in relief measures that will at least prevent starvation and distress sales of land.

From 1971 to 1973, severe drought, similar to that in the Sahel, affected much of India. In the face of widespread crop failures, human and animal migration and the very real threat of starvation, the government of the State of Maharashtra organized massive food-for-work programs in its rural areas. At its peak over 5 million people were employed on relief work, building roads, dams and canals. Subramanian (15) claims that there were no deaths directly attributable to starvation in those years. Many critical and perceptive reporters visited Maharashtra and came away full of praises for the state effort. Wolf Ladejinsky, in the Economic and Political Weekly (Bombay) of 17 February 1973, called the Maharashtra drought "a disaster of unprecedented dimension" and complimented the state government for organizing relief and employment works on such a large scale. John Pilger, in the New Statesman (London) of 8 June 1973, commented, "the Herculean relief efforts of both the Indian and State Governments have undoubtedly prevented [the drought] from becoming a famine in the classic sense with people dying in their tracks."

COMPONENTS OF TECHNOLOGIES TO IMPROVE FOOD PRODUCTION

Past approaches to resource development to increase agricultural production in the semi-arid tropics have achieved only limited success because they have not recognized the basic climatological and soil characteristics of the region nor utilized natural watershed and drainage systems (16). Better technologies are now being developed to ameliorate the effects of drought, increase food production per unit of land and capital, assure stability, and contribute directly to improving the quality of life.

Improved Soil Management Practices

    Effective soil management practices in the rainfed semi-arid
tropics must produce a suitable seed bed, ensure the proper placement of seed and fertilizer, destroy weeds, conserve soil moisture
and minimise runoff and erosion.

    Dry season primary tillage is now a common practice on the
Vertisols of semi-arid India. The soils are ploughed in March
or early April after the post-rainy season crop is harvested.
Because there is still some moisture in the soil, the power
required for tillage is less than it would be later in the dry
season, and the draft animals are well-fed and strong. Some weeds
do grow, but these are removed in the final preparation of the
seed bed which is done nearer to the onset of the rains. Pre-
monsoon showers, which are nearly certain climatic events, soften
the soil for the final land preparation.

    Crust formation on the surface is a problem with many sandy
textured soils in the semi-arid tropics. The power required to
break the crust is low for most of these soils [see, however,
Nicou and Charreau (17)] but the crust reforms after rains and
impedes seedling emergence. Incorporating organic matter into
the soil helps to decrease the strength of the crust - probably
by increasing soil moisture near the surface - but no satisfactory
technology exists so far for dealing permanently with this problem.

    Time of planting is important. Planting as early as the
rains will permit will generally ensure good yields in most years,
but research shows clearly that highest yields are obtained if
planting occurs about two weeks later. [Table 8, Randhawa and
Venkateswarlu (18)]. Probabilities based on climatological
evidence can now be used to predict optimum planting dates.

    Accurate placement of seed and fertilizers ensures high
seedling densities, vigorous early growth and resistance to
drought, but is seldom attained in traditional agriculture.
Practicable and economic new technologies to ensure accurate
placement in rainfed semi-arid tropical agriculture have yet to
be developed. It is probably the area where greatest gains can
be made from increased research efforts.

    Timely weed control measures are important to conserve
moisture and to avoid competition to the standing crop. Poor
weed control can reduce yields by 50% or more. The use of
herbicides is uneconomic or uncommon except on some cash crops,
and weeding is done manually or by using animal drawn implements.
Weeding is a major source of employment for landless labor in
India, and a major bottleneck to increased production in labor-
scarce areas of Africa.

Since the amount of water infiltrating into the soil is a function of the infiltration opportunity time and soil surface conditions, vegetative cover and land slope can be suitably modified to retain most of the rainfall on the ground surface. Reduced tillage maintains crop residues on the surface and contributes to improved infiltration and reduced evaporation and erosion. Land shaping in various forms has similar effects.

Use of Fertilizers

Most of the soils of the semi-arid tropics are of low fertility, being almost universally deficient in nitrogen and phosphorous (19, 20). Sulphur deficiency is common in Africa where the annual rainfall exceeds 600 mm.

Fertilizers are not commonly used in rainfed agriculture. Unirrigated districts in semi-arid India use an average of 18 kg/ha of fertilizers ($N + P_2O_5 + K_2O$) per hectare of cropped area compared to 57 kg/ha in irrigated districts (21). Most of the fertilizer used in the unirrigated areas is used on cash crops such as cotton, tobacco and groundnut.

There is much evidence to show that fertilizer use is economic on the staple semi-arid cereals. Several hundred experiments on cultivator's fields with sorghum, maize and pearl millet in semi-arid India have given average gains of 14 kg of grain per kg of N and 7 kg of grain per kg of P_2O_5. Benefit to cost ratios are 4 or greater (22).

Soil type and particularly water-holding capacity have significant effects on the efficiency of fertilizer use (Fig.3).

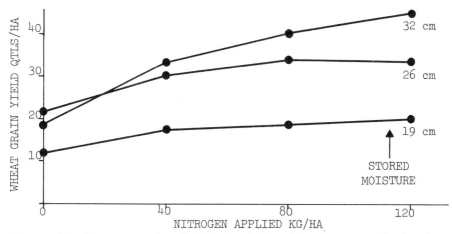

Figure 3.   Response of rainfed wheat to nitrogen on soils having different stored moisture (22).

Crops grown on the same soil in the rainy season will usually
have higher fertilizer use efficiencies than crops grown in the
dry season on receding stored moisture.

Good plant nutrition stimulates early plant growth and root
proliferation into the subsoil. The fertilized crop is able to
draw effectively on subsoil resources of water and is better
protected against drought (23).

Use of Improved and Appropriate Seeds

The combination of genes that reduced plant height and
susceptibility to lodging with genes for responsiveness to added
nutrients resulted in quantum jumps in the yields of wheat and
rice in irrigated agriculture. Similar approaches are proving
successful with maize, sorghum and pearl millet grown in rainfed
conditions. Improved varieties and hybrids often outyield local
varieties. Hybrids that have satisfactory grain quality and
levels of disease resistance combined with high yields are
attractive to farmers. Although year to year variability in
yield is higher for hybrids than for local varieties, the yield
gains and other characteristics of good hybrids are sufficient
to persuade farmers to take the higher risks involved. In
Maharashtra State hybrids now make up 35% of all sorghum planted,
and more than 70% in the postrainy season. Average yields in
the state are 50% higher than they were ten years ago.

The best crops to use in any particular district depend upon
climate, soils and agricultural and socioeconomic traditions.
Considerable efforts by the All India Coordinated Research Project
for Dryland Agriculture have now determined the most appropriate
crops for most of the semi-arid regions of India. For example,
Randhawa and Venkateswarlu (18) give tables for most suitable
crops and cultivars, for six districts with growing seasons
usually less than 20 weeks, seven districts with growing seasons
between 20 and 30 weeks and eight districts with growing seasons
exceeding 30 weeks. The crop combinations recommended are the
results of 3 to 7 years research work.

Intercropping

Farmers in the semi-arid tropics commonly intercrop their
land. There is considerable scope for improving the usefulness
and productivity of intercropping as can be illustrated by considering two contrasting intercropping systems, sorghum/pigeonpea
and millet/groundnut. The first is typical of those systems in
which an early cereal (maize, sorghum or pearl millet) is
combined with a slow-growing, reasonably tall long-season crop

(pigeonpea, cotton, castor or cassava). The second is representative of systems using a tall cereal combined with a quick-
growing low legume (beans, soybean, groundnut or cowpeas).

Sorghum/Pigeonpea. Pigeonpea is a long season crop that
matures in 6 months but its early establishment is characterized
by prolonged slow growth for as long as 2 months. During this
period the crop is inefficient in utilizing resources, e.g. the
sole crop during this period intercepts just 50% of available
light and produces only 20% of its total dry matter. Intercropping with earlier maturing crops such as sorghum improves
the use of natural resources for the total crop period. This
combination is very important in the Deccan plateau of India
covering Maharashtra, parts of Madhya Pradesh, Andhra Pradesh
and Karnataka where sorghum is the staple food. It shows promise
in similar environments in West Africa. Its value as a hedge
against disaster has already been mentioned.

In the traditional system, farmers grow several rows of
sorghum (6-12) with one or two rows of pigeonpea. This cropping
pattern is chosen deliberately to ensure a full yield of the
staple cereal. The yield of pigeonpea is considered as a bonus.
The pattern is inefficient in the use of late-season resources
because the pigeonpea is sparsely distributed and cannot cover
the ground. Studies by ICRISAT and the All India Coordinated
Research Project on Dryland Agriculture have shown that the
proportion of pigeonpea can be increased to 2 sorghum : 1 pigeonpea without seriously affecting sorghum yield provided the sorghum
population is maintained equivalent to the sole crop optimum
(24, 25). The sorghum in this combination has a similar growth
pattern to the sole crop and produces 90-95% of the sole crop yield.
The pigeonpea, after sorghum harvest, compensates, i.e. it spreads
to make use of the entire field space, and can produce 70% of sole
crop yield. The combination gives a 60-70% advantage over sole
cropping in terms of land productivity. Furthermore, the intercropped pigeonpea is efficient in the sense that a greater
proportion of dry matter is harvested as seed yield (30%) compared
to the sole crop (20%). Vertisols, with their better water-
holding capacity than Alfisols enable intercropped pigeonpea to
produce much better growth in the postrainy season resulting
in higher relative yields and greater yield advantages (Table 4).

Millet/Groundnut. This combination is adapted to soils
with moderate water-holding capacities. Both crops usually
mature less than 3 weeks apart, so the intercropping advantage
is not as great as for sorghum/pigeonpea. Because the groundnut
is an important cash crop, the farmer usually requires quite
a high proportion of this crop and a good row arrangement
appears to be 1 millet : 3 groundnut with the same within-row

Table 4. Grain yields and land equivalent ratios in sorghum/pigeonpea intercropping on two soil types (Average of 3 years)

	Vertisols					Alfisols				
	Yield		Land equivalent			Yield		Land equivalent		
	Sole	Inter-crop	LES*	LEPP**	Total	Sole	Inter-crop	LES*	LEPP**	Total
Sole crops										
Sorghum	4500	–	1.00	–	1.00	4573	–	1.0	–	1.00
Pigeonpea	–	1314	–	1.00	1.00	–	1770	–	1.0	1.00
Intercropping										
1:2 Sorghum/ Pigeonpea	4240	945	0.94	0.72	1.66	3641	1140	0.80	0.64	1.44

*LES (Land equivalent ratio for Sorghum) is the relative land area required by a sole crop of sorghum to achieve the yields produced by the same component in the intercrop.

**LEPP (Land equivalent ratio for Pigeonpea) is the relative land area required by a sole crop of pigeonpea to achieve the yields produced by the same component in the intercrop.

population as in the respective sole crops (26). This system
has given a proportionate yield of 50% of pearl millet and 77% of
groundnut giving a 27% yield advantage over sole cropping
(Table 5). The increase comes mainly from the pearl millet
which compensates for the low density by increased yield per
plant.

Table 5.   Grain or pod yields and land equivalent ratios in
           pearl millet/groundnut intercropping (Average of
           3 years)

	Yields (kg/ha)		Land equivalents		
	Millet	Groundnut	LEM*	LEG**	Total
Sole Crops					
Pearl millet	2370	–	1.0	–	1.00
Groundnut	–	2332	–	1.0	1.00
Intercropping					
1:3 millet/G'nut	1177	1796	0.50	0.77	1.27

*LEM (Land equivalent ratio for millet) is the relative land
 area required by a sole crop of millet to achieve the yields
 produced by the same component in the intercrop.

**LEG (Land equivalent ratio for groundnut) is the relative
  land area required by a sole crop of groundnut to achieve
  the yields produced by the same component in the intercrop.

     The advantages obtained in the above combinations are the
result of the complementary use of resources over time and space
without additional cost. The advantage of intercropping, in fact,
tends to be higher at low fertility or in low moisture conditions
(27); but this does not mean that this practice is valid only in
poor situations. Although the relative advantage over sole cropping decreases with higher fertility or better moisture supplies,
the absolute advantage in total crop yields is usually increased.

Other Cropping Systems

     Many of the Vertisols in India are cropped in the postrainy
season after a rainy season fallow. ICRISAT research has shown
that some of these soils can be cropped during the rainy season

without affecting the postrainy season crop resulting in substantial increases in yields and profits. The rainy season crop
must be planted early in dry soils before the onset of the rains,
which also avoids the problem of working in wet, sticky soils.
Postrainy season crops can be established either sequentially
(e.g. chickpea or safflower), or - with some difficulty - by
relay planting (e.g. sorghum and pigeonpea). Intercropping of
sorghum/chickpea or safflower/chickpea in the postrainy season
has been examined but the intercropping advantage is only about
20%. Double cropping is less profitable than intercropping. For
example, intercropped maize/pigeonpea is 73% more profitable and
34% less variable than maize followed by chickpea. The problems
with the double crop are the additional cost involved in establishing the second crop and the poor crop stands that may sometimes result due to the early cessation of rains.

On soils such as Alfisols the shorter growing period
precludes double cropping with two full season crops. But
efforts have been made to extend cropping beyond a single
season sole crop system, either by intercropping or by using
additional short season crops. Sole castor, an industrial
crop, or combinations involving it, have given highest net
returns (Table 6). Averaged over two years, a sorghum/pigeon

Table 6. Cropping Systems on Alfisols 1978-79, 1979-80

Cropping Systems	Grain yield (kg/ha)		Net monetary returns (Rs/ha)*	
	1978-79	1979-80	1978-79	1979-80
Sole castor	1462	1144	2039	2814
Sole groundnut	1236	1173	1133	2211
Sole sorghum	2516	2241	1402	1406
Sole millet	1940	2099	1158	1915
Sorhum/pigeonpea intercrop	2169/417	1680/831	1873	2855
Millet/groundnut intercrop	849/869	1063/881	1293	2681
Mungbean + Relay castor	634+885	603+737	2075	3393
Mungbean + Sequential castor	593+672	569+613	1766	2928
Millet + Relay Horsegram	1866+594	2099+536	1412	2181

*Gross returns less the cost of seeds, fertilizers and pesticides.
Prevailing market prices (Rupees/100 kg) one month after the
harvest of the respective crops used for calculating gross returns
for 1978-79 and 1979-80 were, respectively, Castor 170 and 185,
Groundnut 150 and 250, Sorghum 80 and 90, Millet 80 and 110,
Pigeonpea 230 and 260, Mungbean 200 and 330, Horsegram 100 and
102, Cowpea 180 and 222.

pea intercrop has proved almost as good as sole castor and the extended cropping systems have proved more profitable than sole food crops.

Proportionate cropping, i.e. allocating land resources to crops based on formulae relating crop durations to the probabilities of adequate soil moisture can help decrease the risk of loss and increase overall farm productivity. Research conducted at Haryana Agricultural University in northern India showed that allocation of 40% of the land to guar, a very drought resistant crop, with a 120 day growing season, 40% to pearl millet, a drought resistant crop, with a 70 day growing season and 20% of land to mung bean with a 50 day growing season allows the farmer to harvest all the three crops in the best years and at least two crops in all but severe drought years.

Crop rotations are used particularly in semi-arid West Africa to increase production. In low intensity cropping, the cropping sequences include fallow. Charreau (28) recommends one year of fallow or green manure plowed under followed by groundnut by cereal and finally by groundnut or cowpea. The cereal is usually long season pearl millet or sorghum. The effects on soil physical properties of incorporating green manure are well known resulting in beneficial effects on rooting and production of the succeeding crop. Where rainfall is fairly low and irregular as in northern and central Senegal, a four year rotation of fallow or pearl millet as a green manure followed by groundnut followed by cereal followed by groundnut is recommended or a five year rotation such as fallow or green manure - pearl millet - groundnut - pearl millet - groundnut.

In intensive cropping, the fallow period is eliminated and plowed under grass fallow or green manure is replaced by plowed under straw of short-season cereals (29).

Watershed Management

In rainfed agriculture the main source of available water is rain, but many of the soils of the region have poor infiltration characteristics and excess runoff and erosion or waterlogging can be serious problems at various times during the rainy season. The solution lies in developing technologies that make use of the natural topography and drainage patterns. The small watershed is a natural framework for resource development aimed simultaneously at stabilizing and increasing crop production through more effective use of available water and at resource conservation through insitu conservation measures.

At ICRISAT over the last eight years and in operational-scale village-level studies over the last three years, a technology for land and water management on deep Vertisols using graded broad-beds and furrows within small watershed units has proved successful in achieving both the above objectives (30).

The 150 cm wide beds are graded across the contour to a 0.6% slope and are separated by furrows that drain into grassed water-ways. The broad-beds are not likely to be breached in heavy rain-fall, and allow a flexible planting pattern in rows spaced at 30 cm, 45 cm, 75 cm or 150 cm. They reduce runoff under both fallow and cropped conditions and greatly reduce soil erosion in comparison to ungraded fallow soils (31). The use of graded broad-beds and furrows gives higher gross returns and profits. They can be established successfully within existing field boundaries at some loss in profits.

Dry seeding of the crop about two weeks before the onset of the rains is possible on these soils if the early rainfall is fairly reliable (32) and in most years will enable a postrainy season crop to be taken thereby increasing greatly the gross returns, profits and the rainfall use efficiency.

Supplementary 'life-saving' Irrigation

Dry periods within the monsoon are typical of many semi-arid tropical regions even when normal or above normal seasonal rain-fall occurs. The result is usually a reduction in crop yield, especially on soils which have relatively low water holding capacities. The availability of water for supplemental irrigation is an important means to reduce risk and improve production.

During the 1974 rainy season at ICRISAT Center, most of the runoff storage reservoirs were partially (50-70%) filled during the early part of the rainy season, thus providing water, if required, for 'life saving' irrigation during drought. The results of supplemental irrigation to crops on Alfisols during a 30-day drought in late August and early September were spectacular. Yields of sorghum and maize were approximately doubled by the application of one 5 cm irrigation. At product prices prevailing at the time of harvest, gross values of the increase due to the use of 5 cm of water were 3,120, 2,780, 1,085 and 650 Rs/ha for maize, sorghum, pearl millet and sunflower respectively (33).

During the postrainy season, supplemental irrigation can substantially boost crop yields because residual soil moisture is usually insufficient to prevent some drought stress at important physiological stages of growth. Sivakumar et al (34) showed that supplemental irrigations given at the time of panicle emergence and flowering of grain sorghum grown on a deep Vertisol gave an additional yield of 3,560 kg/ha over the control treatment (2430 kg/ha). The net benefit accruing from the supplemental irrigations was Rs.2,500/ha.

CONTRIBUTIONS OF THE RAINFED SEMI-ARID TROPICS TO FOOD PRODUCTION

Most of the cropping in the semi-arid tropics will continue to be done in rainfed conditions by small farmers with meagre resources producing staple cereals of low value for which few price incentives to production exist. Currently yields are low and production is unstable due to aberrant weather and the high incidence of pests and diseases. The semi-arid tropics with 13% of the world's land and 15% of its people produces only 11% of its food.

Food production can be increased in the rainfed semi-arid tropics; in India mainly through improved technology, in Africa mainly by increasing the land area under cultivation. Average yield levels in India for the major rainfed cereals and grain legumes are between 300 and 800 kg/ha. They can be doubled by the use of improved technologies. Taking sorghum as an example, the average yield is 800 kg/ha and the potential under rainfed conditions is at least 2500 kg/ha. More importantly, better farmers are already obtaining 1500 to 2000 kg/ha in rainfed conditions using improved seeds and some fertilizers. The highest yields are obtained on fields where the farmer can use supplementary irrigation. For pearl millet the average is around 600 kg/ha, the potential in rainfed agriculture is about 1500 kg/ha and better farmers' yields are 1000 to 1200 kg/ha. Lifting the averages to the level of better farmers' yields will mean more food production, more agricultural employment or higher returns to labor, more food for the farmers and their dependents, and significantly increased gross returns and profits.

The situation in the Sahel and other regions of semi-arid Africa is less promising. Improved technologies, and, particularly, improved seeds, are not so available and the institutions and infrastructure necessary for successful diffusion of new technologies are less developed (35). The differences between average yields, best farmer practices and potentials appear to be much narrower than in India from the limited evidence available.

There is scope to increase the area under cultivation in Africa without environmental loss. FAO in developing a food plan for Africa considered that for increased production of staple cereals in the Sahel, improvements in cropping patterns would contribute 17%, improved yields 36% and increased acreage 48% (36). Even if the necessary improvements and increases are made, the combined self sufficiency ratio in the Sahel for maize, sorghum and pearl millet will still be below 100 by 1990, with a deficit in excess of 200,000 tons of grain. The total deficit in food grains in the Sahel in 1990, including irrigated crops, is projected to be more than 700,000 tons.

For the semi-arid tropics as a whole, yield and production of sorghum, pearl millet and the two major grain legumes, are increasing slowly (Table 7). Only for sorghum does the increase in production exceed the average annual increase in population for the region, which is approximately 2.5%. Total food grain production in the semi-arid tropics, including irrigated crops in developed countries, is actually in excess and the region is

Table 7. Annual compound growth rates of five major crops in the less developed countries of the SAT during 1964-74 (4)

Crop	Area (%)	Yield (%)	Production (%)
Sorghum	0.71	2.08	2.81
Millets[a]	0.04	1.20	1.24
Chickpea	-1.29	1.46	0.15
Pigeonpea	-0.08	0.83	0.75
Groundnut	-0.40	0.02	-0.38

[a]Includes pearl millet (Pennisetum americanum), as well as the minor millets such as Setarias, Panicums, and Eleusines.

in a position to export food (Table 8). The figures in the table allow mild optimism in the long-term about the prospects for food production in the region, except perhaps for drier areas in the Sahel and except in the driest years. If successful technologies can be transferred within the region, and the necessary institutions and infrastructure provided, some aspects of the variability in climate can be overcome. Sustained increases in production, employment and incomes are possible, but buffer stocks and food aid will always be needed to ameliorate the effects of severe drought.

Table 8.   Estimated food requirement and food production in the
           semi-arid tropics*

Parameters	Based on countries in SAT as a whole [1] (a)	Based on percentage area of each country in SAT (b)
1. Population (persons)	1375.4	625.1
2. Consuming population (persons)[2] [(1) x .85]	1169.1	531.3
3. Estimated food requirements for adult population (MT) [3]	233.8	106.3
4. Total food grain production (MT)	316.8	135.2
5. Surplus/deficit of food grains [(4) - (3)] (MT)	+ 83.0	+ 28.9
6. Surplus/deficit (Exports-Imports) of food grains (MT)	+ 11.9	+ 1.7

*Data is based on FAO (Publication Yearbook 1978) results on
 cereal production in 1976, 1977 and 1978 years. Food refers to
 cereals only.

[1] Semi-arid areas and populations were estimated by using Troll's
    map (37) on climatic regions of the world by estimating the
    regions in V_3 and V_4 (dry Savanna and Thorne Savanna) in each
    country.

[2] Consuming population was derived by decreasing 15% of the total
    population under the assumption that this constitutes the popu-
    lation under the age of five years who do not consume much food
    grains (Osmania University, Personal Communication).

[3] Food requirements were estimated on the basis of 200 kg per
    adult per year by C. Gopalan et al. (38).

ASSUMPTIONS

(1) In these data no provision is made to separate the cereal
    yields under irrigated and rainfed conditions; these will
    be substantially different.

(2) Population in the semi-arid tropics is computed simply by
    multiplying the population of the country by the percentage
    area of the country in the region, without giving weightage
    to the density of population.

REFERENCES

1. IFPRI (International Food Policy Research Institute): 1977,
    *Food Needs of Developing Countries: Projections of Produc-
    tion and Consumption to 1990*, Research Report 3, Inter-
    national Food Policy Research Institute, Washington, D.C.,
    U.S.A.

2. Ryan, J.G.: 1974, *Socio-economic Aspects of Agricultural
    Development in the Semi-Arid Tropics*, Occasional Paper 6,
    Economics Program, International Crops Research Institute
    for the Semi-Arid Tropics (ICRISAT), Patancheru, A.P.,
    India.

3. Virmani, S.M., Sivakumar, M.V.K., and Reddy, S.J.: 1979,
    *Climatological Features of the Semi-Arid Tropics in
    Relation to Farming Systems Research Program*, ICRISAT,
    Presented at the International Workshop on the Agroclimat-
    ological Research Needs of the Semi-Arid Tropics, 22-24
    November 1978, ICRISAT, Hyderabad, India.

4. Ryan, J.G., and Binswanger, H.P.: 1980, "Socioeconomic
    Constraints to Agricultural Development in the Semi-Arid
    Tropics and ICRISAT's Approach," in *Proceedings of the
    International Symposium on Development and Transfer of
    Technology for Rainfed Agriculture and the SAT Farmer,
    28 August-1 September 1979, ICRISAT, Patancheru, India*,
    International Crops Research Institute for the Semi-Arid
    Tropics (ICRISAT), Patancheru, A.P., India, pp. 57-67.

5. Jodha, N.S., Asokan, M., and Ryan, J.G.: 1977, *Village
    Study Methodology and Resource Endowments of the Selected
    Villages in ICRISAT's Village Level Studies*, Occasional
    Paper 16, Economics Program, International Crops Research
    Institute for the Semi-Arid Tropics (ICRISAT), Patancheru,
    A.P., India.

6. Jodha, N.S.: 1980, "Some Dimensions of Traditional Farming
    Systems in Semi-Arid Tropical India," in *Proceedings of
    the International Workshop on Socioeconomic Constraints
    to Development of Semi-Arid Tropical Agriculture, 19-23
    February 1979, ICRISAT, Hyderabad, India*, International
    Crops Research Institute for the Semi-Arid Tropics (ICRISAT),
    Patancheru, A.P., India, pp. 11-24.

7. Rastogi, B.K.: 1979, "Cropping Patterns, Farming Practices and Economics of Major Crops in Selected Dryland Farming Regions of India," in *Proceedings of the International Workshop on Socioeconomic Constraints to Development of Semi-Arid Tropical Agriculture, 19-23 February 1979, ICRISAT, Hyderabad, India,* International Crops Research Institute for the Semi-Arid Tropics (ICRISAT), Patancheru, A.P., India, pp. 25-36.

8. Rao, M.R., and Willey, R.W.: 1980, *Evaluation of Yield Stability in Intercropping: Studies on Sorghum/Pigeonpea,* Experimental Agriculture 16, 105-116.

9. World Bank: 1979, "World Development Indicators," Annex to *World Development Report, 1979,* The World Bank, Washington, D.C., U.S.A.

10. Newman, M., Oudedraogo, I., and Norman, D.: 1980, "Farm-Level Studies in the Semi-Arid Tropics of West Africa," in *Proceedings of the International Workshop on Socioeconomic Constraints to Development of Semi-Arid Tropical Agriculture, 19-23 February 1979, ICRISAT, Hyderabad, India,* International Crops Research Institute for the Semi-Arid Tropics (ICRISAT), Patancheru, A.P., India, pp. 241-261.

11. Jodha, N.S.: 1975, *Famine and Famine Policies: Some Empirical Evidence,* Economic and Political Weekly 10(41), pp. 1609-1623.

12. Jodha, N.S.: 1978, *Effectiveness of Farmers' Adjustment to Risk,* Economic and Political Weekly 13(25), pp. A38-A48.

13. Kloth, T.I.: 1974, *Sahel Nutrition Survey, 1974,* U.S. Public Health Service, Center for Disease Control, Atlanta, Ga., U.S.A.

14. Faulkingham, R.H., and Thorbahn, P.F.: 1975, *Population Dynamics and Drought: a Village in Niger,* Population Studies 29, pp. 463-477.

15. Subramanian,V.: 1975, *Parched Earth: the Maharashtra Drought 1970-73,* Orient Longman, Bombay, India.

16. Kampen, J., and Burford, J.R.: 1980, "Production Systems, Soil-Related Constraints, and Potentials in the Semi-arid Tropics, with Special Reference to India," in *Priorities for Alleviating Soil-Related Constraints to Food Production in the Tropics,* International Rice Research Institute, Los Banos, Philippines, pp. 141-165.

17. Nicou, R., and Charreau, C.: 1980, "Mechanical Impedance Related to Land Preparation as a Constraint to Food Production in the Tropics (with Special Reference to Fine Sandy Soils in West Africa), in *Priorities for Alleviating Soil-Related Constraints to Food Production in the Tropics,* International Rice Research Institute, Los Banos, Philippines, pp. 371-388.

18. Randhawa, N.S., and Venkateswarlu, J.: 1980, "Indian Experience in the Semi-Arid Tropics: Prospect and Retrospect," in *Proceedings of the International Symposium on Development and Transfer of Technology for Rainfed Agriculture and the SAT Farmer, 28 August-1 September 1979,* ICRISAT, Patancheru, India, International Crops Research Institute for the Semi-Arid Tropics (ICRISAT), Patancheru, A.P., India, pp. 207-220.

19. Jones, M.J., and Wild, A.W.: 1975, *Soils of the West African Savanna. The Maintenance and Improvement of their Fertility, Technical Communication* Number 55, Commonwealth Bureau of Soils, Harpenden, U.K.

20. Kanwar, J.S.: 1976, *Soil Fertility - Theory and Practice,* Indian Council of Agricultural Research, New Delhi, India.

21. Jha, D., and Sarin, R.: 1980, *Fertilizer Consumption and Growth in Semi-Arid Tropical India. A District-Level Analysis,* Progress Report 10, Economics Program, International Crops Research Institute for the Semi-Arid Tropics (ICRISAT), Patancheru, A.P., India.

22. European Nitrogen Service Program: 1980, *Fertilizer Use in Dryland Agriculture,* Fertilizer Information Bulletin Number 13, European Nitrogen Service Programme, New Delhi, India.

23. Norum, E.B.: 1963, *Fertilized Grain Stretches Moisture,* Better Crops with Plant Food 47, pp. 40-44.

24. Natarajan, M., and Willey, R.W.: 1980, *Sorghum-Pigeonpea Intercropping and the Effects of Plant Population Density. 1. Growth and Yield,* Journal of Agricultural Science, Cambridge 95, pp. 51-58.

25. AICRPDA (All India Coordinated Research Project for Dryland Agriculture): 1976, *Achievements for the Period 1972-1975,* All India Coordinated Research Project for Dryland Agriculture, Hyderabad, India.

26. Reddy, M.S., and Willey, R.W.: 1980, *Growth and Resource Use Studies in an Intercrop of Pearl Millet/Groundnut,* Field Crops Research (in press).

27. Reddy, M.S. and Willey, R.W.: 1980, The Relative Importance of above-and below-ground resource use in determining yield advantage in pearl millet/groundnut intercropping. 2nd symposium on Intercropping in Semi-Arid Areas, University of Dar-es-Salaam, Morogoro, Tanzania, Aug. 4-7.

28. Charreau, C.: 1974, *Soils of Tropical Dry and Dry-wet Climatic Areas of West Africa and their Use and Management,* Agronomy Mimeo 74-26, Department of Agronomy, Cornell University, Ithaca, New York, U.S.A.

29. Charreau, C., and Nicou, R.: 1971, *L'Amelioration du Profil Cultural dans les Sols Sableux et Sablo - Argileux de la Zone Tropicale Seche Ouest-Africaine et ses Incidences Agronomiques,* Agronomie Tropicale 26, pp. 903-978; 1184-1247.

30. Kampen, J.: 1980, "Farming Systems Research and Technology for the Semi-Arid Tropics," in *Proceedings of the International Symposium on Development and Transfer of Technology for Rainfed Agriculture and the SAT Farmer, 28 August-1 September 1979, ICRISAT, Patancheru, India,* International Crops Research Institute for the Semi-Arid Tropics (ICRISAT), Patancheru, A.P., India, pp. 39-56.

31. Binswanger, H.P., Virmani, S.M., and Kampen, J.: 1980, *Farming Systems Components for Selected Areas in India: Evidence from ICRISAT,* Research Bulletin Number 2, International Crops Research Institute for the Semi-Arid Tropics (ICRISAT), Patancheru, A.P., India.

32. Virmani, S.M.: 1980, "Climatic approach to Transfer
     of Farming Systems Technology in the Semi-Arid Tropics,"
     in *Proceedings of the International Symposium on Develop-
     ment and Transfer of Technology for Rainfed Agriculture
     and the SAT Farmer, 28 August-1 September 1979, ICRISAT,
     Patancheru, India,* International Crops Research Institute
     for the Semi-Arid Tropics (ICRISAT), Patancheru, A.P.,
     India, pp. 93-101.

33. Krantz, B.A., Kampen, J., and Virmani, S.M.: 1978, *Soil
     and Water Conservation for Increased Food Production in
     the Semi-Arid Tropics,* Presented at the 11th Congress of
     the International Society of Soil Science, 19-27 June
     1978, Edmonton, Canada.

34. Sivakumar, M.V.K., Seetharama, N., Sardar Singh, and
     Bidinger, F.R.: 1979, *Water Relations, Growth and Dry-
     matter Accumulation of Sorghum Under Post-rainy Season
     Conditions,* Agronomy Journal 71, pp. 843-847.

35. OECD (Organization for Economic Cooperation and Develop-
     ment): 1976, *Analysis and Synthesis of Long Term Develop-
     ment Strategies for the Sahel,* Organization for Economic
     Cooperation and Development, Paris, France.

36. FAO (Food and Agriculture Organization): 1978, "Regional
     Food Plan for Africa," in *The State of Food and Agriculture
     1978,* FAO Agriculture Series Number 9, Food and Agriculture
     Organization, Rome, pp. 2-3 to 2-17.

37. Troll, D.: 1966, "Seasonal Climates of the Earth," in
     E. Rodenwaldt and H.J. Jusatz, eds., *World Maps of
     Climatology,* Springer-Verlag, Berlin, pp. 19-25.

38. Gopalan, C., Ramasastri, B.V., and Balasubramanian, S.C.:
     1971, Nutritive Value of Indian Foods. National Institute
     of Nutrition, Indian Council of Medical Research, Hyderabad,
     India.

CLIMATIC VARIABILITY AND SUSTAINABILITY OF CROP YIELD IN THE
MOIST TROPICS

Charles F. Cooper

Department of Biology
San Diego State University
San Diego, California 92182 U.S.A.

ABSTRACT. Except on the most fertile and level sites, production
systems that attempt to mimic natural ecological processes are
more likely to yield sustained crops of food and fiber in the
humid tropics than are high technology systems introduced from
ecologically very different developed countries. The climates
and soils of the humid tropics are more diverse and variable than
is commonly realized. The same biological processes that occur
in tropical forests after natural disturbance occur in cultivated
areas. This fact should be taken advantage of in design of sustainable agricultural systems. Row crops and small grains are
ecologically better adapted to temperate than to tropical regions.
Although the extreme view of the causal relationship between
diversity and stability in ecosystems has been discredited, monocultures are more susceptible to disease and insect attack than
are multi-species stands. Increasing atmospheric CO_2 is unlikely
to have a great effect on tropical climate, but there could be an
important fertilizing impact. Production of biomass fuel should
be considered as part of a multiple cropping system. Implementation of an ecologically based multiple cropping system will
require site-specific agricultural research, substantial skill on
the part of farmers, and overcoming serious social, economic, and
institutional barriers.

W. Bach, J. Pankrath, and S. H. Schneider (eds.), Food-Climate Interactions, 167–186.
Copyright © 1981 by D. Reidel Publishing Company.

Nature, to be commanded, must be obeyed.
Sir Francis Bacon, Novum organum, 1620

Most of the humid tropics was originally forested. Indeed,
that can be taken as a definition: the humid tropics comprise
those areas between the Tropics of Cancer and Capricorn where the
natural vegetation of well drained sites is predominantly closed
evergreen forest. This accounts for less than a third of the
total land surface in the tropics (5). Humid tropical forests
have characteristics which differentiate them, in varying degree,
from forests of the temperate zone: year-round growth, high
species diversity, tightly coupled cycling of nutrients between
plant and soil, rapid weathering and high rates of soil biological
activity, and, except where soil fertility is inherently low,
rapid biomass accumulation after disturbance. These features
have permitted most moist tropical forests to maintain productivity and stability despite climatic variability.

Much of this forest has been converted to agriculture,
although the fraction of the land in cultivation is smaller than
in the drier tropics. Systems employed range from large industrial farms based on heavy mechanical equipment to traditional
subsistence cropping.

If the land is to be of maximum use to its people, it must
produce ample crops not just for a few years but in perpetuity.
I contend that, except in a few especially favored sites,
production systems that endeavor to mimic natural ecological
processes are likely to yield greater sustained crops of food
and fiber for a given level of productive input than are high
technology systems introduced from ecologically very different
regions. Furthermore, such ecologically based systems will be
better able to prosper under natural climatic variation.

These systems are likely to include a mixture, in space and
in time, of species and of life forms--annual crops, shrubs,
trees, and perennial herbs. Such mixtures often characterize
traditional agriculture. Native peoples have learned by experience much about the ecological capabilities of their landscape.
As the distinguished tropical ecologist M. E. D. Poore has put
it (22), "It is always worth studying the detailed operations of
traditional stable systems of subsistence agriculture in the
region to determine what may be learnt from them for new agricultural development." Modern research may point the way toward
agricultural development that is socially, economically, and
ecologically sound.

The goals of land management in the humid tropics should be
to manipulate ecosystems without increasing soil loss, exhausting
fertility, increasing stream sediment, or accelerating other

degrading processes. Systems should be designed so that if
errors are made in management, correction can allow recovery to
an acceptable state. In the long run, these goals can best be
attained by utilizing natural processes so as to retain their
conservative impact on system degradation, leading to adequate
and sustained production for man's use (10).

It is a commonplace that much of the tropical world is poor.
The immediate barriers to increased crop production in poor
countries are primarily social and economic, not biological, as
Theodore Schultz forcefully emphasized in accepting the 1979
Nobel Prize in Economics (27). Schultz stated, "What matters
most in the case of farmland are the incentives and associated
opportunities that farm people have to augment the effective supply of land by means of investments that include the contributions
of agricultural research and the [improvement] of human skills."
Agricultural research and improvement of human skills will be
most successful in augmenting production of food and fiber,
however, if they are guided by ecological imperatives.

VARIABILITY OF CLIMATE IN THE HUMID TROPICS

Temperate zone residents seldom recognize the diversity of
tropical climates. Once having divided tropical climates into
"wet" and "dry", authoritative spokesmen from temperate regions
commonly regard climate as of little further significance in
development of land resources. The prevalent popular view
fostered by press, cinema, and fiction is that the tropics are
seasonally monotonous and somewhat too warm and humid for comfort-
able living by people of European extraction, but otherwise
benign (29).

Actually, there are many more combinations of rainfall and
temperature regimes in the tropics than in temperate regions.
The widely accepted Holdridge World Life Zone System (flawed
though it may be) distinguishes more than 100 different bio-
climates worldwide. The tropics include 39 of these life zones.
The warm temperate region has 23 life zones and the cool temperate
only 16--and yet the tropics are commonly said to be climatically
the least complex of the world's regions (29).

Despite this regional diversity, certain general climatic
characteristics can be recognized in the humid tropics. Rainfall
is often torrential, leading to accelerated erosion of unprotected
soil. Soil is often at or near saturation, making it hard to
work. Year-round warm temperatures stimulate soil biological
activity and chemical weathering as well as plant growth. Heavy
cloudiness sometimes reduces available solar radiation.

Benign as the tropical climate may be thought, its year-to-year variability nevertheless often leads to significant human deprivation. Despite the adjective "humid", most tropical regions so designated have at least a two- to three-month period of greatly diminished rainfall. Weather variations can accentuate the length or severity of the dry season. Agriculturally significant drought has occurred in places as humid as Sarawak (3). Or there may be no dry season in some years, encouraging growth of plant pests. Probably most devastating to food production and to human welfare in the tropics, however, are flooding of river valleys after torrential rainfall and of coastal lowlands as a result of storm surges. Both types of flood are brought about by severe tropical storms. Paradoxically, though, these storms also bring the moisture needed for good crop production in the uplands.

Historical and paleoclimatic studies have shown that the year-to-year variations in rainfall and other climatic factors in the tropics are not random fluctuations about a constant climatic mean, but that long-term climate itself varies (8). However, the existing data are insufficient to define past climates in the humid tropics with much reliability. The prospect of long-term climatic change, either natural or man-induced, should be taken into account in long-term planning of food production systems. Man's addition of carbon dioxide and other substances to the atmosphere may now be altering global climate.

SOIL DIVERSITY

Diversity of soils in the humid tropics is at least as great as the diversity of climate. Soils range from the ancient, deeply weathered and heavily leached Oxisols (Latosols) of central Africa and the western Amazon Basin to the recent volcanic or alluvial soils of tropical Asia, Central America, and the Caribbean. The former have a low mineral reserve for supplying nutrients to plants, are largely dominated by kaolinite in the clay fraction, and are high in iron and aluminum oxides. Low cation absorption capacity makes them particularly susceptible to leaching. By contrast, immature soils of recent volcanic or alluvial origin (Inceptisols) can be extremely rich.

The natural vegetation of the humid tropics is the result of interaction between climate and soil. Of the two, climate is the more important. From a review of the literature and from his own field work, Knight (15) concluded that forest composition and structure tends to be relatively insensitive to soil characteristics in the tropics, but by no means completely so. The principal exceptions are sites markedly deficient in specific nutrient elements. Certainly, tropical forests are often lush and dense even on infertile soils. These ecosystems have evolved to

establish tight nutrient cycling and nutrient retention; most of
the nutrient capital is contained in the living biomass. When
forests are removed from infertile soils to permit crop production, there is usually rapid deterioration of the site.

Forests are equally lush on better sites, but inherent
fertility permits sustained production of food and fiber on these
sites under careful management. Similarity of physiognomic
appearance of forests on productive and unproductive soils, and
poor understanding of forest taxonomy and of ecological requirements of species make superficial appearance of the native forest
in undeveloped regions a poor guide to its sustainable potential.

DISTURBANCE AND SUCCESSION

Wherever economies have progressed beyond the primitive
hunting and gathering stage, food and fiber production in the
tropics has led to manipulation of natural ecosystems. Many
tropical forest ecosystems are also subject to natural disturbances such as hurricanes and volcanism which can be qualitatively
similar to those produced by man.

Ecologists have devoted much attention to the response of
tropical forests to disturbance (6). A number of biological
processes are usually emphasized in these studies: seed dispersal
and germination, rates of photosynthesis and of biomass development, species composition, development of three-dimensional
structure, and nutrient loss and immobilization, among others.
Ecologists tend to prefer to study "natural" succession; however,
the same processes occur in ecosystems manipulated by man. An
important management objective should be to work with, not
against, these processes.

Seed dispersal, germination, and vegetative reproduction in
managed stands are strongly influenced by man, but natural dispersal, particularly of weeds, also goes on. Studies of
reproductive patterning can thus be an effective aid to ecologically based weed control.

Regrowth after moderate disturbance is usually rapid in the
humid tropics. Unless tree felling or burning is followed by
plowing or mechanical land clearing, the soil is normally protected in less than three months by a monolayer of greenery. More
leaf layers are added as height growth proceeds. Leaf area index
is normally three or more by the end of a year, and approaches
that of a mature forest by the time the regenerating stand is six
years old (5).

I observed similar recovery in second growth forest

extensively defoliated by ash from the eruption of Volcan Arenal,
Costa Rica. Six months after the eruption it was virtually
impossible to walk through the defoliated stands, whereas similar
unaffected forests were easily passable. Shading by canopy
leaves inhibited the understory in the intact forest; defoliation
permitted the residual vegetation to grow exuberantly.

Mature tropical forests commonly have an extensive system of
fine roots and associated mycorrhizae, which take up soluble
nutrients that might otherwise be lost through leaching. These
fine roots have a very short lifespan, and those of non-coppicing
species die almost immediately when the parent tree is cut. However, if secondary vegetation is permitted to develop, regrowth
of fine roots is rapid (23). In one study in Costa Rica, a
system of fine roots developed that was nearly complete only one
year after forest cutting; this presumably had a nutrient uptake
capacity similar to that of the mature forest, at least near the
surface (Table 1).

Table 1. Biomass of roots less than 2 mm
          in diameter in a mature forest
          and one-year-old regrowth at La
          Selva, Costa Rica. After Raich
          (24).

Depth (cm)	Fine root biomass $(g/m^2$ dry weight)	
	Mature forest	Regrowth
Litter	2.5	8.0
0–5	95.5	85.6
5–10	50.1	65.5
10–15	31.9	47.0
15–30	55.8	46.1
30–50	53.6	13.6
Total	289.4	265.8

Erosion and nutrient loss can indeed be severe where
mechanical treatment or other measures leave the soil exposed
and prevent regrowth. If an early vegetation cover is allowed
to develop, however, nutrient and soil loss is not as great as
is sometimes thought. Evidence, in addition to that cited above,
includes the absence, or at least the scarcity, of documented
cases of permanent deflection of succession or change in vegetation type where there has been only intermittent, as opposed to
continuous, human intervention (11).

Nutrient loss likewise need not be severe under agricultural cropping if a continuous vegetation cover is maintained. For example, in an area of Costa Rica where sugar cane, coffee, and managed pasture are grown on inherently fertile volcanic soil which was formerly forested, the soil does not appear to have deteriorated seriously after 22 years of cultivation. There have been some losses of essential minerals, however. Future farmers may have to pay more attention to soil management and maintenance of fertility than has so far been necessary in this region (17).

Growth of productive second growth forests can be rapid in the humid tropics. Ewel (5) describes naturally regenerated Trema micrantha in Costa Rica which attains a height of 9 m at one year of age and more than 30 m at eight years. Second growth forest 30 to 50 years old is structurally very similar to the pre-disturbance forest, but usually differs significantly in species composition.

The wood density of successional species is commonly low. Thin cell walls are perhaps a penalty for fast growth. These fast-growing, low-density trees will constitute the wood resource of the future as mature forests are replaced. A utilization technology directed toward the high density woods of mature tropical forests will not be long useful in the humid lowlands (5). This has significant implications for the design of tropical agricultural systems that include wood production as an important component.

ROLE OF FIELD CROPS IN THE HUMID TROPICS

Although some annual field crops are native to the humid tropics, large-scale commercial field crop production ecologically derives from temperate climates. The great natural grasslands of the mid-latitudes are well adapted to row crops and small grains; the humid tropics are less so. High intensity of rainfall, intense sunlight, soils of poor structure, rapid weed growth, and heavy pest infestation all mean that greater care is needed than elsewhere to avoid soil deterioration and excessive erosion in cropped fields.

The level of managerial ability and capital investment required for successful sustained production of field crops in the humid tropics can usually be justified only on good soils where there is easy access to markets and a suitable skilled labor force (22). These conditions are seldom met; mostly they occur on level alluvial or volcanic soils. The latter, although often highly fertile, are rarely level enough for successful large-scale commercial row crop production. They may be well

suited, however, to incorporation of row crops in the mixed
cropping systems advocated here.

Even where soil and labor conditions are suitable, grain
yields in the humid tropics are typically less than in comparable
warm temperate regions. According to Chang (3), the yield potential for rice in the tropics "is much lower than in temperate
regions because of the lower radiation, shorter day length, and
higher temperature during the ripening period in the tropics.
Compared with the temperate zone, maize yield in the tropics is
even lower than rice yield because of the additional effect of
high night temperatures."

Despite its lower yield potential, paddy rice still affords
the best opportunity for sustained grain production in the
tropics, relatively insulated from climatic variability.
Indonesian rice paddies have maintained production, at modest but
acceptable yields, for hundreds of years with little or no artificial nutrients (9). The fertility of Indonesian rice paddies
is maintained by nitrogen fixation by blue-green algae and by
weathering of volcanic soil on mountain watersheds.

Modern agricultural research has shown that yields can be
improved and maintained by transformation of rice paddies to a
totally man-dominated ecosystem--a clear exception to the generalization of the superiority of ecologically based cropping
systems in the tropics. As Fukui points out (8), lowland rice
cultivation is the only agricultural system that has historically
supported great concentrations of people on a sustained basis in
the humid tropics. The reason is presumably the high and stable
productivity of lowland rice, productivity which has the potential
for substantial improvement under proper conditions (Table 2).

High intensity rice cultivation requires much more than just
better seeds. It requires total control of the hydrologic regime
of the paddy fields. As Fukui (8) puts it, this involves not
merely the replacement of original vegetation by rice, but also
the creation of a new environment which does not exist in nature.
This demands cooperation from the entire community. According to
Fukui, "Water control can be likened to a road system which
begins with foot paths and is improved to make cart roads, then
single lane roads, double lane roads, and finally highways. Thus,
such a system can be called an infrastructure rather than one of
the inputs needed for production." When this infrastructure is
fully developed, the crop is almost fully insulated from the
vagaries of climate. Rice in less fully transformed localities
is more sensitive to climatic variability.

Because sustained production of upland crops is much more
difficult in the humid tropics than in the temperate zone, Fukui

Table 2. Rice yield in selected Asian countries. After Fukui (8). Differences are presumably more a reflection of technology than of climate or soil.

Country	Average rice yield (tonnes/ha)
Laos	1.4
Thailand	1.8
Burma	1.9
Bangladesh	2.0
Sri Lanka	2.2
Indonesia	2.7
Malaysia	3.0
Taiwan	4.3
South Korea	5.4
Japan	5.9

(8) believes that "Perhaps rice cropping will be the only alternative in large parts of the humid tropics, including those countries where rice cultivation is not common at present." It is surely not appropriate to use as a model for such cropping, though, the capital-intensive, low labor input cultivation systems that characterize rice production in California (26). The advanced technology, including laser-controlled giant land levelling machines and heavy crawler tractors, and the heavy chemical inputs used there are wholly unsuited to social and economic conditions in the tropics.

PESTS AND PLANT DIVERSITY

Although the broad generalization that postulated a causal link between stability and diversity in natural ecosystems has largely been exploded, there is still abundant evidence that pest attacks tend to be more frequent and virulent in monocultures than in more diverse communities. This is especially true in the humid tropics, where growth of both crops and pests can continue all year.

In temperate regions, the life cycle of insects, and to a lesser extent of diseases, is typically interrupted by the cold of winter. In the semi-arid tropics, there is a similar interruption during the long dry season. There is ordinarily no such interruption in the humid tropics.

Unlike in drier regions, dispersal of plant diseases is relatively inefficient in the humid tropics. Long distance wind transport of spores is inhibited by climate. Somewhat the same is true of insect pests. Most agricultural pests are relatively species-specific. Widely separated plants of a given species are thus less likely to be attacked than are those growing close together. Species diversity in tropical forests may in part be an evolutionary adaptation to the continually favorable growth conditions but poor dispersal mechanisms of pest species. It is an adaptation that can be taken advantage of in designing sustainable crop production systems.

The etiology of pest attack is quite different in annual and perennial crops. In annuals, the important distinction between a severe disease outbreak and a mild one is time. An unusually severe epidemic is one that reaches a high level of disease a week or two earlier than usual in relation to the time when the crop ripens. Early development usually arises from infection early in the growing season (30). Diseases of long-lived perennials, on the other hand, involve continued growth of the pathogen or repeated reinoculation until it eventually overwhelms the host. This difference needs to be taken into account in designing cropping systems that involve both annuals and perennials.

Tropical monocultures on a commercial scale are commonly possible only with heavy application of pesticides. Not only do pesticides at the levels regularly applied have adverse effects on adjacent ecosystems, but purchase of these petroleum-based chemicals represents an increasingly severe drain on foreign exchange in many developing countries. Chemicals will surely remain a part of any viable pest control strategy in the tropics, but a dispersed, multi-species crop design will make possible the more effective application of integrated pest management systems that minimize pesticide use while still protecting the crop. This is a field that requires intensive local research involving specific crops, insects and pathogens. A major drawback to integrated pest control in temperate zone commercial agriculture is that it is relatively labor-intensive in comparison with broad-scale application of pesticides. This could become an advantage in some labor-surplus regions of the tropics.

Pollination is another biological process in the tropics in which species diversity plays a role. Attempts to cultivate Brazil nuts (Bertholletia excelsa), which grow wild throughout the Amazon Basin and which yield a valuable cash crop, have given disappointing results so far. A large plantation established near Manaus about 25 years ago has been practically abandoned because of low yields. It is suspected that monocrop plantations of this species lack breeding sites for the pollinating bees (Bombus spp.), which occur naturally in the forest, but this hypothesis has not been experimentally demonstrated (1). It may well be that interplanting of improved varieties of this species with other tree crops would materially improve nut harvest. It has been demonstrated that selected clones propagated by grafting not only increase yields but reduce the time between planting and first harvest from about 15 years for natural seedlings to 4-6 years for the grafted plants (1), making Brazil nuts a potentially useful cash crop.

CARBON DIOXIDE, CLIMATE, AND PLANT GROWTH

     There is incontrovertible evidence that burning of fossil fuel and, perhaps, clearing of tropical forests is increasing the concentration of carbon dioxide in the atmosphere, and that this increase will continue into the next century. (For a general discussion of the CO_2 question, see ref. 20). There is general consensus that doubling of atmospheric CO_2, now expected sometime in the latter half of the next century (25), will increase global mean temperature about 3°C, a consequence of the well known "greenhouse effect". Global climate models indicate that this warming will be greatest near the poles and least in the tropics (7). Indeed, best estimates are that a 3°C global warming would mean an increase of 1°C or less near the equator. Global precipitation is also expected to increase, but the relationship between precipitation and evaporation is predicted to change hardly at all from the equator to about 15° north latitude, the zone that basically defines the humid tropics. Evaporation is expected to increase more than precipitation between 15° and 50°N (Fig. 1). It can therefore be concluded that any worldwide climate change due to increased atmospheric CO_2 will be felt less in the humid tropics than in other climatic zones. It does not appear that there will be a significant impact on agriculture in this part of the world except to the extent that climatic shifts alter worldwide crop production patterns and the terms of trade.

     Direct effects of CO_2 itself are another matter. CO_2 can limit crop growth when other nutrients and water are optimal; indeed CO_2 fertilization is a common means of increasing crop production in temperate latitude greenhouses. How crops and non-cultivated ecosystems will respond to increased atmospheric CO_2 under field conditions is not known, either in temperate regions or in the tropics. Whatever effect does occur is expected to be either beneficial or neutral; significant adverse effects on plants have not been noted until CO_2 concentrations exceed 1000 ppm--three times present levels.

     Plants with different photosynthetic systems respond differently to increased CO_2. Species whose pathway of photosynthetic carbon fixation is through 3-carbon carboxylic acid (C_3 plants)

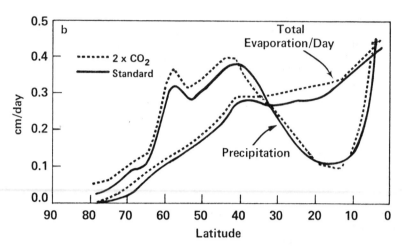

Figure 1.   Zonally averaged global evaporation and
             precipitation under standard conditions and
             under a doubling of atmospheric CO_2, as pre-
             dicted by computer simulation models of the
             general circulation. From Watts (31).

tend to be more responsive to CO_2 than those having a 4-carbon
pathway (C_4 plants) (20).

    C_4 plants are adapted to high temperatures and intense sun-
light. Therefore, C_4 plants are generally of tropical affinity,
whereas C_3 plants are more common in temperate regions. This
generalization is by no means universal, however. Among important
tropical crop plants, maize and sugar cane are C_4 species, whereas
rice and cassava have C_3 metabolism. The nature of the photosynthetic pathway in most tropical trees has not been investigated,
but virtually all temperate climate trees are C_3.

    The adaptation of C_4 plants to intense sunlight is of course
not of much importance in those humid regions where cloudiness
limits insolation. C_3 plants, with their responsiveness to CO_2,
may be the best choice in cloudy locations.

    Research to develop optimal site-specific cropping patterns
in the humid tropics should take account of the fact that continued increase of atmospheric CO_2 in coming decades is a virtual
certainty. To the extent that there is an effect on plant growth,
it will generally be greater in C_3 plants than in C_4. The effect
is also expected to be greater under conditions of optimal soil
nutrient and water status than when either is limiting. There is
a significant lack of hard research data on interactions of CO_2
with other plant stresses, particularly in the tropics. A major

international research program has been proposed by the American Association for the Advancement of Science (2) to define the environmental and societal consequences of increased atmospheric CO_2. It is important that a significant portion of this global effort be directed toward the humid tropics. Specific suggestions to that end were made by a workshop that was part of the AAAS effort (28).

BIOMASS FUELS

Per capita energy consumption in tropical developing countries is far less than in industrialized nations. Although energy use in the tropics will not--indeed, probably should not--catch up with that in temperate regions, increased development implies more energy consumption. Better matching of energy sources to end-use requirements may reduce the need for large-scale industrial energy conversion (18), but will not eliminate it.

Biomass fuels already play a large role in the energy budget of tropical countries. Integrated agroforestry development may permit more efficient exploitation of biomass resources.

Biomass fuels--wood, charcoal, agricultural residues, and cattle dung--provide an amount of energy in south Asia about equivalent to that supplied by commercial sources, chiefly electricity and liquid fuels (24). The picture is roughly similar in the American tropics. Most of this non-commercial energy is devoted to cooking and other household uses, and to agriculture.

Fuel wood is obtained almost entirely from native trees, often very slow growing. Higher energy yields can be obtained from fast growing trees introduced from other parts of the world. A highly touted example is the evergreen leguminous tree Leucaena leucocephala, which is well adapted to humid and semi-humid environments. In dense plantations in the Philippines, this tall, branchless tree can grow to a height of 20 m in 6 to 8 years, and yield 25 or more tons of wood per hectare per year. It is a good nitrogen fixer, and the protein-rich foliage is eagerly sought by livestock. It does not, however, grow well on the acid Oxisols high in aluminum oxides which are characteristic of many of the poorer tropical regions (24).

It is notable that most proposals for cultivation of Leucaena and similar fast growing trees published in temperate regions advocate that they be planted in monoculture energy plantations. Better, it would seem, would be development of planting schemes which intersperse these trees with other species, including both food and cash crop species and other fuel woods. Advantage could then be taken of Leucaena's nitrogen fixing ability and forage production capability. Such a cropping system, mimicking in part the natural ecological processes of the humid tropics, would be truly successful only on the basis of site-specific local research and experimentation. Unfortunately, it is simpler to transfer to the tropics forest practices ecologically suited to temperate regions rather than to make the more difficult transition to locally adapted practices.

Although most biomass fuel is now burned directly for household use, there are good prospects for transforming both wood and agricultural residues into methane or methanol, the latter usable for transportation fuel. This can be done now; the problem is to make the processes economic. Microbiological research is needed to develop more efficient strains of anaerobic bacteria adapted to local plants.

Competition for land between fuel and food production could lead to increased food prices and reduce the real incomes of the poor under some conditions, particularly in the more densely populated countries (24). This problem might be in part alleviated if effective locally adapted agroforestry systems were developed to produce both food and fuel from the same hectare. This is a significant challenge to research and development.

IMPLEMENTATION

Achieving ecologically based crop production requires a labor-intensive agrarian system incorporating a high degree of managerial skill on the part of the individual farmer. The social and institutional obstacles, including land tenure, should not be minimized. Yet I submit that in much of the humid tropics, systems which so far as possible mimic natural ecological processes are the only way of maintaining agricultural productivity in the face of varying climates and deteriorating soils.

The deleterious consequences of applying temperature zone agricultural practices to the humid tropics have often been described. Not only is soil depletion and subsequent land abandonment common, but the local population is often inadequately supported (21). The monocultures or moderately mixed stands that characterize modern temperate zone agriculture are a much greater departure from normal ecological conditions in the tropics than they are in their region of origin (14).

The shifting cultivation widely practiced by primitive people in the tropics, though, is not a good model for improving the living conditions of present day rural people. For one thing, it requires too much land per person. However, site-adapted agricultural research can develop cropping systems that retain the heterogeneity of shifting cultivation.

Hart (12) has used a natural ecosystem analog to design a tropical crop system in Costa Rica which is structurally and functionally similar to a tropical forest ecosystem (Fig. 2). Instead of planting a perennial crop on bare soil and continuously removing weeds until the crops reach maturity, annual crops are substituted for the weeds, increasing production from the crop system and somewhat reducing weeding costs. Planting proceeds in two phases. The first includes beans, maize, cassava and plantain. The second phase begins when coconut, cacao and rubber are interspersed with the plantain. This two-phase arrangement is consistent with reports that during tropical forest succession, seedlings of climax species are almost never present in pioneer stages, and with existing evidence of apparent decline in diversity between herbaceous and woody phases of natural succession (12). The system designed by Hart is specific to a particular site. On wetter sites in the same region, cowpea could be substituted for common bean, and African oil palm for coconut. Similar locally adapted schemes could be devised for other parts of the world.

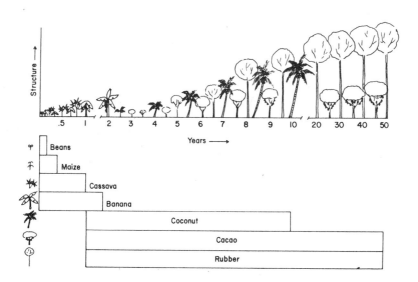

Figure 2.   The chronological arrangement of crop components in a hypothetical successional crop system on a well drained site in the humid tropics of Costa Rica. From Hart (12).

Both native and domestic animals can be introduced into a mixed cropping system to provide a much needed source of high quality protein. Again, practices for doing so are highly site specific. Nigh and Nations (21) describe how the Lacandon Maya of Chiapas, Mexico, plant specific crops for the express purpose of attracting specific animals to the fields, where they can be harvested for food.

In ecological cropping systems, particular attention should be given to ensuring that proper mycorrhizae are present. Most vascular plants depend to some extent on vesicular-arbuscular mycorrhizae for mineral uptake, though a few species do not (13).

Aquaculture--culture of freshwater fish in managed ponds-- is a particularly suitable form of protein production in the humid tropics, and is well adapted to a mixed cropping system. Many crop residues can be used aş fish food, and the harvested fish are a good cash crop if suitable marketing and transportation facilities are available.

Marketing and transportation are indeed one of the principal barriers to mixed cropping in the tropics at anything beyond a local subsistence level. Many products are bulky and perishable. Some such as cassava are difficult if not impossible for urban residents to prepare for safe consumption, yet cassava is poten- tially a very important and productive component of most such systems. A solution may be local cooperatives in which the raw crop is converted to usable flour for urban consumption.

Practical experiments with ecological farming have been described from other tropical localities, including Africa (16) and Venezuela (4). In the latter case, "It would appear that the high-diversity approach to landscape management offers a socially, economically and ecologically viable alternative or complement to existing land use systems" (4). In this approach, fertilizers are not used as a massive energy supplement to achieve high yield and associated high leaching rate, but rather as a replacement of minerals lost through harvest in a system with tight mineral cycles.

Kock (16), working in Africa, in a similar vein advocates a "Minimum Input Strategy" that is the opposite of conventional high input techniques. He says, "The objective of farming in the trop- ics thus moves away from pursuing short-term goals of achieving the highest production with the highest input use (at present price relationships). A long-term strategy is needed which mobi- lizes natural factors to a maximum in order to ensure a high and secure production level with minimum use of non-renewable resources and energy. Ecological adjustment helps to bring this about."

CONCLUSIONS

, A principal thrust of this paper is that sustained production of food and fiber in the presence of climatic variability will usually best be attained in the humid tropics through a multi-species, multi-layered cultivation system that mimics natural successional processes. Ecologically sound though such a scheme may be, there are serious political, social and économic barriers to its large scale implementation.

A provocative review presented in April 1980 at an International Conference on Amazon Agriculture and Land Use Development in Cali, Colombia (1), is important as much for who said it as for what was said. The author, Paulo de T. Alvim, is a respected Brazilian agronomist and plant physiologist who is closely acquainted with both social and ecological conditions in his own country and elsewhere in the humid tropics.

Alvim concluded that perennial tree crops are the most appropriate land use for tropical regions where rainfall is high and soils are predominantly poor. Their important ecological advantages over annual crops include better protection of the soil against erosion, leaching and compaction; lower demand for soil nutrients; and higher tolerance for soil acidity and aluminum toxicity, all of which are problems in much of tropical America. Many species, including those already commercially cultivated and potentially useful new crops, appear suitable for plantation agriculture. Emphasis throughout Alvim's review is on cultivation of old and new tree crops in essentially pure stands.

Alvim concludes with a discussion of agroforestry, which he recognizes as ecologically superior to conventional forms of agriculture. He points out significant obstacles to its widespread adoption, however. The main problem is the scarcity of hard data from actual research to use in implementing or recommending specific practices to farmers. Experiments in agroforestry require an integrated approach; they are much more complex than conventional field experiments with single crops. Interactions among different plant species are usually site specific, making it difficult to generalize conclusions from isolated studies. In short, research in agroforestry does not fit the temperate region Agricultural Experiment Station model of high-yield monocrop development and demonstration which developed countries have attempted to impose on the tropical world. Alvim concludes that as the problem stands today, agroforestry is to be regarded mainly as a promising field of research to be strongly encouraged and supported in the wet tropics, but not as a system widely recommended for promoting agricultural development.

Another set of obstacles arise from the marketing and

distribution infrastructure of most tropical countries. History
shows that scientific agriculture in the tropics has almost always
been started by industrialized countries for production of export
crops in their tropical colonies or in other tropical countries
where a high return on investment appeared possible. Many of
these export crops have been based on tree cultivation. Food and
fiber have often not been produced in adequate quantities for the
local market. Local economies have thus become dependent on
fluctuations in commodity prices in developed countries. The
result has often been imposition of a capital-intensive mechanized
production system on regions short of capital but with a surplus
of labor. This mismatch has in turn led to much social and
economic disruption.

How these barriers can be overcome is far from clear. The
problem is social rather than ecological. Present high intensity
export-based systems often lead to short-term advantage at the
expense of long run stability. I suggest that continued research
on ecological processes in the tropics and agricultural research
on site-specific multi-crop production systems for food and fiber,
coupled with a political will to alter current social conditions,
can lead to sustainable yields in the humid tropics even in the
face of climatic variability.

REFERENCES

1. Alvim, P. de T.: 1980, "A perspective appraisal of perennial
    crops in the Amazon Basin." Paper presented at International
    Conference on Amazon Agricultural and Land Use Development,
    Cali, Colombia, April 16-18, 1980.
2. American Association for the Advancement of Science: 1980,
    "Environmental and societal effects of a possible CO_2-induced
    climate change: A research agenda, Volume I." Rept. DOE/EV/
    10019-01. U.S. Dept. of Energy, Washington, D.C.
3. Chang, J-h.: 1980, Some aspects of climatology in southeast
    Asia and New Guinea. GeoJournal 4, pp. 437-445.
4. Dickinson, J.C. III: 1972, Alternatives to monoculture in
    the humid tropics of Latin America. Professional Geographer
    24, pp. 217-222.
5. Ewel, J.: 1980, Tropical succession: Manifold routes to
    maturity. Biotropica 12(Supp.), pp. 2-7.
6. Ewel, J. (Ed.): 1980, "Special issue on tropical succession."
    Biotropica 12 (Supp.).
7. Flohn, H.: 1980, Possible climatic consequences of a man-made
    global warming. Intern. Inst. Appl. Systems Analysis,
    Research Report 80-30.
8. Fukui, H.: 1979, Climatic variability and agriculture in
    tropical moist regions. pp. 223-245, In: World Meteorological
    Organization, "Extended summaries of papers presented at the
    World Climate Conference." WMO, Geneva.

9. Geertz, C.: 1963, "Agricultural involution: The process of ecological change in Indonesia." University of California Press, Berkeley.

10. Golley, F.B., and E. Medina: 1975, Ecological research in the tropics. pp. 1-4, In: F.B. Golley and E. Medina (Eds.), "Tropical ecological systems," Springer-Verlag, New York.

11. Harcombe, P.A.: 1980, Soil nutrient loss as a factor in early tropical secondary succession. Biotropica 12 (Supp.), pp. 8-15.

12. Hart, R.D.: 1980, A natural ecosystem analog approach to the design of a successional crop system for tropical forest environments. Biotropica 12 (Supp.), pp. 73-82.

13. Janos, D.P.: 1980, Mycorrhizae influence tropical succession. Biotropica 12 (Supp.), pp. 56-64.

14. Janzen, D.H.: 1973, Tropical agroecosystems. Science 182, pp. 1212-1219.

15. Knight, D.H.: 1975, A phytosociological analysis of species-rich tropical forest on Barro Colorado Island, Panama. Ecol. Monogr. 45, pp. 259-284.

16. Kock, W.: 1979, Ecological farming in the tropics. Food and Climate Review 1979, pp. 69-75.

17. Krebs, J.E.: 1975, A comparison of soils under agriculture and forests in San Carlos, Costa Rica. pp. 381-390, In: F.B. Golley and E. Medina (Eds.), "Tropical ecological systems." Springer-Verlag, New York.

18. Lovins, A.: 1977, "Soft energy paths: Toward a durable peace." Ballinger, Cambridge, Massachussetts.

19. National Research Council, Geophysics Study Committee: 1977, "Energy and climate." National Academy of Sciences, Washington, D.C.

20. Newman, J.E.: In press, Impacts of rising atmospheric carbon dioxide levels on agricultural growing seasons and crop water use efficiencies. In: "Environmental and societal consequences of a possible CO_2-induced climate change, Volume II." U.S. Dept. of Energy, Washington, D.C.

21. Nigh, R.B., and J.D. Nations: 1980, Tropical rainforests. Bull. Atomic Scientists 36(9), pp. 12-19.

22. Poore, D.: 1975, Ecological guidelines for the development of the American humid tropics. pp. 225-247, In: "The use of ecological guidelines for development in the American humid tropics." International Union for Conservation of Nature and Natural Resources, Publication (New Series) No. 31.

23. Raich, J.W.: 1980, Fine roots grow rapidly after forest felling. Biotropica 12, pp. 231-232.

24. Revelle, R.: 1980, Energy dilemmas in Asia: The needs for research and development. Science 209, pp. 164-174.

25. Rotty, R., and G. Marland: 1980, Constraints on carbon dioxide production from fossil fuel use. Institute for Energy Analysis, Oak Ridge Associated Universities, Research memorandum ORAU/IEA-80-9(M).

26. Rutger, J.N., and D.M. Brandon: 1981, California rice culture. Scientific American 244(2), pp. 42-51.

27. Schultz, T.W.: 1980, The economics of being poor. Bull. Atomic Scientists 36(9), pp. 32-37.

28. Slater, L.S. (Ed.): In Press, Climate change and agricultural production in non-industrial countries. In: "Environmental and societal consequences of a possible CO_2-induced climate change, Volume II." U.S. Dept. of Energy, Washington, D.C.

29. Tosi, J.A., Jr.: 1975, Some relationships of climate to economic development. pp. 41-58, In: "The use of ecological guidelines for development in the American humid tropics." International Union for Conservation of Nature and Natural Resources, Publication (New Series) No. 31.

30. van der Plank, J.E.: 1963, "Plant diseases: Epidemics and control." Academic Press, New York and London.

31. Watts, R.G.: 1980, Climate models and CO_2-induced climatic changes. Climatic Change 2, pp. 387-406.

CLIMATE AND AQUATIC FOOD PRODUCTION

John E. Bardach and Regina Miranda Santerre

Resource Systems Institute, East-West Center,
    Honolulu, Hawaii, USA
Department of Geography, University of Hawaii,
    Honolulu, Hawaii, USA

ABSTRACT. Ever since man started to gather marine and freshwater
animals, the level of harvest has been influenced by the climate.
In this report we consider the effects of climate on aquatic food
production by aquaculture and by major fisheries in the high
seas, in regions of upwelling, and in tropical riverine areas.
Our main findings are:

1.   The literature dealing with the relations between climate
and aquatic food production is incidental to rather than eluci-
dative of those relations.

2.   Long-term climatic changes and short-term deviations from
the climatic norm affect fishing activities and levels of aquatic
productivity.

3.   The most important overall effect of climate on aquatic food
production is through variations in temperature, which influence
the distribution, migration, growth, and availability of food for
aquatic organisms.

4.   Other important climate-induced changes in the environment
affect ocean currents, wind patterns, and the onset of seasonal
rains such as the monsoons. The changes may act through variations in food, or malfunction of physiological, or behavioral
signals.

5.   Enhancement of aquatic food production through climatic
changes is difficult to measure but deleterious effects could

W. Bach, J. Pankrath, and S. H. Schneider (eds.), Food-Climate Interactions, 187–233.

reduce the annual global aquatic food production by more than 10 percent.

6.   Aquaculture production is less affected by vagaries of climate than is fishing.

7.   Improvements should be made in the methods of collecting fishery statistics to permit better resolution of the effects of climate and exploitation on fish stocks. As is true for other types of food production, more reliable and longer-range capabilities for predicting impending changes of climate are also desirable.

8.   Based on very preliminary analyses, an increase in atmospheric carbon dioxide might result in slight increases in the natural occurrence of fishes and hence fish production.

INTRODUCTORY OVERVIEW

Man has gathered food from the sea and the lakes and rivers of the world even before the dawn of history. Now this harvest, predominantly from untended, wild stocks fluctuates around 70 million metric tons (MMT).

The oceans, a main spring of climatic change, cover 70.8 percent of the earth's surface, while the various bodies of fresh water occupy barely one percent of the globe. All the waters of our planet furnish about 18 percent of the world's average daily intake of directly consumed animal protein (1) not considering spoilage and discards. In 1977 the food fish catch was 52.9 MMT in the round; in addition to it one must also mention the take of 20.6 MMT of small schooling fishes which are turned into fish meal and into oil for industrial purposes (1), amounting to a total harvest of 73.5 MMT of fish and shellfish.

It is noteworthy that in 1977 freshwater bodies produced about 11.5 MMT, virtually all fish, while mollusks and crustaceans contributed 6.8 MMT of the total marine catch of 62 MMT (1). The relatively large production from freshwaters is in part due to the greater fertility of lakes and rivers, compared to the sea, especially in the tropics. Also, fishing in freshwater is easier than in the sea with lesser expenditures for boats, equipment, and fuel. Another factor that contributed to relatively faster growth in freshwater, than in marine fisheries, is the recent emphasis on aquaculture. Defined as the husbandry of aquatic organisms, it now contributes ca. 5.3 MMT to total aquatic food production, an amount equal to 10 percent of the entire marine food fish catch (2). Mainly for the reasons given above, namely relative ease of control, aquaculture is still predominantly pursued in fresh and brackish waters.

The overall potential of conventional marine foodfish resources is probably under 100 MMT (3), while freshwater fish production could be increased to 15 MMT and aquaculture yields could at least be doubled, to produce a total of between 120 and 130 MMT of aquatic food. In addition there are unconventional resources such as the antarctic krill, squid from the high seas and even small mid-depth schooling fishes. These sources are believed capable of furnishing at least the same tonnage of animal protein as conventional marine fisheries albeit only with as yet unassessed but certainly high fossil fuel expenditures, and possible adverse ecological effects, accompanied by sub-stantial changes in processing technologies and marketing practices (3).

The most productive and, by and large, also the most heavily fished areas in the world seas are on both sides of the North Atlantic, prominently the North Sea and its environs, the reaches between Northeast Asia (Korea, Japan and Russia) and North America (Oregon, Washington, British Columbia, and Alaska). Also highly productive, and the sites of intensive fisheries, are the upwelling areas off the coasts of Peru and Northern Chile and those along the west coast of tropical Africa extending southward through Angola, even into the Republic of South Africa proper (Figure 1). There is also intensive fishing in the shallow Southeast Asian seas with some leeway for expansion to occur where an upgrading of the fishing technology can be employed. Expansion potentional also exists for the harvest of tuna stocks in the tropical Pacific and Indian Oceans.

Another substantial unused fishery potential is believed to be near the tip of South America, especially in the Atlantic and nearby in the Westwind drift region of open seas between Antarctica and South America, Australia, and South Africa, respectively. Its living resources are virtually unused, except for the whales now hunted to near extinction. Obviously, only available during the southern summer, they consist primarily of krill, that is, vast swarms of plankton-feeding shrimp that are the food of the baleen whales. Their potential sustained harvest has been estimated to be as high as 50 million metric tons (4); but both gathering the krill as well as preserving it will require high technology and arrangements for long-distance trans-portation of the fishery products. Japan, the USSR, and a few European nations have, so far, engaged in krill exploitation.

The distribution of fish and shellfish in the seas (except for those produced through aquaculture) depends on the prevalence of organisms lower in the food chain and the abundance of the latter is great only in some regions of the oceans. Productive

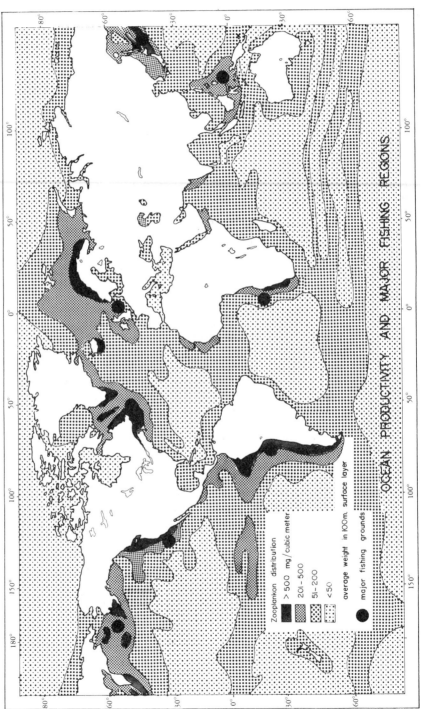

Figure 1. Ocean Productivity and Major Fishing Regions of the World. (Redrawn from FAO – Atlas of Living Resources of the Sea – 1972.)

waters lie, by and large, near the shores rather than in the high seas and in the temperate and subtropical rather than the tropical zones (Figure 1). In fact, with the recent promulgation by most nations of the 200 mile Extended Economic Zones (EEZ) as the result of seven-year negotiations in the third United Nations Conference on the Law of the Sea (UNCLOS III), about 99 percent of all fishing activities will take place in national sea spaces, generally closer to shore than 200 miles or over the continental shelf. Incidentally, with developed nations of the Northern Hemisphere having established preeminence in global fishing (mainly Japan, the USSR, Western European nations, and the Eastern European socialist countries), the re-ordering of marine property rights embodied in the 200 mile EEZs promises to change the pattern of exploitation of living marine resources.

Proteins of aquatic provenance differ in their role as contributors to overall nutrition. The role of fish in the Japanese diet is well known, but they are also important in the USSR and in South and Southeast Asia, where the populations are clustered in the valleys of large rivers, for example, the Indus, the Ganges, the Irrawaddy, the Chao Phraya, and the Mekong; that is, in the nations of India, Bangladesh, Burma, Thailand, Cambodia, and Vietnam. Thus fish may make up a major portion of the total, albeit often deficient, animal protein intake of more than a quarter of mankind (Table 1). In Africa, certain nations, for example, Senegal, Liberia, and the Ivory Coast, obtain 50 to 60 percent of their total animal protein intake from fish; and in South America along the Amazon fish are of comparable importance. Emphasis will thus be placed in the later, more detailed discussion on how fluctuations in rainfall can affect the supply of these locally important food commodities.

Fish have a different economic, but not less important, status in other regions: Peru and Chile, close to the Humboldt Current, and several African nations in the vicinity of the Benguela Current depend to a large extent on fish meal for their national incomes. Small schooling fishes, anchovies, or sardines teem in offshore waters made highly fertile by upwelling associated with these currents. These fish populations have made up as much as one-third of the total marine catches in the 1960s. However, they can be greatly depressed by surface incursions of warm water (the El Niño phenomenon) which, in turn, are engendered by pressure anomalies in distant parts of the global sea-air interactive system. In short, the economic effects of these short-term climatic phenomena can be ruinous to fishing countries. These resources are also easily over-fished and because fish meal is an important ingredient of compounded animal feed, their status has affected the food economies of Japan, North America, and Western Europe (Table 2).

Table 1 (A&B). Nutritional Significance of Fish and Seafoods in the Diets of People in Asian Countries.

Table 1.a. ASIA AS A WHOLE

	1961-63	1966	1970	1974
Population, 10^6	1696.9 (100.0)	1845.1 (108.7)	2006.9 (118.3)	2184.3 (128.7)
Calorie, Person/Day	2032 (100.0)	2055 (101.1)	2200 (108.3)	2220 (109.7)
Vegetable Products	1872	1889	2024	2049
Animal Products	160	167	176	182
Protein, g/Person/Day	52.8 (100.0)	53.4 (101.1)	56.9 (107.8)	57.2 (108.9)
Vegetable Products	43.3	43.2		
Animal Products	9.4 (100.0)	10.2 (108.5)	10.9 (116.0)	11.4 (121.3)
Fish and Seafoods	2.9 (100.0)	3.5 (120.7)	3.8 (131.0)	4.0 (137.9)
In % of Animal Product	30.9	34.3	34.8	35.1

Prepared by Y.H. Yang, RSI/EWC, From: 1. Provisional Food Balance Sheets 1972-74 Average. FAO Rome 1977.

2. World Development Report 1978. The World Bank, D.C.

Table 1.b. INDIVIDUAL COUNTRIES (Population and Nutrients Availability 1972-74 Average, Per Capita GNP 1976 Mid-Year)

COUNTRY	POPULATION, 10^6	PER CAPITA G.N.P. US$	CALORIE	PROTEIN, g	DAILY PER CAPITA AVAILABILITY		
					PROTEIN OF ANIMAL PRODUCTS, g	FISH AND SEAFOODS, g	IN % OF ANIMAL PROTEIN
East Asia							
China, People's Republic	796.6	410	2,278	61.7	11.8	3.8	32.2
China, Taiwan	15.6	1,070	2,757	74.2	24.7	11.1	44.9
Korea, Republic of	32.6	670	2,749	72.0	13.1	9.0	68.7
Korea, DPR	15.1	470	2,636	76.4	12.1	8.0	66.1
Japan	108.3	4,910	2,832	85.5	40.1	22.1	55.1
Hong Kong	4.1	2,110	2,641	78.6	44.4	14.8	33.3
Mongolia	1.4	860	2,477	92.8	60.9	.2	.3
Southeast Asia							
Singapore	2.2	2,700	2,823	74.6	37.6	13.7	36.4
Malaysia, Peninsula	9.6	860	2,539	45.1	8.6	1.0	11.6
Malaysia, Sarawak	1.1	...	2,518	51.9	15.5	9.3	60.0
Malaysia, Sabah	.7	...	2,776	59.7	22.1	11.3	51.1
Indonesia	129.2	240	2,031	42.0	5.3	3.6	67.9
Philippines	41.6	410	1,957	45.5	16.7	8.9	52.7
Thailand	39.4	380	2,302	49.5	13.2	6.8	51.5
Burma	29.8	120	2,125	55.2	8.8	4.0	45.5
Kampuchea	7.7	...	2,081	48.4	7.5	3.2	42.7
Lao	3.2	90	2,064	56.2	9.5	1.9	20.0
South Asia							
India	584.0	150	1,967	48.5	5.3	.8	15.1
Bangladesh	71.1	110	1,949	43.2	6.7	3.5	52.2
Pakistan	66.3	170	2,128	57.2	12.7	.3	2.3
Sri Lanka	13.4	200	2,075	40.9	6.6	2.6	39.4
Nepal	12.0	120	2,019	49.7	7.5	.1	.1
Afghanistan	18.3	160	2,001	61.5	6.9	---	---
Iran	31.0	1,930	2,319	60.8	11.8	.1	.1

Table 2. Catches of Peruvian Anchovy (Millions of Metric Tons) and Prices of Fish Meal, Soybean Meal, and Pork Meat (Current $/Ton).

Product \ Year	1965	1966	1967	1968	1969	1970	1971	1972	1973	1974	1975	1976	1977	1978
anchovy catch[1]	7.23	8.53	9.82	10.26	8.96	12.27	10.28	4.45	1.78	4.0	3.3	4.3	0.8	0.5
fish meal[2]	190	160	134	129	172	197	167	239	542	372	245	376	454	410
soybean meal[2]	94	101	98	98	95	103	102	129	302	184	155	198	230	185
pork meat[3]	677	809	694	811	940	916	885	1046	1543	1570	2023	1939	1796	2286

Sources:
1. Colin W. Clark, "Bioeconomics of the Ocean," BioScience, in press.
2. Price Prospects for Major Primary Commodities Fats and Oils 1979, World Bank, for meals. (Soybean meal U.S. 44%, fish meal any origin).
3. Trade Year-Book 1979, FAO, for pork meat. (U.S. price).

In the following section of this paper, emphasis will be on more detailed description of stages in the life cycles of aquatic organisms as they are prone to be influenced, by various climatic anomalies and/or fluctuations. Historic examples will be cited in some detail concerning the influence of the climate on fisheries. Prognosis with mention on the special status of aquaculture as well as the possible future of fisheries in relation to the carbon dioxide problem are discussed in the last section.

THE CLIMATE AND THE LIFE CYCLE OF FISHES

Reproduction

To begin with, spawning (as the reproductive act is called in aquatic animals) is strongly under the influence of temperature: in most non-tropical species spawning occurs once or intermittently throughout a season when the water has warmed or cooled to a certain level; but changes in the duration of the day length also affect the onset of spawning (5). In certain of them, for example, herring (Clupea harengus), there are, however, spring- and fall-spawning stocks with the former depositing eggs on gravel or seaweed progressively later in the year as one goes from more southern to more northern spawning sites. Large cod-fish of over 20 kilograms scatter over 20 million eggs during the late winter and early spring, albeit in multiple spawning bouts. The onset of their spawning appears to be tied to temperature; hence climate induced warming and cooling of surface waters influence the range of their occurence (see section on "Ocean Fisheries" below). Exact timing for repeated spawning is often triggered by stages of the tides or phases of the moon, that is by astronomic rather than climatic signals. Thus the Pacific threadfin (Polydactylus sexfilis) will begin to spawn in Hawaii when the water has warmed to 24 degrees C. and then repeat the spawning episodes faithfully throughout the summer on the outgoing night tides of 5 to 6 days bracketing the last-quarter phase of the moon (6). Moon-phase determined spawning is, in fact, generally more prevalent in tropical rather than temperate seas.

The behavioral characteristics of the various species are such that a high rate of fertilization is favored if not assured; observers of herring spawning on the Dogger Banks tell of milky seas when the males, extruding milt, follow the females which stick their egg masses on to the substrate. Once the eggs are laid, the perils truly begin with their peak reached after the onset of feeding and throughout the larval period.

Spawning is timed in such a fashion that survival at hatching is ensured. Thus, if the eggs are laid in the fall, they may be near dormant during the winter as in certain races of salmon. In fact, compared to birds and mammals, the lack of internal temperature control, such as prevails in aquatic animals (except for tunas), makes ambient temperature exert overriding and continuous accelerating or decelerating influences on metabolic functions and hence on growth of aquatic animals at whatever stage.

Larval and Post-Larval Stages

Most species of importance to world fisheries have long free swimming larval and/or juvenile stages [cod (<u>Gadus</u> <u>morhua</u>), haddock (<u>Melanogrammus</u> <u>aeglefinus</u>), sea-basses (family Serranidae)], if they are not free swimming or pelagic all their lives, [herring, sardines (family Clupeidae), anchovies (family Engraulidae), mackerel (<u>Scomber</u> <u>scombrus</u>), tuna, squid]. They rely on free swimming minute food organisms, that is plant (phyto) and/or animal (zoo) plankton or small schooling fishes. We should note though that before feeding can begin, variations in the onset of vernal warming or autumnal cooling may cause stresses at various stages of egg and very early larval development and make for mass mortalities (5). With the exception of prominent areas of upwelling waters throughout the year, the temperate regions over the continental shelf where most fisheries are located show pronounced seasonal variations in plankton productivity. The spring plankton bloom over George's Bank for instance, nourishes the larvae of late winter or early spring-spawning fishes (e.g. herring) (Figure 2). The matching in time of the onset of feeding of the larvae (when their yolk sac is absorbed after hatching) and the spring bloom which in turn depends on climatic events for its timely occurrence will in fact determine the sizes of year classes (7). Thus, with intervals of one or more years, profitable or precarious fishing seasons occur (Figure 3).

There are also instances where feeding or spawning migrations in the mature adult organism are triggered by temperature or currents or the onset of seasonal rains (e.g., <u>Hilsa</u>, an Indian Ocean herring which migrates up-river to spawn, or <u>Rastrelliger</u>, a small mackerel in the Gulf of Thailand, and Pacific salmon <u>Oncorhynchus</u> spp. which depend on thermal clues for their upstream spawning migrations). By and large, however, temperature and other climatic factors have substantially less influence on later life history stages of aquatic animals than they have during their larval periods.

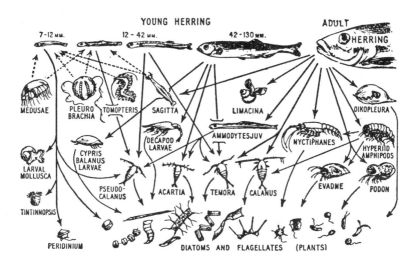

Figure 2. A Sketch Showing the Feeding Relationships
Between the Herring of Different Ages and Plankton,
Redrawn from Hardy (1924). The Arrows Point in the
Direction of Predation. (Source: The Ocean Sea: Its
Natural History. Part II. -- Sir Alister Hardy.
Boston 1965).

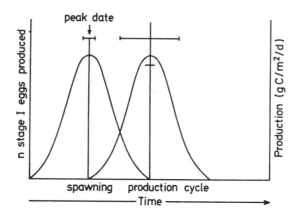

Figure 3. The Match/Mismatch Hypothesis: The Overlap
Between Spawning Time and the Time of Plankton
Production Might Indicate the Magnitude of the Subsequent Year Class. (Source: Cushing & Dickson, 1976)
(8).

OCEAN FISHERIES

General Considerations: Climatic Background

The earth's climatic history is comparatively well recorded
for the past 1,000 years or so, while detailed records of
fishery catches and fish stock densities are available only since
the beginning of the twentieth century. Earlier and somewhat
detailed statistics are primarily from the North Temperate
latitudes; but more recently information from other fishing
regions is being documented. This section outlines the climatic
events of the past one hundred years and will not attempt to
describe in detail the complex climatic processes at work.[2] The
climatic elements which seem to be responsible for the observed
changes in midlatitude climate are mentioned since the corresponding fisheries events discussed are predominantly from these
latitudes.

The first four decades of the 20th century were marked by a
sustained tendency of warming (Figure 4) causing an increase in
mean annual surface air temperature, resulting in the reduction
of sea ice to its recorded minimum extent in 1938, and an
increase in the strength of global atmospheric circulations (8).
Since the late 1940s a reversal of this warming trend occurred
(Figure 4) accompanied by a southward extension of the northern
sea ice, slight equatorial shifts in the wind tracks and a
weakening of the main westerly winds (8). The reflective
surfaces of the sea ice further intensified the cooling effect
in the northern latitudes. These changes may be related to an
apparent reduction in incoming solar radiation (10).

The climatic shifts experienced in the mid-latitudes are
mainly influenced by the so-called Rossby Waves - or the Midlatitude Westerlies formed as a result of the latitudinal
pressure gradient (8,11). Bounded equatorward by the subtropical
high pressure belt, the Westerly belts widen from the surface and
are strongest towards the upper troposphere (8,11). Due to the
earth's rotation these Westerlies are deformed and form undulat-
ions, each poleward extension generating a warm ridge and an
extension towards the equator forming a cold trough (Figure 5).
This whole system can be modified by the earth's topography
because ridges tend to form over warm surfaces and troughs over
cold ones (8).

Very recent evidence seems to suggest that the cooling
experienced in the last few decades might again be reversing.
Since 1970 there has been a general tendency for rising surface
and air temperatures in the Arctic basin, as well as a strengthening of the Westerlies and a series of mild winters in Europe.
An increase in temperature and salinity of surface layers of the

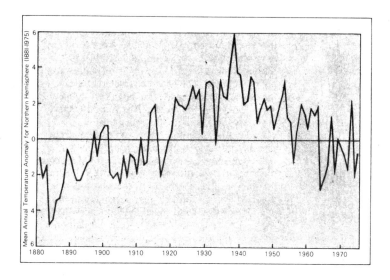

Figure 4.   Mean Annual Surface Air Temperature as
Deviating from a Period Mean of 0° C. over the Northern
Hemisphere from 1880 to 1975. (Source: Mason, 1980)
(9).

Northern Icelandic waters led to another reduction in sea ice in
these areas (12).

Biological Events

     An examination of the records of temperate fisheries reveals
the reaction of the harvested fish populations to the climatic
changes mentioned above.   An illustration of such a response to
the overall warming and cooling tendency experienced this century
is provided by the change in the mean date of capture of the cod
fishery at Lofoten from 1900 to 1970.   This change indirectly
reflects the changes in the mean date of spawning, a temperature-
dependent event, as fish stocks seem to be adjusting to a warming
and subsequent cooling of the oceans (Figure 6).

     Another noteworthy effect of long-term climatic variations
was the establishment of the cod fishery in West Greenland.
Landings of cod which were almost nonexistent at the beginning of
the twentieth century increased from 23.5 metric tons in 1912 to
243.5 metric tons in 1917 and to about 600 metric tons per year
between 1919 and 1923.   Correlated with the warming of the
Atlantic Ocean, the cod fishery advanced progressively to the
north, reaching as far as 73 degrees North latitude in the 1930s.
With the onset of the cooling trend in the late 1940s, the

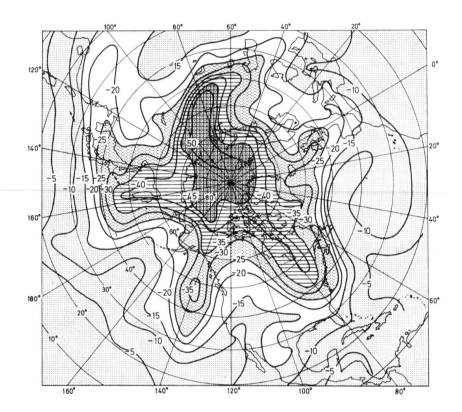

Figure 5.   Rossby Waves in Circumpolar Circulation.
Shown in the Temperature Distribution at the 500 mb
Level in the Atmosphere, Isotherms are Drawn at 5°C.
Intervals.   Rossby, 1959.   Source: Cushing & Dickson,
1976) (8).

fishery moved southward with only small quantities of cod being
caught north of 69 degrees North latitude by the 1950s (13).
While the decline in sea temperatures reduced catches in West
Greenland, the cooler climate favored the survival of the cod in
the North Sea which is the southern limit of its range.   The
annual average landings of cod from the North Sea more than
doubled between the period of 1950-1963 (average of 94,000 metric
tons per year) and 1964-1971 (average of 247,000 metric tons per
year (14).

During the post 1940 cooling period, the southward extension
of sea ice in the northern hemisphere reached its maximum
recorded extent in 1968 (Figure 7.a), bringing with it a drastic
change to the Icelandic herring fishery.   Catches of herring

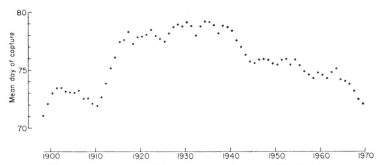

10-year running means of the mean date of capture of the Lofoten spawning cod fishery 1894–1974 (1 January = day 1 etc.).

Figure 6. Change in Dates of Cod Spawning in Response to Long-Term Climate Changes. (Source: Cushing & Dickson, 1976) (8).

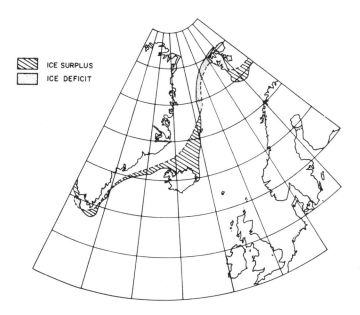

Figure 7.a. Extent of Ice in the Northern Atlantic, in 1968 (solid line) compared to normal (1911–1950) (dashed line). (Source: Johnson, 1976) (15).

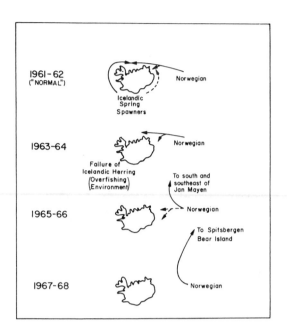

Figure 7.b. Summary of Recent Changes in the Feeding
Migration of Herring at Iceland. (Source: Johnson,
1976) (15).

dropped from 750,000 tons in the mid-1960s to 50,000 tons in the
1970s. The ice cover north of Iceland had altered the annual
feeding migrations of the herring stock, moving them out of reach
of the main fishing fleets (15) (Figure 7.b).

Events in the North Sea

    Although fishery statistics could be used as indicators of
deviations in climate in and around the fishing grounds, the
figures also reflect changes in techniques of exploitation and
utilization of fish catches, exemplified by the North Sea fishery
landings since the beginning of the twentieth century.

    From 1919 to 1938 total landings from the North Sea averaged
1.2 MMT, with catches increasing to an average of 1.7 MMT in the
period between 1946 to 1961. Since 1961 there was a rapid
increase, with landings exceeding 2 MMT every year from 1964
onward and reaching the maximum level of 3.4 MMT in 1968 (16).
Overall climatic conditions in the North Sea during the afore-
mentioned period were in keeping with conditions around the
globe. A general warming in the North Sea was recorded from 1905

to 1954, with the highest temperatures being experienced in the
1940s. This was followed in the next two decades by a strong
cooling trend, especially pronounced in the northern regions of
the North Sea and in the winter. Yet, notwithstanding the
reversal in climatic conditions, landings from the North Sea
have increased, apparently for the following reasons.

Harvesting fish from the North Sea was chiefly by drift
netting and trawling in the 1950s. The increased exploitation of
fish stocks was made possible primarily by the introduction of
purse seiners[3] in the 1960s, together with improved intervessel
communications and the development of sonic devices for locating
fish, greatly enhancing the fishing efficiency of the fleets.
Interestingly, it was a few species of fish that contributed to
these increasing yields. The initiation of industrial fishing
for the production of fish meal and fish oil in the early 1950s
more than doubled the output from the North Sea. The percentage
of fish being converted to fish meal and oil grew from a mere 2
percent in 1950, to almost 22 percent in 1960 and to approximately one-half of the total catch in 1970. This flourishing
fishery, being directed at species with pelagic (i.e. open sea)
feeding habits, such as the Norway pout (Trisopterus esmarkii),
sand-eel (family Ammodytidae), sprat (Sprattus sprattus), and
horse mackerel (Trachurus trachurus) might have caused a
reduction in the predation on pelagic fish larvae and competition
for food supply. This in turn could have led to increased
recruitment of commercial species, such as cod, plaice
(Pleuronectes platessa), haddock, and whiting (Merlangus
merlangus) during the sixties resulting in very high yields of
these demersal species in certain years (17).

At the same time, the exceptionally cold years between 1963
and 1966 seemed associated with high year-class strength of cod
plaice, and haddock, apparently cold loving species; while
relatively warm conditions in their spawning grounds seem to
produce good whiting year-classes (18). There also appears to
have been a northward shift in maximum densities of several
species from the English Channel, such as the plaice, sole (Solea
solea), turbot (Scophthalmus maximus), anchovy, and horse
mackerel, in the early 1960s, thereby increasing their availability in the North Sea (16). Postuma (19) records the
immigration of southern subtropical and warm temperate species
into the North Sea. The pilchard Sardina pilchardus was recorded
every year since 1936, with peak catches in 1959, but declining
in 1969-70. Similarly catches of Atlantic sea breams (Pagellus
centrodontus) increased in the 1950s and 1960s with maximum
catches during 1967-1969, but they dropped to very low levels in
1970-73. It is not clear if these fish populations immigrated
to the north in response to a cooler climate or if expanding

fishing efforts simply captured these species in larger numbers.
It is also documented that human activities (effluents) have
increased nutrient levels in the North Sea since the 1940s (20).

Thus, despite the strong post-1940 cooling trend the overall
picture has been one of increasing catches from the North Sea area.
However, it is important to note that during this period, three
major fisheries of this region collapsed, that of herring, mackerel, and dogfish (mainly Squalus acanthias). Depletion of fish
stocks by increasing fishing pressure is considered the prime
reason for the failure of these fisheries and not climatic
influence (16). At the same time, it must be remembered that
reduced and stressed by intense fishing these natural populations
might be less prone to withstand climatic stresses, hence decreasing their chances of survival.

A direct relationship is observed between the fluctuating
landings of sole from the North Sea and winters with abnormally
low temperatures caused by high pressure regions over Scandinavia
driving very cold winds out to the sea. Such winters lead to
high mortality, some fish being killed directly by the 3 degrees
C. waters, others becoming moribund and more vulnerable to fishing.
In subsequent years catches were very small and adult stocks very
low. However, such winters are followed by high year-class
strengths in the fisheries two or three years later when catches
begin to increase and continue to do so until another severe
winter kills the stock or the year-class is fished out. This
effect was most marked in the winters of 1924-25, 1928-29,
1946-47, and 1966 (16).

Some Events from the Pacific Fisheries

As was mentioned earlier, the North Atlantic Ocean seems to
be warming since the 1970s, but the northern waters of the
Pacific experienced anomalous cold air and surface water temperatures, and in the summers of 1971-75 colder bottom temperatures. These cold temperatures in the Bering Sea, the Gulf
of Alaska and the west coast of North America were due to the
extended sea ice cover in the eastern Bering Sea and persistant
cold winds from the north. The colder climate has adversely
affected the fishery for sockeye salmon (Oncorphynchus nerka)
and halibut (Hippoglossus stenolepis) in these waters (21).
In fact, the landings from the Bristol Bay sockeye salmon
fishery in 1973-74 were the lowest since the 1800s. The cold
waters delayed the spawning migrations of the adults into
streams and had negative effects on the hatching of eggs. As
as result, the survival of juveniles and their return to the
sea was very poor, the 1971 seaward migration into Bristol
Bay being the smallest on record. The cold temperatures also
disrupted the distribution of halibut in these waters. Normally,
two year olds are a significant proportion of the catch inshore

in shallow waters near the northern Alaska peninsula, while three
year olds dominate the catch in Bristol Bay. The colder waters
have inhibited the migration of the older fish in to the bay so
that they were found in inshore waters with the younger fish
(21).

Teleconnections

    The difference in climatic conditions between North Atlantic
and North Pacific as discussed above may well be the result of a
warm ridge of the Rossby Wave over the North Atlantic and a cold
trough over the Pacific. By the same token, comparable biological events, in this case, the collapse or appearance of
certain fisheries, occurring in regions separated by considerable
distances seem to be linked by common climatic events i.e.
teleconnections. The Rossby Waves, which are a strong influence
on climatic change in these latitudes, seem to control these
remotely linked events. For example, the formation or amplification of a stationary cold trough at a given point in the
westerly wavebelt will lead to the formation or amplification of
a response wave immediately downstream at a distance compatible
with the strength of the westerlies (8). Thus, in these mid
latitudes teleconnected centers tend to be spaced according to
the scales of the Rossby Waves. Therefore it is not surprising
that biological events appear to be correlated even across
considerable distances. Cushing and Dickson (8) recorded the
common periodicities of the North-West Atlantic, Pacific and
Mediterranean sardine stocks. Records of these fisheries extend
back a few centuries, although events documented in the earlier
periods may be less reliable; however, if the collapse or
appearance of the fishery was a major event, it was recorded.

    The progress of the fishery for overwintering North Sea
herring on the western coast of Sweden, the Bohuslan coast has
been well documented since 1400. The fishery has appeared and
disappeared five times since the fifteenth century. Just as
changes in the weather, harvests, locust swarms, etc., had been
related in the mid-nineteenth century to the 55 year sunspot
cycle that Schwabe had discovered in 1943 (22), the herring
periods became associated with it. Yet Petersson (1922), [in
Höglund (22)] related them to a 111 year cyclic phenomenon in
tidal variation. Now the causality of regular cyclical periodicity of herring abundance has been disproved, while climatic
variation has been adduced for the changes in abundance of the
fish (22).

    The alternation between the Swedish Bohuslan fishery and the
Norwegian spring fishery for Atlanto-Scandian herring is well
known to fishermen of the region. It was generally believed that
climatic conditions cause the herring to move their spawning

grounds alternately between the Norwegian coast and the Swedish coast farther south. It has since been realized that the inefficiency of fishing methods to reach the herring in certain years was the main reason for the periodic failure of these fisheries, and the alternation theory between the Swedish Bohuslan fishery and the Norwegian spring fishery as suggested by Cushing and Dickson (8) has been more or less abandoned (22).

Nevertheless, the periods of the Norwegian spring herring fishery appear to be correlated with the period of ice cover retreat north of Iceland and the general warming of the North Atlantic. This periodicity seems to conform also to highs in the Japanese sardine (Sardinops melanosticta) fishery in the East Pacific. Cushing and Dickson (8) also report a correspondence in the California sardine (Sardinops caerulea) periods to that of the Japanese and Norwegian herring periods. In addition, examination of the records since the sixteenth century of the Adriatic sardine fishery of Yugoslavia suggest a correspondence between "good" (1718-1725, 1830, 1929-40) and "bad" (1730-1780, 1884, 1946-55) year-classes of the Adriatic sardine to that of the Japanese sardine. Thus, Cushing and Dickson (8) conclude that the periodicities of Norwegian, Californian, Japanese, and Adriatic fisheries are related to conjoint warm periods in the Atlantic, Pacific, and Mediterranean waters. Indeed the development of either a warm meridional ridge or a cold meridional trough over the North Sea area and responses to it could well be paralleled by similar developments in the North Pacific and Mediterranean.

Fluctuations in fisheries yields and variations in climate are less well documented from the tropics and subtropics than they are from the temperate seas. Yet, tropical marine fisheries have steadily increased in importance with the tunas being the most important in value.

The Climate and Tuna Fisheries

World-wide hunting of tunas is mainly pursued by Japan, although Taiwan and Korea are also engaged in it. The development of the high seas fishery for tuna dates from the 1950s; it depended on advances in purse seining, long-lining[4], brine freezing, and canning techniques. The fishery is so widespread that there are not yet adequate synoptic data on tuna distribution and climate to permit a thorough discussion on global dynamics of tuna stocks. Some facts are known, however, about the relation of tuna distribution to ocean conditions (23). In view of their unit value and the apparent expansion potential, mainly in the fishery of one species, the skipjack, a few remarks on tuna seem to be in order here. Six species of tuna are of commercial importance with yellowfin (Thunnus albacares) and

skipjack (<u>Katsuwonus</u> <u>pelamis</u>) in the lead. They are active
predators resorting even to cannibalism on their own young;
because of their fast swimming habits they need ample oxygen,
and their capacity to regulate their body temperature above
those of the surrounding seas imposes on them a relatively narrow
temperature range (24).

Thus tuna are relatively scarce in certain portions of the
Eastern tropical Pacific where the mixed layer is shallow and the
thermocline occurs closer to the surface than in the Western
Pacific (see also section on "Upwelling Fisheries"). Tuna are
believed to rely on major currents for their extensive feeding
and spawning migrations (25). Thus both yellowfin and skipjack
follow the shifts of the 20 degrees C. surface isotherm in the
California Current. Yellowfin occurrence, in abundance, is
bounded by the 20 and 30 degrees C. isotherms, areally and bathy-
metrically, with skipjack having only a slightly narrower temperature range. They are abundant between 20 degrees and 29
degrees C., except off the coast of Eastern Australia where their
lower temperature boundary falls to 15 degrees C. Bigeye tuna
(<u>Thunnus</u> <u>obesus</u>) occur in deeper waters since they and albacore
(<u>Thunnus</u> <u>alalunga</u>) can stand lower temperature with a range of
14-23 degrees C. Bluefin tuna (<u>Thunnus</u> <u>thynnus</u> <u>orientalis</u> and
<u>Th.</u> <u>maccoyli</u>), which are the largest and most cold adapted
species, roam between 14 and 19 degrees C. in the northern and
between 10 and 21 degrees C. in the southern seas respectively.
Within these bounds of temperature tuna make long migrations; the
skipjack which wander farther than several other tuna species are
believed to rely on some kind of active or passive orientation to
surface currents and to some extent on temperature variations (25).
Field observations are still inadequate to explain mechanisms of
orientation and behaviour in response to the environmental variables that would be followed, but laboratory experiments have
established high sensitivity by tuna to rates of change in
ambient temperature (26).

It was observed that skipjack larval abundance was highly
positively correlated with high sea surface temperature. This
led to the hypothesis that warm years in the Central Pacific
might lead to higher abundances of adults in the Eastern tropical
Pacific fishery. There is an apparent contradiction, however,
that concerns the differential temperature tolerance levels of
adults and larvae. Adult skipjack spawners are adversely
affected by high temperatures, while skipjack eggs and larvae can
well withstand them. Thus spawning seems to occur in deeper
waters and the eggs then rise to float in the surface layer where
their survival is enhanced (25).

The importance of the upwelling of cold and nutritious water
in the larval survival of other tuna species has been indicated

by several studies. Investigations of the recruitment of yellowfin in the Western Pacific (27) and in the Indian Ocean and
albacore recruitment in the Southwest Pacific around New Caledonia
[Nakagone, unpublished, in (31)] indicate high year-class
strengths when the water was cool. A similar finding for the
North Pacific bluefin suggests that a strong year-class enters
the fishery a few years after an upwelling occurrence off the
eastern Philippines (28).

The summary of observations does permit one to state,
however, that climate-induced shifts in currents and surface
temperatures would change tuna distribution and regional
abundance. The tuna temperature relationship has already influenced redeployment in fishing fleets as demonstrated both in the
Pacific and in the Indian Ocean, where fishing centers moved
parallel with the isothermal lines (29,30). For instance, in
1972 an El Niño year in the Eastern Pacific with abnormally high
temperatures, the tropical Western Pacific waters were abnormally
cool and the center of the yellowfin fishery in the area was
located about 4 degrees south of the average (31).

Another example of temperature conditions affecting the
distribution of fish species comes from the albacore fishery of
the Northwest Pacific. Albacore tuna catches decreased off
southern California as the fish moved northward in response to
the overall global warming that was occurring in the early
twentieth century. An interesting similarity between tree ring
widths of conifers in western North America and albacore distribution along the west coast is documented by Clark, et al., (32)
(Figure 8). Below-normal tree growth during dry conditions is
associated with cyclonic activity in the fishing season.
Favorable fishing conditions occur in sunny weather when excess
stored heat in the ocean increases precipitation which, in turn,
results in greater tree growth in the following spring.

A similar association has been recognized between mackerel
landings along the eastern coast of the U.S. and the tree ring
widths of the Canadian larch, another conifer (33). These
results suggest that although the different organisms are
responding to their respective environments, they are reacting
to the same climatic fluctuation influenced by air-sea interactions.

UPWELLING FISHERIES

The Peruvian Anchovy Fisheries and El Niño

In 1970, Peru's anchovy harvest amounted to 12.3 MMT or more
than 20 percent of the total world marine finfish catch for that

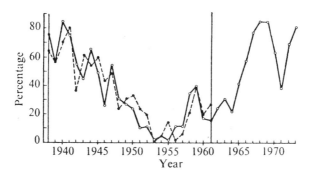

Figure 8.   Correlations of Albacore Catches and Tree
Ring Widths. (Source: Clark, et al, 1975) (32)

———————   Percentage Landings Data
————————   Derived from Tree-Growth Data

year. These fishes rely on plankton blooms that result from the
surfacing of colder nutrient rich water. The fisheries for ancho-
veta (Engraulis ringens) in Peru and Sardinella aurita off the
coast of Africa where there is similar upwelling, developed after
World War II as the demand arose for fish meal as an ingredient
of animal feed.

In 1972, the Peruvian catch plummeted to 4 MMT to fall
farther in 1973 to 1.8 MMT. The primary cause of this drastic
collapse was the atmospheric and oceanographic event known as
El Niño, although heavy fishing greatly aggravated the situation.
This episodic climatic occurrence is characterized by a far
greater than normal influx of warm tropical water along the coasts
of Equador, Peru, and Chile around December. (El Niño, is the
Spanish name for the Christ child.) The incursion of this water
mass, poorer in nutrients than the upwelling waters is accompanied
by heavy rains along the usually dry coasts.

Cushing (34) noted that very strong offshore winds precede
the El Niño episodes. They depress surface plankton production
possibly even to contribute to year-class failures. If a failure
such as the one noted in the fall of 1971 when the hatch was only
one-tenth that of the average of previous four to five years (35)
were to occur before poor growing conditions for juveniles--
i.e. an El Niño--the collapse of the anchovy fishery that follows
(1972-73) can be largely explained (8). Added to these climatic
effects on the biology of the species is the fact that the
anchovy schools occur deeper during El Niño episodes than in
normal years--even in the thermocline and somewhat below it--

so that the fishermen cannot reach the fish with their surface
purse seines.

Such events where the climate and the oceans conspire
against the fishermen, as it were, have occurred in other years
and always with severe negative effects on the economy of the
coastal states. Documented El Niños in 1957-58 mainly affected
the guano birds, anchovy fishing having barely begun at that
time; but in 1965 and especially 1972-73 the economic effects on
the fishery were catastrophic. A shortfall of fish meal in the
world's markets had severe ripple effects on other animal feeds
and on livestock prices as illustrated in Table 2.

The stronger than normal--usually they are slight--incursions
of warm water are explained by occurrences of weak trade winds
permitting the South Equatorial Current to cause a massive
southward invasion of warmer North Equatorial waters (36).
Winters with warm surface water off the coasts of Peru and Chile
have far reaching oceanographic and hydrographic connections
than previously postulated. The warmer water is not only on the
surface but the thermocline, which is normally at a depth of
about 20 meters along the coasts of Peru, dips during El Niño to
a 100 or even 200 meters. Because of the tendency of oceanic
water masses to be in hydrodynamic equilibrium, this phenomenon
pointed to oscillations of the entire warm water layer between
Peru and the Western Equatorial Pacific (37).

Time series of water level measurement in the Solomon
Islands, Guam, Galapagos, and Peru corroborated the theory that
strong trades cause the water to pile up in the Western Pacific
and that these waters actually flow back, in a pivoting mixed
layer, above the thermocline, when the trades weaken below normal
strength. Average tradewinds cause the surface water level to be
40 cm. higher in the Western Pacific, than in Peru. With strong
trades there are added to these levels 10 cm. of additional water
while the level at Galapagos falls 5 cm. below normal. After
three months of weaker winds in early 1972, the Galapagos level
rose by 15 cm. and the Solomon level fell by 20 cm. Corroboration
of this apparent see-saw movement was found in the strengthening
and southward displacement of the Equatorial Countercurrent with
subsequent further southward diversion when the returning water
mass reached the South American coast. As a result of a modelling
exercise, the speed of propagation of an internal "Kelvin" wave
was found to agree with the observed behavior of the thermocline
in Peru. Observations and models also suggested that the time-
volume relations of the hydrodynamic phenomena permitted the
prediction of an El Niño from four to six months before it occurs.
That is, the event may be predicted after the onset of certain
atmospheric changes (weakening trades) but, one should note that
El Niño's do not resemble one another completely, though they

have certain characteristics in common, and predicting an El Niño
and its approximate strength would in itself be very important (37).

An El Niño-like Phenomenon in the Atlantic Ocean

The fisheries that rely on the upwelling along the African
coast now produce about 2.5 MMT of fish meal fish annually. The
general wind patterns around the equator are broadly comparable
there to those in South America. However, the configuration of
the coast, such as the Gulf of Guinea, and the lesser expanse of
the Atlantic Ocean do not permit close comparisons with the
Pacific.

There are nevertheless, reports of El Niño-like phenomena
prevailing along the coast of the Gulf of Guinea (38,39), and it
is reported that the Guinea Current regularly transports some
warmer water along the coast but that this transport is far
stronger in certain years, e.g., 1968, than in others, and that
the wider occurrence of warmer water then gives rise to prey
fish mortalities and declines in the production of tunas. The
phenomenon thus seems to resemble the Pacific El Niño rather
strongly, but the long-range predictions of the phenomenon in
either ocean still elude us (38).

Assuming the recovery of upwelling fisheries after El Niño
effects through very strict control of fishing levels (a possibility but by no means a certainty, even though Peru has nation-
alized its anchovy fisheries), longer-range predictions of
upwelling anomalies would be of major importance for the world
food trade as indicated in the relationship between anchovy
harvest, soybeans, and pork prices (Table 2). If these fisheries,
which make up a substantial portion of the world fish catch, were
eventually used to supply human nutrition directly rather than be
channeled through domestic animals, the latter use still seeming
to be predicated by economic and technical considerations (40),
predictions of impending fluctuations in them would be equally,
if not more, desirable.

Upwellings Along the Eastern Edge of Continents

The last decade has brought evidence that major ocean
currents are the key agents nourishing the rich ecosystems along
the eastern as well as the western edges of continents. For
instance, menhaden (Brevoortia tyrannus) fisheries prevail from
Florida to North Carolina (annual harvest on the average being
1.2 MMT). The fish feed on the rich plankton pastures that are
the result of upwelling induced by the Gulf Stream, which has
changed in strength and patterns due to changes in climate during
the last few thousand years. Its meanderings deflect some of its

waters onto the continental shelf and this action forces
nutrients to the surface. Upwelling in another area (off
Charleston) is triggered by the bottom topography, which deflects
the Gulf Stream eastward (41).

Similarly, changes in nutrients with monsoon direction of
the waters off Somalia are indeed striking. During the northeast
monsoon (in the northern winter months) when the water comes
across the Indian Ocean, fertility approaches very low levels,
while the southwest monsoon (in the northern summer months)
elevates nitrate and phosphate levels to the heights of other
very fertile stretches of upwelling (42). Very strong climate-
induced bouts of the latter monsoon can bring a great richness
of fish; they are, however, prone to suffer cold-shock mortalities
and they may not be accessible because of foul weather.

Fisheries for Alternating Species Pairs

North of the equator in the Pacific Ocean, in California,
there is another fishery for pelagic offshore schooling fishes,
the anchovy (Engraulis mordax) and the Californian sardine. The
occurrence of this species pair in a more temperate latitude
than the Peruvian anchovy is favored by some upwelling caused by
the interaction of the California Current and the inflow of
deeper water from the south. The fishes grow more slowly than
those of Peru and their populations include several year-classes.

The anchovy competes for its plankton food with the sardine
which is, however, less fecund. Sardines were selectively sought
by fishermen from California, where in the 1920s an intensive
fishery was developed for them. In 1945 the sardine fishery
collapsed while the anchovies increased in numbers since then (43).

The selective fishing pressure on the sardine, which was
heavily sought by a fishery that almost neglected the anchovy,
was perhaps the prime reason for the decline of the sardine
population. Furthermore, the beginning of the major decline in
sardine stocks coincided with the start of an anomalous period in
the California Current, from 1943 to 1957, of unusually cold
conditions and of unusually strong and regular upwelling. These
conditions might be expected to favor the anchovy at the expense
of the sardine since the anchovy will spawn in waters above 11
degrees C. while the sardine requires the water to warm up above
13 degrees C. for successful reproduction, and it is unlikely
that abundant year-classes of the sardine would occur during this
period (43).

There is also the occurrence of scales of anchovies, sardines,
and other pelagic species in the varved mud sediments of several
of the anaerobic basins in the Santa Barbara Channel off South-

west California (44). They show similar shifts in biomass
dominance of cold water species associated with the anchovy and
warm water species associated with the sardine. Apparently,
these two main species in the fishery has alternated during the
many thousands of years during which the varved sediments were
being deposited.

The existence of another species pair, possibly differen-
tially affected by climatic changes has been documented (45).
The fate of the Japanese sardine appears to have been similar to
that of the California sardine. There is a coexisting anchovy
population, and while there is now a significant fishery for the
anchovy, it is not quite known whether there was an increase in
the anchovy landings following the decline of the sardine,
although the evidence suggests this. The present anchovy fishery
is only one-fourth as large as the peak sardine fishery,
suggesting that the anchovy has not yet completely filled the
void left by the sardine (45).

In Japan and in California, the falling sardine catches--
generated considerable scientific interest and research. In
both places, scientists tended to be divided into two camps.
Environmentalists maintained that the changes in the ocean
climate were responsible for the observed qualitative and
quantitative changes in the populations. The second camp
maintained that overfishing was the basic reason for all the
problems (45).

El Niño episodes leading to severe fluctuations in upwelling
fisheries and alternations between species pairs are occurrences
in a complex web of climatic perturbations. These latter also
find expression in timing and intensity of the monsoons which
influence both biologic and economic events in coastal fisheries;
they have even greater relevance for agriculture and freshwater
fisheries in South and Southeast Asia.

FLOODPLAIN FISHERIES AND THE MONSOONS

The Large River Valleys

The names of the rivers Indus, Ganges, Brahmaputra, Irrawaddy,
and Mekong also conjure up the vision of teeming millions of
people, many of them undernourished if not starving. Fish from
the rivers themselves but more prominently from their vast floodplains have traditionally been important in supplying the animal
protein component in the food of these people. In inland Bangladesh, for instance, riverfish make up more than 50 percent of this
supply, such as it is; and in Cambodia, fish from the inundated
areas of the Mekong River are equally important. It is perhaps
less well known that fish is the staff of life also in inland
tropical Africa, where smoked fish from the shallow lakes and

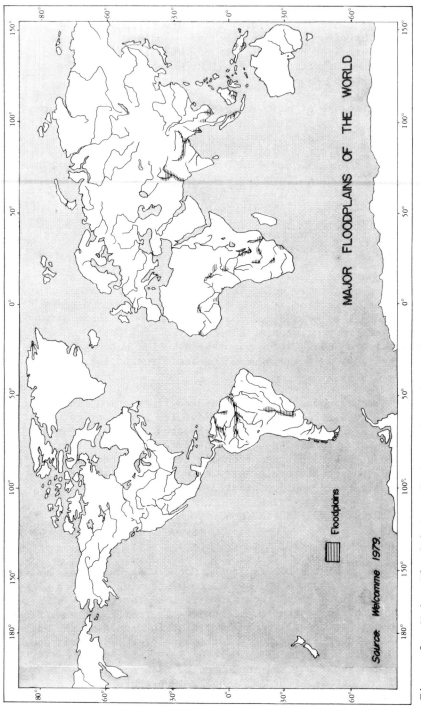

Figure 9.   Major Floodplains of the World.   (Redrawn from Welcomme, 1979) (46).

swamps of the floodplains of the Senegal, the upper Niger, and Quémé (in Dahomey), the Kafue and the Volta Rivers, among others, supply much of the animal protein.

Large South and Southeast Asian rivers, as well as the Niger, other African rivers, and the Amazon (Figure 9) are in climatic zones characterized by intense rains. In tropical Asia, this

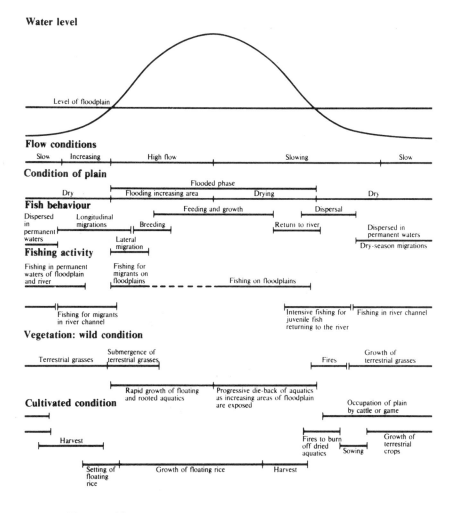

Figure 10.   Summary of the Major Activities on a Floodplain Throughout the Year.   (Source: Welcomme, 1979)   (46).

phenomenon, ultimately caused by the differential warming and
cooling of land and sea, gives rise to seasonal winds that blow
from the Indian Ocean onto the land for ca. six months, then
reverse their direction toward the sea. The winds from the sea,
the original monsoons (from the Arabic word "mausim" meaning
time or season) are moisture-laden and a good southwest monsoon
means sufficient rainfall from ca. June to September or October.
"Sufficient" here refers to the needs of agriculture and,
incidentally, also to enough water in the shallow overflow areas
of the rivers and often beyond to make the fishes grow (Figure 10).
Differences of water areas in the floodplain during the wet and
the dry seasons can be fifty to a hundred fold, and mean annual
maximum and minimum discharge rates vary accordingly (46).

Both late and early rains, as well as excessive amounts
thereof, can affect the levels of harvest of staple crops as well
as fish while a drought often means famine (47), with the lack
not only of carbohydrate staples, but also of such little animal
flesh as is usually available. Market prices often indicate
levels of supply of food commodities, as is illustrated in
Figure 11, depicting that a climatic anomaly--a deficiency in
rainfall-pushed fish out of the reach of many. A stronger but
similar point is made by a look at fish landings of the Senegal
and the Niger over many years--including those of the Sahel
drought (1969-73) (Table 3); only slightly over one-half the
tonnage previously available reached the markets at its peak.

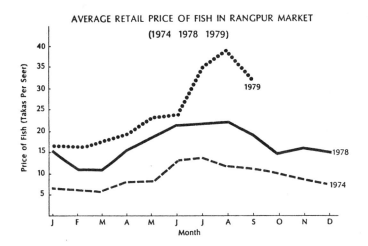

Figure 11. Rise in Fish Prices from Floodplain Fisheries
(Bangladesh) in Response to Drought Conditions. (Source:
Eusuf and Currey, 1979) (48).

Table 3. Trends in Catch in Senegal River and the Central Delta
of the Niger. (Source: Welcomme, 1979) (46).

River	Catch tonnes x 1000								
	1967	1968	1969	1970	1971	1972	1973	1974	1975
Senegal	30	25	20	18	18	15	12	21	25
Niger*	9.5	10.8	11.1	11.2	8.8	7.8	4.2	3.6	7.6

*Smoked fish at Mopti landing only.

Some species of fish spawn in the rivers proper, while
others wait to lay their eggs until the waters rise into sometimes forested lowlands with vast shallow lakes, ponds, and
oxbows that are warm, well-lit, and fertile; in particular,
these abound in plankton, supplying abundant food for fish
larvae. An important species, the Indo-Pacific anadromous
herring Hilsa ilisha, takes its signal for entering the rivers
from the sea by the onset of monsoonal rains. Hilsa occurs from
the Arabian Sea to the South China Sea, including the East Indian
Archipelago, where adult and maturing stocks are found in lower
estuaries and inshore continental shelf regions. The stimulus
for the upstream spawning migration is the swelling of the river
due to flood waters of the southwest monsoon and the spring thaw
in the northern mountains. Fishing occurs in the lower parts of
the rivers during the early part of the monsoon flood (43).

Juveniles of most riverine species spend at least a portion
of their lives in the inundated areas. When the waters begin to
recede, the main fishing activities begin (Figure 10). Fishermen place obstructions with varied types of collecting gear in
hydrologically strategic positions. A seasonal glut of fishes is
the rule in floodplain fisheries and thus many traditional
methods of fish preservation were developed. The fish are
preserved in various ways, smoked, salted, or rendered into a
cheese-like consistency by bacterial action. The latter products
are not only prominent in Southeast Asia where they are best
known, but also they occur in Africa and in South America. In
large rivers, there may be many dozens if not a hundred species
that enter a floodplain fishery; since most of the fish are
caught during their first year, they are small on the average (46).

Monsoon rains also influence the productivity of coastal
waters: shoals of the Indian oil sardine (Sardinella longiceps)
come inshore to spawn early during the monsoon. The final
ripening of the gonads of these fish seem to be tied to the temperature drop in surface waters that occurs at the onset of the
monsoon rains, usually in June, along the coasts of India (49).

There follow blooms of plankton when the larvae and juveniles appear, illustrating the match between the genetically governed timing of events in the life cycle of the fish with environmental signals. Simply put, the fish are programmed to fully ripen their gonads and spawn when the temperature drops due to the onset of the monsoon rains. The coincident winds stir the water and cause blooms just at the time when the eggs hatch and the larvae need ample food.

As the fishery relies on several growth stages; it can be influenced by anomalies in the timing and the strength of the monsoons. Raja (49) reported a significant correlation for the period of 1960 to 1969 between the abundance of 0 age group juveniles on which part of the fishery depends and the rainfall during the peak spawning period, which is June to August. He also noted that the vagaries of the southwest monsoon appear to be responsible for heavy pre- and post-ovulation atresia (i.e., the eggs that do not hatch or die soon after they are laid), affecting the spawning potential of the population and, in turn, the recruitment to the fishery.

The intensity of the monsoon has been reported to affect the fishery in yet another way: Murty and Edelman (50) demonstrated the existence of a critical value of monsoon intensity above which catches improve as the wind becomes stronger. This positive effect is thought to rely both on the increased oxygenation of surface waters as they are mixed by the wind and on the local upwelling similarly generated.

As to the volume of these river fisheries, Welcomme (46) estimated the maximum sustainable yields from relatively unmodified (see below) floodplains at 40 to 60 kilograms per hectare or 4 to 6 tons per square kilometer. Assuming it to be 5 tons and using global, albeit highly deficient and rough estimates for areas at maximum flood in the unregulated tropical floodplains of Asia, Africa, and South America of about 750,000 square kilometers, one arrives at a fishery potential from them of over 3 MMT, or more than one-quarter of the present catches from freshwater worldwide. Incidentially, this potentially available tonnage is also more than what is presently caught in the sea by any of the nations which rely so prominently on these floodplain fisheries.

A shortfall due to drought in this often crucial supply of animal protein where no water simply meant no fish was in the past only due to the vagaries of the climate. More recently, one may note two important opposing trends in floodplain fisheries, namely, intensification of relatively rarely practiced culture operations and water regime controls for intensified agriculture and related uses of the floodplain, even to include urbanization.

Floodplain control implies first some attempts at regularization of inflow and drainage, with increasingly large and widespread installations. Ultimately, this means barrages and dams with a change from a fluctuating aquatic environment with much water to a stable one with far less seasonal flood water. Only rarely has the planning of the overall development of a river basin and hence its floodplain taken into account the living aquatic resources. As a result, valuable fisheries such as the sardine fishery off the Nile Delta have disappeared. During the Nile flood (from August to December), the sardines (Sardinella aurita) moved to the region in front of the delta to feed, greatly increasing in body weight during that period. After the flood receded, the standing crop of zooplankton diminished and the fish moved into deeper waters. The Egyptian fishery which was based on this annual flood-caused boon in ocean fertility is now virtually dead, as there is a dam across the Nile and hardly any flood water from it reaches the sea (43).

The hard-learned lesson of the Nile now makes for careful consideration of plans for river development. Teams which plan integrated river development, preferably with many small rather than few large dams, are confident that it can be done in such a way that the fish yield of the basin is not only unimpaired, but also increased. It is especially important to note that droughts and anomalies in the monsoon regime will have far worse consequences on the fisheries of unregulated tropical rivers than they would on the often extensively practiced aquacultural fish production of large and small reservoirs. After all, in the latter, some modicum of water level control is possible and water can be retained even in very poor years, such control having been the purpose for which the dams and barrages were installed. But flood control structures are ephemeral; and even though large dams may endure for some centuries, inundated floodplains may well arise again and climate-induced droughts may once more decimate fish populations.

Where aquaculture can be practised on these floodplains the production per hectare of such ponds can be a multiple of that available to the unmanaged fishery; 200 kilograms per hectare are easily reached and even 500 or even 1,000 kilograms are reported (51). In the Chao Phraya river basin, for instance, the control of the water regime over much of floodplain has decreased natural fish production. To compensate for this loss, many state and private fish farmers rear various species of cyprinid and siluroid as well as freshwater prawns (Macrobrachium rosenbergii). The impervious nature of the alluvial soils, the flat terrain, and the ready access to water supplies through irrigation canals make this type of modified floodplain particularly suitable for aquaculture in ponds (46). Thus, the potential of fish protein production of managed floodplains may well be two

or three times the amount (3 MMT) previously mentioned; in fact,
with appropriate inputs, 6 to 10 MMT might be achieved on a
global scale.

WORLDWIDE AQUATIC PLANT AND ANIMAL HUSBANDRY

    The rearing under more or less controlled conditions of
freshwater or marine animals and plants holds substantial promise
of increasing contribution to world protein supplies, especially
in the tropics. The intensity of aquaculture is reflected in
the amounts of inputs to aquafarming operations and in the level
of control over the stocked species at various phases of their
life cycles. It can range from heated multilevel silos with
complicated filters and the feeding of high-quality rations to
walled or fenced enclosures in bays where fish or shellfish are
simply contained for a period (52). In between, lie the rearing
schemes for oysters and mussels, which mostly rely on natural
seeding, but supply means of attachment extending throughout the
water column, or the ranching of salmon where hatchery-reared
juveniles are released for later capture after they have foraged
for food in nature. The culture of marine algae, copiously
practiced in Asia (China, Japan, Korea), is also based on man's
control over the early life history stages of the cultivars;
that is algal spores or very young thalli are collected and
treated with care until they can be set out into nature in places
that favor their growth.

    Fresh and brackish water aquaculture is at present more
varied, more sophisticated, and more advanced than mariculture
proper; the former now supplies ca. 2.5 MMT of finfish worldwide[5] while the latter is estimated at 2.8 MMT of yield of which
seaweeds makeup 39 percent (2). Except for many, but by no means
all, variants of mollusc culture, aquaculture practices now
depend on full or partial control over reproduction and the
larval and juvenile stages of the cultivars; it goes without
saying that such control holds the promise of genetic improvement in growth, disease resistence, etc. Earlier maricultural
practices were based on the hatching of eggs simply to larval
stages and the pouring into the sea of myriads of larvae that
had barely absorbed their yolk-sacs. It was recognized in the
last few decades that such practices were ineffective and now
protection from predation and feeding extend well into the
juvenile stages of the few marine species that are set out
(e.g., sea bream Sparus major in Japan, flatfishes, mainly sole,
in Great Britain).

    Man has no influence on climate and what climatic variations
do to the stocks of fish and shellfish in nature. If he cultures
them it is only natural that he would try to save both the stock

and his investment in the husbanding of it. This is to say that larval rearing tanks may be heated and that plankton production for larval food is practiced in such a manner that it furnishes continuous and ample food to the hatchlings. Also, under conditions of drought, an aquafarmer would tend to consolidate his adult stock from several of fewer ponds, as necessary.

Thus, the influence of climate on the stock of aquaculture is generally less than on aquatic organisms in nature. Although in brackish water and marine culture operations, where there is generally less control than in pond or raceway culture of freshwater animals, vagaries of nature may still often bedevil the mariculturist. Aquatic animals are cold-blooded (except for tuna which are not reared, at present) and the Q_{10} of their growth is generally around two,[6] indicating, for instance that a rainbow trout (Salmo gairdneri) at 17 degrees C. would grow twice as fast as at 7 degrees C. The first mentioned temperature, incidentally, is the optimum upper limit for the species and also that of the thermal springs of the Snake River Valley in Idaho, U.S.A., an important center of the trout rearing industry where the trout farmers need not worry about the climate. At this point, another caveat is in order. While aquaculturists have certain control over their production facilities, climatic variations will influence their yields. This is especially true in temperate zone landlocked nations which rely substantially on aquaculture for their fish supply (e.g., Austria, Hungary). Common carp (Cyprinus carpio) is important there as elsewhere in Eurasia and being a warm water fish it is especially prone to experience a slowing in growth during years that are colder than normal. The onset of its spawning readiness may also be affected.

It then follows that temperate zone aquaculture, just like fisheries can be sensitive to climate-induced temperature variations but that it is not likely to suffer from a mismatch in peak larval growth periods with available food and that it is not likely to suffer from drought. That climate-related contingency is of more concern to the tropical aquafarmer, especially in freshwater, as it was in floodplain fisheries of tropical rivers.

SUMMARY AND CONCLUDING REMARKS

When estimates of potential marine harvest were made without attention to economics, as was often the case, they ranged from more than three to four to ten times the present take. More recent extrapolations which take into consideration economic demands and per annum production increases of the last decade arrive at around 100 MMT of likely annual world fish production from all sources in the year 2000 (53). It is stressed in all those estimates that the actual catches now and in the future are

likely to rely on far fewer species than could in fact be
harvested in theory.

Cases in point here are the several millions of metric tons
of so-called by-catches and the harvesting of krill in Antarctic
waters. The former consist of species of miscellaneous fishes
caught in trawls with shrimp. They are of low value, in spite of
the fact that the people living near the shore from which the
shrimp are taken could well use additional animal protein in any
form. Economic incentive for their utilization simply does not
appear to exist even though all or nearly all the technology components are extant for their utilization as fish meal, fish
silage, or fish paste (40). Krill is now harvested on a small
scale by technologically advanced nations, naturally with very
large inputs in fuel. Whether or not krill harvest will expand
significantly will depend on advances in technology that would
make capture and on-board processing cheaper than it is now,
probably on the conclusion of an Antarctic treaty and, last but
not least, on the future cost of fuel.

In the recent past, Quantum jumps have occurred in the
development of fishing. First came steel vessels and steam
engines, then diesel motors, synthetic yarns for nets, and electronic fish finding devices. These contributed to the tendency
to overcapitalize the fisheries of developed nations and with it
to overfishing. In the developing countries in spite of the lack
of capital and of technologies and technical skills, sheer numbers
of fishermen led to the overexploitation of the very nearshore
areas. Under those circumstances, climate-induced variations of
fish stocks, apparently quite natural events, but severely aggravated by fishing can cause grave economic havoc.

If El Niño-like phenomena were to occur simultaneously off
South America and off Africa in the same year and if there was
also a weak and late southwest monsoon over South Asia (true, not
a likely occurrence), the global shortfall of fish harvest might
well reach 4 or 5 MMT; in fact a deficiency of 7 to 9 MMT is
possible. Depending on the global catch at the time, this could
be significantly more than ten percent of the global fish harvest.
Predictions of these events are not yet possible with much
certainty, although we have mentioned that an El Niño may be
presaged a few months before it can occur.

Other relief from such a catastrophy might be thought by
some to reside in greater reliance on aquaculture. While it is
true that aquaculture is less vulnerable to suffer from short-
term climatic variations than are fisheries, it should be noted
that aquaculture cannot be cranked up overnight, whatever the
level of inputs. Hence it has only a limited capacity to make up
for sudden nutritional shortfalls aside from the fact that in the

foreseeable future the products of aquaculture will be more
expensive than most fishery commodities.

Fish stocks suffer most from the vagaries of climate through
the effects that changes in ambient conditions have on the larvae.
The biological dynamics at the larval stages in the life cycles
of fishes and aquatic invertebrates have a short-term horizon and
are highly variable. Instant mortality coefficients[7] during the
larval stages range between 300 and 400 while they are 0.2 to 1.5
in the juvenile and adult stages (53). The most catastrophic and
not all too rare event that can befall larvae is a climate-
induced mismatch in time between their food needs and the prevalence of their food (7). Larval stages last in the order of
days while later stages live at least a year but mostly years.
Thus the range of variability of events in the larval stage can
be several hundred fold while it is probably no more than four
to five fold later on. We should state as a corollary that small
as the biomass of fishes is during very early life, their future
numbers are largely determined then and not later in life when
the biomass is large and when the fishery takes its toll.

Spawning is an important mark in the adult life history of
aquatic animals; in some cases the event terminates their life
(e.g., Pacific salmon). Spawning is often triggered by environmental signals such as temperature levels, rainfall, day length,
and stages in the lunar cycle. Some of these can be shifted out
of their norms by climatic influences and lead to biological
catastrophies for certain species. For instance, the anadromous
river-spawning herring of South Asia receive their spawning
"Zeitgeber," which evolution has programmed them to heed, from
the wind and the rains of the monsoon. Since climate-induced
anomalies can indirectly influence the numbers of larvae by
throwing spawning events out of kilter, they are probably next in
importance to those that affect larval food supply directly.

Post-larval stages of aquatic animals are strongly influenced
by temperature; long-range changes in climate such as the global
warming trend of the 1930s and 1940s have extended the range of
temperate and tropical fishes to the north or to the south, as
the case may be. This is superimposed on the possibility that a
more benign climate affords to fishermen the opportunity of
more frequent outings and of work under less strenuous conditions;
thus it can increase catches. The receding ice cover in the
vicinity of Iceland, mentioned earlier, is a case in point.

In fact, Bell and Pruter (54) strongly suggest a re-examina-
tion of the reported temperature-fish productivity relationships
as discussed earlier in this paper, since in most cases adequate
consideration has not been paid to changes in fishing effort, in
economic conditions, and in the efficiency of fishing technology.

As they pointed out:

> The warming of waters may have played a far less
> important role than is generally accepted in
> affecting the pattern of stock changes and annual
> yields. Also, while it is true that warming of
> the seas could increase recruitment or migra-
> tion of adults - it is also true that more
> clement weather would facilitate fishing.

Man's influence on or in the environment may well go hand-in-
hand with climatic factors: for instance, the addition of industrial effluents and increased sewage disposal to the North Sea
noticeably enhanced nitrite and phosphate levels in the southern
part of the sea over the last 15 years (20). This coincided with
improved year-class strengths of cod, plaice, and haddock from
this area, generally attributed to cooling conditions in their
spawning grounds (18). Eutrophication increased fertility of the
waters and productivity at primary and secondary levels, while
climatic factors influenced survival and growth of the animals.
A similar situation where human and climatic influences interacted
can be seen in Lake Erie, where the fish community has changed
dramatically over the past century. Eutrophication resulting
from nutrient loading on the lake, intense exploitation, and the
general cooling of the early twentieth century followed by warming
in the second half of the century has caused cold water species
such as lake herring Coregonus artedii, lake whitefish
C. clupeaformis, and lake trout Salvelinus namaycush to virtually
disappear (55).

Thus far in history, man's influence on climate has not been
noticeable; the next few decades may well present a different
picture in this regard. Several man-caused interferences are
presaged among which there looms largest an increase in
atmospheric carbon dioxide due to deforestation and the burning
of fossil fuels. Present, albeit speculative and highly
tentative, estimates of man-induced warming due to more carbon
dioxide in the atmosphere in the next century range from an
increase in global annual average surface air temperature of 2 to
3 degrees C., with a maximum warming even of up to 7 degrees C.
or more in polar regions in winter. The rise in temperature there
will increase annual precipitation by 5 to 7 percent, but
maximum warming as well as about 20 percent increase in precipitation is predicted to occur in the mid-latitudes (9).

A number of scenarios have been painted about the effects of
increases in atmospheric carbon dioxide on various phases of
human endeavor. If the maximum level of warming were indeed to
prevail, ice covers would recede everywhere and uncover new
fishing grounds. Leaving aside whether these might be highly or

only moderately productive, the fact remains that catches from
them would likely be good for a while, as they are in any newly
opened fishery. That the earlier mentioned extension of species
would also take place goes without saying.

If there is some reduction, due to the differential warming,
in zonal temperature differences, trade winds might weaken somewhat. In consequence one could postulate that the intensity and
frequency of El Niño episodes might be reduced since they do
seem to occur in response to strong bouts of the trades which
displace surface water toward the west (see also section on
"Upwelling Fisheries"). The upwelling fisheries might possibly
become more stabilized and perhaps be easier to manage in the long
range. Effects of the warming on the monsoons are even more
speculative; less fluctuation in them could mean fewer severe
floods thus enabling better river management, including the
establishment of large extensive aquaculture operations in the
floodplains of tropical rivers. At the same time other consequences of carbon dioxide increases may affect fisheries
adversely: the carbonate balance may shift to decrease the pH of
oceanic waters leading to perhaps deleterious effects in the algal
community. Likewise the higher carbon dioxide-induced "greenhouse" effect may lead to increased photosynthesis by virtue of
decreased ultraviolet radiation. The former may give rise to as
yet unpredictable changes in pelagic and benthic (i.e. bottom)
community structures (56).

Carbon dioxide enhancement may affect floodplains and
estuaries more than the open sea because their waters are relatively poorly buffered. Stresses of fluctuating pH values on
the fresh- and brackish-water biota can be further aggravated
by variation in continental run-off with its load of high
nutrients and, occasionally toxic substances. But it should be
remembered that estuarine species are adapted to a highly
variable environment, a mitigating factor in the impact of
carbon dioxide on these communities (56).

Added to these negative effects of global warming due to
more carbon dioxide is the possible flooding of ports and cities
by the sea; one can now make no judgement, and there might be more
positive than adverse effects as far as fisheries per se are concerned. Comparable considerations of the effects of rising levels
of carbon dioxide on agriculture raise the possibility that these
higher concentrations may have beneficial effects on crop yields
(57). On the same subject, Bach (58) indicates that there may
be regionally beneficial effects of CO_2 increases. He also
stresses, however, that other regions may be adversely affected
and that there is need for more research before definite statements can be made.

The above discussion bears out that a great deal more knowledge and careful investigations of aquatic populations, and
climate-related events must be available prior to establishing
truly valid relationships between climatological changes and
aquatic food production.

When considering aquatic populations, care must be taken to
account for changes in fishing technology, fishing effort, and
the extent of exploitation, in other words--the overfishing
problem. Since deficient data may perhaps be used to judge
climatic effects on the fisheries. There exists a need,
especially in developing countries to obtain and record proper
fishery statistics.

The study of the impact of climate on food resources princi-
pally involves climatic observations and monitoring, and field
studies need to be carried out describing the "natural" state of
the climate and its variations. There are significant elements
of climate about which little is known, such as ocean atmospheric
interactions in remote areas which have local consequences elsewhere. Some development of technology is also required, for
example, more cost-effective remote sensing techniques, particularly to interpret low-level winds and energy flux over the open
oceans (59).

The aquatic environment has long been considered the "logical"
dumping grounds for both industrial and domestic wastes. This has
led to drastic changes in nutrient levels of enclosed freshwater bodies, rivers, estuaries, and marine environments, the
effects of which often override climatic influences. Such addi-
tions have to be carefully monitored and kept under control, for
often their ultimate results, namely, eutrophication can be quite
disastrous to food production. Similarly, warming of the climate
may hasten eutrophication.

Climate modelling and prediction research is also important
since they can sharpen the predictions of year-class strengths
in a fishery. These, in turn, are the bases for catch quota
regulations and fishing seasons. The fate of entire economies
can depend on such prediction capability (e.g., the Peruvian
anchovy and the El Niño phonomenon). The problem of anticipating
regional or global climate can be approached either from deterministic modelling or by empirical or statistical studies based
on observed past climatic behavior (59).

Whatever path one chooses, one must remember that marine
ecosystems are highly dynamic, hence, to understand them so as to
manage their resources it is necessary to design a model complete
with all its interactions. Such a dynamical numerical marine
ecosystem model (DYNUMES) as developed by the National Oceanic

and Atmospheric Administration permits simulation of the mechanics of distribution and abundance of various species in space and time as affected by interspecies interactions, for example, predation, competition, climatic factors, and the activities of man. The main objective of such a model are: 1) evaluation of the effects of exploitation; 2) evaluation of climatic effects and quantitative comparison and man-made and climate-induced changes in the system; and 3) reduction of the data into accessible and reviewable patterns while determining future research priorities (60).

If these are obtained and if prediction capabilities are sharpened of climate- and man-induced changes, singly or in conjunction, technicians, administrator, and policymakers may have the tools for better fisheries management. True, economic considerations and nutritional needs are also important, but without improved statistics from most oceans coupled with enhanced weather and climate prediction capabilities, it is indeed doubtful that the projected six billion occupants of the world in the year 2000 will have at their disposal the hoped-for harvest in excess of 100 MMT of aquatic foods.

FOOTNOTES

1. A shorter version of this material as it pertains to marine fisheries only will be published in a special issue of Bio-Science on Living Marine Resources.

2. The reader is referred to the fall issue of 1978 of OCEANUS titled 'Oceans and the Climate' (v. 21., No. 4) for more detailed elucidation of atmosphere-ocean interactive processes.

3. Purse seines are flat nets used to encircle fish, fitted with floats on top and with a purse-line in the bottom that could be closed around a school of fish.

4. Consists of a longline stretched between floats to which are attached short "dropper" lines at intervals with baited hooks.

5. Information received by the senior author on a very recent trip to China indicates that this figure might be conservative.

6. Typically for every $10^{\circ}C$ increase in temperature the rate of a chemical reaction (metabolic reaction) will increase by a factor of 2.

7. Measures the fraction of a particular age group of fish dying due to fishing as well as natural causes.

REFERENCES

1. FAO: 1979, The State of Food and Agriculture - 1978. FAO
    Agriculture Series, No. 9., Rome., 162 pp.

2. Ryther, J.H.: 1980, More Food from the Sea? II. Mariculture,
    Ocean Ranching, and other Innovative Approaches, BioScience
    (in press).

3. Krone, W.: 1979, Fish as Food - present contribution and
    potential. Food policy. v.4., No. 4, pp. 259-268.

4. Bardach, John E. and Ernst R. Pariser.: (1978), Aquatic
    Proteins, pp. 427-484. Protein Resources and Technology:
    Status and Research Needs. (Edited by M. Milner, N.S.
    Serimshaw and D.I.C. Wang, MIT), Avi Publishing Co., Inc.
    Connecticut.

5. Lagler, Karl F., John E. Bardach, Robert R. Miller, and Dora
    R. May Passino.: 1977, Ichthyology (2nd ed.). John Wiley &
    Sons, Inc., New York. London. Sydney., 506 pp.

6. May, Robert C., Gerald S. Akiyama, and Michael T. Santerre.:
    1979, Lunar Spawning of the Threadfin, Polydactylus sexfilis
    in Hawaii. Fishery Bulletin 76., No. 4., pp. 900-904.

7. Cushing, D.H.: The Production Cycle and the Numbers of Fish.
    Symp. Zool. Soc. Lond., No. 29, pp. 213-232.

8. Cushing, D.H., and R.R. Dickson.: 1976, The Biological
    Response in the Sea to Climatic Changes. Adv. Mar. Biol.
    14., pp. 1-122.

9. Mason, John Sir.: 1980, The Climate of the Future. PHP. -
    a Forum for a Better World. PHP Institute International Inc.,
    Japan. v.11., No. 5, 116., pp. 6-20.

10. Budyko, J.I.: 1968, On the Causes of Climatic Variation.
    Meddelanden fran sveriges meterologiska och hydrologiska
    institut. Serie B28, pp. 6-13.

11. Chang, Jen-hu,: Atmospheric Circulation Systems and Climates.
    The Oriental Publishing Company, Honolulu. pp. 328.

12. Dickson, R.R., H.H. Lamb, S.A. Malmberg, and J.M. Colebrook.:
    1975, Climatic Reversal in the Northern North Atlantic.
    Nature, 256., pp. 479-482.

13. Hansen, Paul M. and Frede Hermann.: 1965, Effect of Long-Term Temperature Trends on Occurrence of Cod at West Greenland. ICNAF Environmental Symposium. Special Publ. No. 6., pp. 817-820.

14. Dickson, R.R., J.G. Pope, and M.J. Holden.: 1974, Environmental Influences on the Survival of the North Sea Cod. In 'The Early Life History of Fish'. (ed. J.H.S. Blaxter). Springer-Verlag, New York - Heidelburg, Berlin, pp. 69-80.

15. Johnson, James H.: 1976, Food Production from the Oceans. In. Proceedings of the Workshop on Climate and Fisheries. U.S. Dept. of Commerce, NOAA/NMFS/EDS. Washington, D.C., pp. 15-36.

16. Holden, M.J.: 1978, Long-Term Changes in Landings of Fish from the North Sea. Rapp. P.-v. Réun. Cons. int. Explor. Mer. 172., pp. 11-26.

17. Madsen, Popp K.: 1978, The Industrial Fisheries in the North Sea. Rapp. P.-v. Réun. Cons. int. Explor. Mer. 172., pp. 27-30.

18. Hill, H.W. and R.R. Dickson.: 1978, Long-Term Changes in North Sea Hydrography. Rapp. P.-v. Réun. Cons. int. Explor. Mer. 172., pp. 310-334.

19. Postuma, K.H.: 1978, Immigration of Southern Fish into the North Sea. Rapp. P.-v. Réun. Cons. int. Explor. Mer. 172., pp. 225-229.

20. Lee, A.: 1978, Effects of Man on the Fish Resources of the North Sea. Rapp. P.-v. Réun. Cons. int. Explor. Mer. 172., pp. 231-235.

21. McLain, Douglas R.: 1976, Anomalously Cold Temperatures in the Southeastern Bering Sea and Along the West Coast of North America. 1971-1975. In. Proceedings of the Workshop on Climate and Fisheries. U.S. Dept. of Commerce, NOAA/NMFS/EDS, Washington, D.C., pp. 149-163.

22. Höglund, H.: 1978, Long Term Variation in the Swedish Herring Fishery off Bohuslan and their Relation to North Sea Herring. Rapp. P.-v. Réun. Cons. int. Explor. Mer. 172., pp. 175-186.

23. Bardach, John E. and Y. Matsuda.: 1980, Fish, Fishing, and Sea Boundaries: Tuna Stocks and Fishing Policies in Southeast Asia and the South Pacific. GeoJournal v.4., No. 5, pp. 467-478.

24. Stevens, E.D., and W.H. Neill.: 1978, Body Temperature Rela-
     tions of Tuna, especially Skipjack. In. Fish Physiology
     (Edited by W.S. Hoar and D.J. Randall), Vol. 7., pp. 316-360.

25. Sund, Paul N., Maurice Blackburn, and Francis Williams.: 1980.
     Tunas and Their Environment in the Pacific Ocean: A Review.
     Oceangr. Mar. Biol. Ann. Rev. (Oct., H. Barnes). Hafner
     Press: N.Y. (in press).

26. Dizon, Andrew E., E. Don Stevens, William H. Neill, and John
     J. Magnuson.: 1974, Sensitivity of Restrained Skipjack Tuna
     (Katsuwonus pelamis) to Abrupt Increases in Temperature.
     Comp. Biochem. Physiol., 49A., pp. 291-299.

27. Yamanaka, I. and H. Yamanaka: 1970, On the Variation of the
     Current Pattern in the Equatorial Western Pacific Ocean and
     its Relationship to the Yellowfin Tuna Stock. Proc. 2nd CSK
     Symposium, Tokyo.

28. Yamanaka, H.: 1970, Relation between the Fluctuation of
     Bluefin Tuna Resource and Oceanographical Conditions.
     Bull. Jap. Soc. Fish. Ocean., 14: pp. 202-207.

29. Uda, M.: 1952, On the Relation between the Variation of the
     Important Fisheries Conditions and the Oceanographical
     Condition in the Adjacent Waters of Japan. J. Tokyo
     Univ. Fish., 38: pp. 363-389.

30. Inoue, M. and Y. Iwasaki.: 1969, Movement of the Thermal
     Equator and the Fishing Grounds Mainly for Yellowfin Tuna in
     the Indian Ocean. Bull. Jap. Soc. Scient. Fish., 35:
     pp. 957-963.

31. Yamanaka, I.: 1978, Oceanography and Tuna Research. Rapp.
     P.-v. Réun. Cons. int. Explor. Mer. 173., pp. 203-211.

32. Clark, N.E., T.J. Blasing and H.C. Fritts.: 1975, Influence
     of Interannual Fluctuations on Biological Systems. Nature,
     256., pp. 302-305.

33. Austin, H.M.: 1980, Tree Rings and Mackerel, Coastal Ocean-
     ography and Climatology News. 2. No. 2., pp. 22-23.

34. Cushing, D.H.: 1971, A Comparison of Production in Temperate
     Seas and the Upwelling Areas. Trans. Royal Soc. South Africa.
     40. M.I., pp. 17-33.

35. Valdivia, J.: 1980, Biological Aspects of the 1972-73 'El
     Niño' Phenomenon. 2: The Anchovy Population. Proceedings
     of the Workshop on the Phenomenon known as 'El Niño'. IDOE
     UNESCO, Guayaquil, Ecuador., pp. 73-81.

36. Bjerknes, J.: 1966, Survey of El Niño, 1957-58 in its rela-
     tion to Tropical Pacific Meteorology. Inter-American Tropical
     Tuna Commission. 12 No. 2., 62 pp.

37. Wyrtki, K.: 1979, 'El Niño'. La Recherche. v. 10., No. 106.,
     pp. 1212-1220. Dec. 1979.

38. Hisard, Ph.: 1980, Observation de résponses de type 'El
     Niño' dans l'Atlantique tropical oriental Golfe de Guinée.
     Oceanol. Acta. 3., No. 1., pp. 69-78.

39. Merle, J.: 1980, Variabilité Thermique annuelle et inter-
     annuelle de l'ocean Atlantique équatorial Est. L'hypothese
     d'un 'El Niño' Atlantique. Oceanol. Acta. 3., No. 2.,
     pp. 209-220.

40. Bardach, John E.: 1977, Aquatic Food Sources. In. World
     Food and Nutrition Study. U.S. National Academy of Sciences,
     Washington, D.C., pp. 251-318.

41. Karplus-Hartline, Beverly.: 1980, Coastal Upwellings:
     Physical Factors Feed Fish. Science 208, No. 4439.,
     pp. 38-40. April 1980.

42. Smith, S.L., and L.A. Codispoti.: 1980, Southwest Monsoon
     of 1979: Chemical and Biological Response of Somali Coastal
     Waters. Science. 209., No. 4456., pp. 597-600. August 1980.

43. Longhurst, Alan R.: 1971, The Clupeiod Resources of Tropical
     Seas. Oceanogr. Mar. Biol. Ann. Rev. (Ed. H. Barnes). 9.,
     pp. 349-385.

44. Soutar, A., and J.D. Isaacs.: 1974, Abundance of Pelagic Fish
     during the 19th and 20th Centuries as recorded in anaerobic
     sediments off California. Fishery Bull. Fish Wildl. Serv.
     U.S. 72., pp. 257-273.

45. Murphy, Garth I.: 1977, Clupeoids. In 'Fish Population Dynam-
     ics'. Edited by J.A. Gulland. Wiley InterScience, pp.
     283-308.

46. Welcomme, Robin L.: 1979, Fisheries Ecology of Floodplain
     Rivers. Longman, London., 317 pp.

47. Currey, Bruce.: 1979, Mapping Areas Liable to Famine in
     Bangladesh. Ph.D. Dissertation. Dept. of Geography,
     University of Hawaii, Honolulu, Hawaii.

48. Eusuf, A.N.M. and B. Currey.: 1979, The Feasibility of a
     Famine Warning System for Bangladesh. Mimeographed Reports.
     Ministry of Relief and Rehabilitation, Dacca, Bangladesh.

49. Raja, B.T. Antony.: 1972, Possible Explanation for the Fluc-
     tuation in Abundance of the Indian Oil Sardine Sardinella
     longiceps Valenciennes. Proceedings of the IPFC Symposium
     on Coastal and High Seas Pelagic Resources., pp. 241-252.

50. Murty, A.V.S., and M.S. Edelman.: 1971, On the Variation
     Between the Intensity of the South-West Monsoon and the Oil
     Sardine Fishery of India. Indian J. Fish. 13, No. 1 & 2.,
     pp. 142-149.

51. Bardach, John E., John H. Ryther, and William D. McLarney.:
     1972, Aquaculture: The Farming and Husbandry of Freshwater
     and Marine Organisms. Wiley-InterScience. 868 pp.

52. Bardach, John E.: 1980, Economic Energy Use in Fish Produc-
     tion. Proceedings of XIV Pacific Science Congress.
     Khabarovsk, Siberia, Academy of Sciences, U.S.S.R.

53. Rothschild, B.J.: 1980, More Food from the Sea. BioScience
     (in press).

54. Bell, Heward F. and Alonzo T. Pruter.: 1958, Climatic
     Temperature Changes and Commercial Yields in Some Marine
     Fisheries. J. Fish. Res. Bd. Canada. 15. No. 4., pp. 625-683.

55. Hartman, Wilbur L.: 1973, Effects of Exploitation, Environ-
     mental Changes and New Species on the Fish Habitats and
     Resources of Lake Erie. Great Lakes Fishery Commission.
     Tech. Rep. No. 22., 43 pp.

56. Carbon Dioxide Effects. Research and Assessment Program in
     the Workshop on Environmental and Societal Consequences of a
     Possible CO_2-induced Climate Change. April 1979. U.S. Dept.
     of Energy. Washington, D.C.

57. Wittwer, Sylvan H.: 1980, Carbon Dioxide and Climatic Change:
     An Agricultural Perspective. Journal of Soil and Water
     Conservation. v. 35, No. 3., pp. 116-120.

58. Bach, W.: 1979, Impact of Increasing Atmospheric CO_2 Concen-
     trations on Global Climate: Potential Consequences and
     Corrective Measures. Environment International v. 2.,
     pp. 215-228.

59. Sprigg, William A.: 1976, Development of a Climate Program,
     an Aid to Resource Management. In. Proceedings of the
     Workshop on Climate and Fisheries. U.S. Dept. of Commerce,
     NOAA/NMFS/EDS, Washington, D.C., pp. 110-117.

60. Laevastu Taivo, and Felix Favorite.: 1976, Summary Review
     of Dynamical Numerical Marine Ecosystem Model (DYNUMES).
     In. Proceedings of the Workshop on Climate and Fisheries.
     U.S. Dept. of Commerce, NOAA/NMFS/EDS, Washington, D.C.,
     pp. 118-148.

ACKNOWLEDGEMENTS

     We are grateful to Profs. Klaus Wyrtki and Chang, Jen-hu
of the University of Hawaii for critical reading of the manuscript
and to Ms. Irene Crossman for coping with the difficulties of
typing.

DISCUSSION

It was commented that one way to divide climatic aspects of food production is to consider climate as a resource and climate as a hazard. Given the long-term climatic statistics of a region, how can we optimize production? Conversely, how can we minimize vulnerability? Technical issues raise value questions. Making up food deficits with food aid creates dependency and may raise political issues. Using technology to maximize climate as a resource may also create dependency and political obligations.

Dr. McQuigg was asked if modern technology has lessened the impact of adverse weather on crops. He said that he has not been able to detect any indication that the impact of drought is less because of modern crop varieties and modern machinery. He does not believe that we have engineered our crops away from the effects of drought in any way. On the contrary the sensitivity of our present agricultural system is greater because of high technology. In 1936, U.S. corn yields were very low, but the system was different. The farmer raised his own seed corn and used manure for fertilizer, investing $ 25 - 30/acre at most. Now a farmer may invest $250/acre of corn. He has to pay high interest on a loan to carry his production costs, and he may or may not be able to get a loan. Thus today's farmer has less ability to absorb a large drop in yield than the farmer of the 1950s had. Many farmers can not make it even when farm prices are high. If we go into the less developed world and put in a higher level of technology and tell the farmers that it will make them less subject to the variability of weather, it will be a cruel hoax. Unless they do a better job of managing, have a higher level of support from industry and government, improve the educational process and raise their perception of agriculture as an important part of the economy, third-world countries may be hurt by the introduction of higher technology. Higher technology does not remove us in any way from the effects of weather.

A participant pointed out that absolute year-to-year variability in crop yields has gone up, although percentage variability has gone down. The variation per year in bushels per acre has increased even though the percentage of the total yield may be less. In the Great Plains, the reduction in the coefficient of variation (the absolute variation divided by the total yield) has dropped during a time when the weather variability was not quite as serious as it had been in the 1930s or 1950s. But it is clear that we need ways to smooth out the shocks that are caused by the effects of weather on crop yields.

W. Bach, J. Pankrath, and S. H. Schneider (eds.), Food-Climate Interactions, 235–238.
Copyright © 1981 by D. Reidel Publishing Company.

Dr. Swindale was asked if the farmers had learned by experience. He said that their principal strategy had been to invest as little as possible on upland areas. He said that the farmers will switch to hybrids if they see an advantage. They will take a risk. A participant pointed out that there is a great deal of indigenous information about drought resistance and similar subjects in Africa, China and other parts of the world. Millet produces the most food for the least water, and people in northern China have produced strains of millet that performed remarkably well. Africans have used a whole variety of crops. The so-called "hunger grass" of West Africa was one of a few crops that even survived very dry weather. It turned out to have a very high protein content. Dr. Swindale said that if there were a true climate change, it might not take as long to come up with appropriate new genotypes as some people have feared. He cited the new genotypes introduced after the U.S. corn blight as an example of how quickly the seed industry can change to meet new requirements.

Dr. Cooper was asked if there are tree crops that might yield better than coconuts. He said that Brazil nuts are suited for agroforestry systems because the bee that pollinates them does not find suitable places to nest in Brazil nut plantations. He suggested a labor-intensive system that could yield three crops simultaneously. The first would be food crops, probably derived from low-growing annuals or short-term grains. A second is biomass fuel and foliage for forage, which should also stimulate meat production. The third product could be an export crop such as Brazil nuts or coconuts. A participant said that FAO has done research that indicates that trees, crops and grazing animals can use the same areas. In response, Dr. Cooper said that the system he had described is based partly on natural ecological processes and partly on traditional agricultural practices. Nevertheless, it involves a substantial technological intensification that will require a higher level of capability from the individual farmer. Such a system will require better agricultural extension work. It was also pointed out that ecologically based systems may not necessarily be the best. They may involve complicated management problems, and pest control may be much more complicated.

Responding to Dr. Bardach's paper, a participant pointed out that the creation of exclusive economic zones has put a large part of the world's fisheries under the jurisdiction of developing countries. This should benefit the world's fisheries in the long run because the main fishing nations will have to enter joint ventures with the coastal countries and stop using the old "hit-and-run" techniques. FAO has been giving legal advice and technical assistance to help countries enter joint ventures. This should help protect resources, maintain a sustained long-term yield in fisheries and increase the fisheries production of developing countries. FAO envisages an increase in fisheries production of deveoping countries from 37.3 million t in 1980 to 51.9 million t in

2000, compared to an increase from 38.0 to 40.6 million t for developed countries. Dr. Bardach agreed that the 200-mile economic zones should lead to an increase in production. He said that there should be a gradual transfer of production capabilities from the joint ventures to the fish-owning nations, but only time will tell whether this actually occurs.

A participant cited a Korean fishing operation in American Samoa. Before the exclusive economic zones the only benefit to Samoa was employment of some Samoans as manual laborers in the fish- processing plant. He feels that this pattern will change, because the immediate economic interest of governments in their own resources will help to improve their infrastructure, marketing and domestic consumption habits for the benefit of their people. Even before the El Nino failure people in the Peruvian highlands did not eat fish. Fish was introduced to them as food aid after the earthquake disaster. He said that this sort of situation, which he described ridiculous, will probably cease.

Dr. Bardach was asked about the possibility of increasing the world's food supply with technology such as artificial upwelling that could bring nutrients up from the deep ocean. Dr. Bardach replied that Ocean Thermal Energy Conversion (OTEC) may prove to be technically and economically feasible in a few cases. But there will be tremendous problems in trying to develop ways to utilize the enriched deep water that such systems would spew out in volumes equivalent to the flow of an Ohio or Rhine River. A participant pointed out that the west coasts of Africa and South America are good fishing areas, while the west coast of Australia is not. Africa and South America have rivers that carry large amounts of phosphate material into the ocean, but Australia does not. Is it possible, he asked, that such areas could be artificially fertilized? Dr. Bardach replied that there may be some possibilities for small areas. But it is difficult to manipulate sea spaces because the sea is an open system with tides and changing currents. Theoretically, you might accomplish something by adding nutrients, but the engineering problems would be quite difficult. A participant pointed out that the world's fisheries could be seriously affected by changes in natural ocean upwelling resulting from increased atmospheric CO_2. A change in the temperature contrast between the poles and the equator could weaken the trade winds and cause a virtual cessation of upwelling.

Dr. Bardach was asked about the relative costs of different kinds of fish. He replied that the least expensive fish are those caught with wood wheels or stationary nets. The next most expensive are those caught with seines close to shore. When you have to go after fish, as you do with tuna, you increase energy costs by a factor of ten. In terms of energy costs, tuna is 20 to 40% more expensive than Wienerwald chicken or American catfish. With shrimp you get another escalation in energy cost by a factor of two. Krill fishing will be possible only for nations with large vessels and substantial investments in preservation techniques. The economy of scale of the krill fishery may save energy, but overall the krill will be at least as expensive as a lobster is today.

A UNEP task force on ozone recently concluded that phytoplankton will be appreciably affected by an expected increase in ultraviolet radiation. Would that affect fisheries significantly? The impact of increased UV radiation would be spread over all the oceans, so it might not have a direct impact on coastal fisheries. However, some coastal fisheries depend on phytoplankton from more remote areas of the ocean. It might be possible to replace the more vulnerable phytoplankton species with less vulnerable ones, but that would also cause a shift in zooplankton and the fish species that depend on the food chain. Replacement would take a long time and might have some very serious implications.

WATER RESOURCES AND FOOD SUPPLY

Gunnar Lindh
Department of Water Resources Engineering,
The University of Lund, Lund, Sweden

ABSTRACT. This paper starts by pointing out the complex role played by water in the human society and the series of water related conflicts that successively will develop. The quantity of water available on a global scale is constant but the quality is gradually decreasing, because of pollution effects. Moreover, the world population is increasing rapidly resulting in a diminishing per capita access to water. On the other side more food is needed. This contributes to a series of conflicts since about eighty per cent of the water resources are used in agricultural food production. Moreover, in an attempt to increase the standard of living, especially in developing countries, industrial production will be increased. Thus more water will be needed and more water will be polluted by industrial processes as well as by agriculture.

THE RELATION BETWEEN FOOD SUPPLY AND AGRICULTURE

The title of this paper is water resources and food supply. One way of approaching this subject is to consider the role that water plays in food production. Water is used for food production in food manufacturing processes but for the most part water has its field of application in crop production. This is of course the reason why agriculture and its water needs play such a central role in man's life. Thus there exists a logical connection between survival, food supply, agriculture, and water.

W. Bach, J. Pankrath, and S. H. Schneider (eds.), Food-Climate Interactions, 239–260.

The importance of water in this case is shown by the fact that about eighty per cent of water used on a global scale is needed for agriculture, including water demand for irrigation. Consequently agriculture plays an important role in the economy, especially in developing countries, and generally it accounts for one to two-thirds of the gross national products of these countries, Radhakrishna (1). As a key sector the output of agriculture will affect changes and improvement of the national product. We also know that for instance in India, about seventy per cent of the working population is engaged in agriculture. However, it is also well known that for much of the developing world, nowhere is the need for a dynamic expansion of production greater than in agricultural activity, UN (2).

There is a lag in agriculture that in fact has impeded the social and economic progress of many developing countries, especially those at the lower end of the income scale. This means that we need a rapid agricultural expansion just to provide food security, to absorb people migrating to urban areas in search for better but in fact more precarious job opportunities, etc. It has been shown to be a difficult task to reduce significantly existing unemployment and underemployment through industrialisation.

THE ROLE OF WATER IN THE TOTAL ENVIRONMENT

As a natural resource water holds a unique position considering its importance in contributing to food production and man's survival. The availability of water is of utmost importance because it will affect the daily life pattern. In regions of the world with abundance of water few problems exist as regards quantities available for domestic use. However, this is more or less an exception. Today nearly half the population of the world, that is, some two billion people, are living without safe access to clean water or water of adequate quantity or quality. Nor are the sanitary facilities acceptable for these people. In order to ameliorate prevailing conditions the initiative has been taken to establish what is called the International Drinking Water and Sanitation Decade 1981-1990. This Decade was originally suggested by the United Nations Conference on Human Settlements (Vancouver, Canada 1976), and later endorsed by the United Nations Water Conference (Mar del Plata, Argentina 1977), and the WHO/UNICEF International Conference on Primary Health Care (Alma-Ata, U.S.S.R. 1978). The goal is that by the end of this decade every person should have access to clean water within a reasonable distance from the home (400 m).

By recognizing the role of water as a factor of human basic need we are indeed touching upon the importance of water in
a broader social context. As an introduction to the comprehen-
sion of the relationship between water resources and food supply
it may be wise to look a little closer at this context. One approach to this subject may be to introduce the concept of the
'total environment', Vlachos and Flack (3). Vlachos (4) suggests
a classification of such an environmental concept by introducing three categories of variables, namely
- a territorial variable which characterizes the physical
  environment
- a sociological variable which depicts the social interaction as well as the organizational and institutional
  network in which man exists
- a cultural variable which has to do with the common ties
  that exist and which refers to the actual, normative
  system.

The role of water should be recognized within this total environment. Considering the use of water within the socio-cultural domain encompassed by the total environment we may distinguish between three different and important uses, namely
- The water use per se, that is the actual versus the symbolic use. When talking about actual use we may for instance view the different ways by which water is used in
  production. As has been said before, water is needed for
  man's subsistence but we may extend this notion to encom-
  pass a much broader class of societal activities. Or we
  may alternatively refer actual use to the use of water for
  communication, etc. Contrary to this, in the symbolic use
  we may for instance perceive water as part of religious,
  mythical, or socio-political expressions. Such concepts
  as Mother Nile or Sacred Ganges are examples of symbolic
  uses of water.
- The possible ways of functional uses of water. Such uses
  may be within agriculture, industry, recreation, land
  building, power, navigation, etc.
- The use of water in relation to social values. This is a
  very important classifying mode that has to do with the
  types of values or goals that we like to achieve by water
  use. For instance, economic development, aesthetic and
  amenity values, health, cultural, and spiritual values,
  may be mentioned as characteristic examples of this specific aspect of water use.

In discussing the importance of water related to a broader
societal context, the role of water as a means of enhancing the

'quality of life' or 'social well-being' has been very much under
debate. Usually the latter term is considered to include predominantly all the structural aspects which determine a person's,
or a group of people's social position in life, Lindh (5). On
the other hand quality of life accounts for some, positive or
negative, internal satisfaction that a given position in life
provides to the individual and it therefore accentuates the socio-
psychological dimensions. We may perhaps call the entire set 'the
good life' which thus may include the spectrum of the notions introduced.

WATER AVAILABILITY AND DEMAND ON A GLOBAL SCALE

It is of great importance to make a survey of the global
water resources since food production is an extremely water de-
manding process, especially because of the large amounts needed
for irrigation purposes. Some eighty per cent of the total global use of water is used in irrigation. In some countries, for
instance in Israel, the amount of water used in agriculture is
about ninety per cent of the total use. As will be shown below,
water for irrigation is in most cases used in a very inefficient
way. Thus, it may be justified to consider the global water situation and to make clear to what extent existing water resources
may be available for various purposes.

Many attempts have been made in order to assess the world's
water resources. Some of the most well-known are those by
Lvovich (6, 7), Kalinin and Shiklomanov (8), and Baumgartner
and Reichel (9). The figures presented in these works are rather
similar. The estimated water resources on the continents, according to Lvovich (7) are shown in Table 1. From this Table the
conclusion may be drawn that the yearly, total available water
resources of the world will amount to 38 800 km^3. Out of these
resources one third is stable, that is these resources are always available. The natural river runoff sometimes gives rise
to very big floods, as for example in India, causing disastrous
inundations. It may often be difficult to take full advantage
of such fluctuating water resources. However, as indicated in
Table 1, already in 1970 reservoirs corresponding to about two
thousand km3 per year runoff had been constructed.

From the Table one may also conclude that as a global average about 4000 m^3 are available per capita and year, that is
some 10 m^3 per capita and day. However, one may also see that
the deviations from the average value are large when comparing
different continents. One may also make a comparison between the

Table 1. Water resources of the continents (1970)

Continents	Natural river runoff km3/year Total	Natural river runoff km3/year Stable	Runoff regulated by Reservoirs	Aggregate stable runoff km3/year	Per capita volume m3/year
Europe	3100	1125	200	1325	2100
Asia	13190	3440	560	4000	1960
Africa	4225	1500	400	1900	5500
North America	5950	1900	500	2400	7640
South America	10380	3740	160	3900	21100
Australia and Oceania	1965	465	35	500	27800
Total	38830	12170	1855	14025	3955

Table 2. Water withdrawals in some countries (1979)

Country	Water withdrawal m3 per person and year
United Kingdom	225
France	456
Germany	537
Japan	707
Belgium	795
Sweden	939
The Netherlands	964
Canada	1230
Australia	1242
United States	2720

figures given in the last column in Table 1 and those presented
by OECD (10) in Table 2, showing the total withdrawals in some
countries in the world. According to OECD, withdrawals are increasing in all countries. However, growth is expected to slow
down for household use, whereas an increase is foreseen in the
amount used for irrigation and cooling purposes.

If now 1200 m^3 per person and year could be considered as
an average value for water demand, an aggregate stable runoff
could allow a world population of about ten billion. This world
population will be reached soon after the year 2000. If, according to Nikitopoulos (11), seventy per cent of the total river
runoff could really be used, this would limit the world population to about 23 billion. This figure may be compared to the
calculations presented by the World Population Conference in
Bucharest 1974. These calculations showed a potential world population of fifty billion people. It must be pointed out that
this was an estimate that was not based on the assumption of
water as a limiting factor.

Lvovich (loc cit) also made some forecasts about the expected withdrawals by the year 2000. His extrapolations showed that
at that time the water withdrawals from lakes and rivers for all
kinds of water uses will amount to more than 7000 km^3 per year.
This corresponds to about twelve times of those today. He also
predicted that all the world's freshwater resources were needed
to dilute the corresponding 6000 km^3 of wastewaters. This is of
course an unacceptable scenario, so Lvovich proposed an alternative that, based on more rational principles, will lead to a
water withdrawal of 1500 km^3 with no wastewater to be diluted.
This second scenario by Lvovich is achieved by using for instance
urban sewage to irrigate fields. New techniques developed could
lead to irrigation quantities corresponding to the wastewater
used through evapotranspiration. This implies that the amount
applied should equal the water-holding capacity of the soil.

Based on the fundamental data presented by authors previously mentioned, some investigators have tried to make some detailed
analyses of available global water resources, see for instance
Falkenmark and Lindh (12, 13) and de Maré (14). In the works by
Falkenmark and Lindh the total water demand from 1965 to 2015 was
estimated by considering the needs within the domestic and industrial water supply and irrigation from a continental point
of view. As regards domestic water supply 200 liters per person
and day were regarded as necessary for rural areas and 400 liters
for urban areas. These figures should be compared to the WHO

recommendations of 150 liters per day and person for sanitary reasons. Statistics on the future development of rural and urban population showed that a 50/50 distribution may be a reasonable assumption. As to the industrial water supply the authors assumed a per capita withdrawal equal to the Swedish industry demand which at that time was 500 m^3 per person and year.

The irrigation demand was calculated from data given by Lvovich. In order to avoid hunger Lvovich counted on a crop production of 800 kilograms per person and year. It was then optimistically assumed that 40 per cent should be produced in irrigated agriculture with a productivity of four tons per hectare. The rest should be produced in rainfed agriculture with a yield of say 1.8 tons per hectare. According to information available at that time it was estimated that future irrigated agriculture should correspond to an increase of irrigated land by 460 million hectares compared to the 190 million hectares at present. If the average need of water for irrigation was assumed to be 2200 m^3 per ton of crops, equivalent to a water need of 9000 m^3 per hectare, a total of 5850 km^3 per year has to be withdrawn from rivers and groundwater reservoirs. Based on such data the total withdrawal needed could be calculated.

Two alternatives were furthermore considered, one without any regeneration of industrial water and another one where regeneration was assumed to be 50 to 90 per cent. Figures 1 and 2 show some of the results. The ordinate gives per cent of total runoff withdrawn and the abscissa shows time, 1970 to 2150. The dotted line indicates the stable runoff as the percentage of the total runoff. Moreover two lines are drawn to show 10 and 20 per cent of total runoff respectively. By this a reference is made to Balcerski (15), who showed the importance of these amounts of withdrawals. Thus, a withdrawal need of 10 to 20 per cent of total runoff indicates that water resources are inadequate. Comprehensive planning and considerable investments are required for solving water problems. If the withdrawal uses are more than 20 per cent of total runoff water is a limiting factor in economic development. The situation for Africa clearly shows that the stable runoff must be increased in some way, for instance by constructing reservoirs or by the aid of water transfers. Figure 2 shows that the water situation may be improved by regeneration of industrial waste waters.

The aim of this chapter has been to show the water resources available on a continental level. Studies on a country level have also been performed. For Europe such a study has been car-

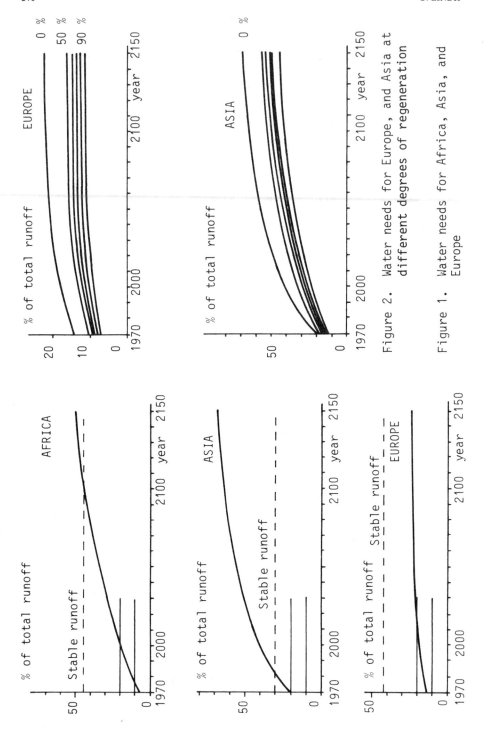

Figure 2. Water needs for Europe, and Asia at different degrees of regeneration

Figure 1. Water needs for Africa, Asia, and Europe

ried out for instance by Bassler (16). In general it may be stated that water is unevenly distributed over the earth's surface.
It is not an overstatement to say that very seldom is water available at the right place and at the right time. This may in some
sense be demonstrated by Figure 3 which is taken from the work
carried out by de Maré (loc cit). This Figure also gives rough
information about regions with severe droughts, scarcity of water, floods, industrial pollution and regions where water quality problems and waterborne diseases are predominant.

THE NEED OF WATER FOR FOOD PRODUCTION

Because of the fact that water is extremely unevenly distributed in space and time in most developing countries, agriculture has to rely on irrigation as an addition to natural rain.
However, the agricultural output is not only determined by quality and quantity of water but it depends also on such factors as
soil, agrotechnique, fertilizers, water application methods etc.
However, there is a clearly documented relation between level of
water control and yield produced. Such relations have been established for rice under experimental field conditions, see for instance FAO (17). The yield successively increases when moving
from the rainfed, uncontrolled flooding system via water table
control to the final step where an advanced management practice
is applied.

But it is also quite evident that increase in yield cannot
be achieved without additional inputs. Such inputs may be a gradual use of fertilizer and an optimal use of agricultural practices. In fact it has been proved that interest and skill of
farmers play a considerable role in attaining maximum yield.
Moreover, it goes without saying that soil characteristics are
important factors. The influence of soil properties on the crop
yield will be dealt with elsewhere in this book. Anyhow, as
this paper deals with water and food relations it might well be
emphasized that the quality of water should be carefully taken
into account. It may otherwise happen that a soil, excellent
for rainfall production may be found to be unusable because, for
instance, the water is too saline. However, the ordinary situation seems to be the one where the physical and chemical properties of the soil will go through a period of change. At a certain time an equilibrium will be established, see for instance
FAO (18). As a consequence the water and soil properties must
be evaluated jointly in order to follow the long-term changes
occuring during irrigation periods.

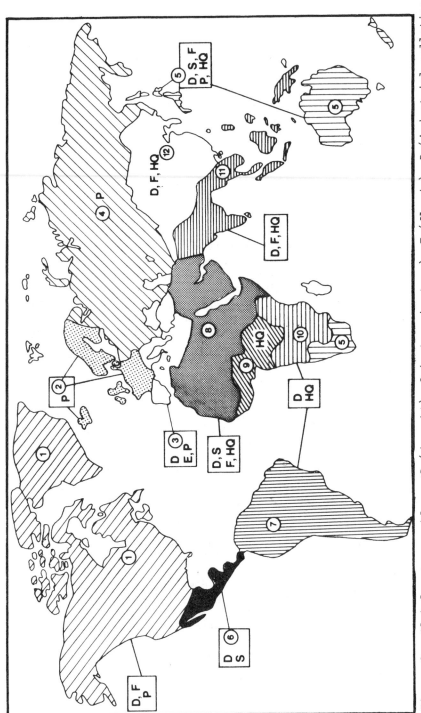

Figure 3. Global water problems: D (drought), S (water shortage), F (floods), P (industrial pollution), HQ (health- and water quality problems)

In order to emphasize the importance of water for irrigation it may be recalled that in 1975 FAO showed that the total irrigated area in the world amounted to 223 million hectares of which 92 million hectares were situated in developing countries. To this may be added that the total area of the world which is under cultivation is about 1400 million hectares. It is remarkable that four-fifths of the irrigated area has been developed in the last 70 to 75 years. In fact, by the end of the 19th century, only 40 million hectares of land were under irrigation and during the last 70 years this area has increased fivefold. Furthermore it may be pointed out that about four-fifths of the present irrigated areas of the world are in the arid and semi-arid regions. Thus about 15 per cent of the world's arable land is irrigated contributing to 30 to 40 per cent of the total agricultural production.

The water used for irrigation, losses included, is in the order of 3000 km^3 per year. This figure corresponds to about 8 per cent of the available global water resources or to 25 per cent of the stable water resources. Through the expansion of irrigation facilities it has been possible, with the aid of an improved technology, to support a much larger population as well as an unprecedented increase in demand for agricultural products which has taken place in the past three or four decades. However, the increased use of water faces very serious ecological constraints. Before such constraints are overcome it may really be doubted if it is possible to go further in many localities. If this is true it may of course affect the accomplishment of the goals set up by FAO for the near future. Thus, according to FAO, by 1990 the total irrigated area has to be increased to 273 million hectares, of which 119 million hectares have to be developed in developing countries. It is also emphasized that this cannot be achieved without strong efforts. For developing countries this may mean
- 45 million hectares of irrigation improvement
- 23 million hectares of new irrigation
- 78 million hectares of drainage improvement,
  including 52 million hectares on irrigated land.

The calculated costs for these activities are about 97 800 million U.S. dollars at 1975 prices. One third of this amount may be needed from external sources. The improvement is expected to require 438 km^3 of additional water. As can be seen, these figures indicate that a vast increase of food production in the Third World will depend upon a quantum jump in irrigation as well as upon an ecological practice that primarily has to preserve the water quality of soils. At the Habitat Conference

in Vancouver 1976, Barbara Ward developed these ideas and it may
be well worthwhile to quote part of her contribution, see Aziz
(19)
-   'In the strangest and yet in the most convincing way,
    nearly every debate these days, on population, on energy,
    on settlements, on food, comes back to water - there is
    no absolute shortage and yet there are increasing risks
    of regional and local crises. There is enough water, but
    not everyone who needs it gets it'.

THE NEED OF WATER RESOURCES ASSESSMENT

     It clearly appears from the figures presented in the pre-
ceding chapters that the demand of water for irrigation is very
high. The total amount is not high in itself but having in mind
the unevenly distributed water resources on earth in time as well
as in space there will be big difficulties in solving the water
problems. On the other side many hydrologists and water resources planners have questioned if the present day demand may be
regarded as a sound starting point in forecasting future need.
The reason is mainly due to the low irrigation efficiencies.
Consequently water demand figures as presented by FAO and other
future projections certainly do not reflect real water needs.
Such observations may, however, give rise to a much more powerful activity in assessing the available water resources on a
national and regional base.

     We may choose Africa as an interesting object in discus-
sing water resources assessment. Then there are two aspects that
have to be considered. The one is the availability of surface
and ground water resources, having in mind that these two resources are hydrologically closely related. The second is water resources available through non conventional means, for instance
desalination.

     Speaking of Africa our first observation may be that in
fact data are available only from a few countries. From only
20 countries out of 47 there are data about water resources.
However, they are not quite reliable because they are incomplete
or because the information is not of direct use. The River Nile
is one example where the runoff is assessed at Aswan including
all flows from the entire catchment, ECA (20). The reason why
data are available from only less than fifty per cent of the
countries has to do with the fact that these missing countries
have not attempted to assess their water resources. However, for

the big rivers there are data available on mean annual river
runoff. There are some very big rivers in Africa but it may nonetheless be stated that river runoff in Africa makes 10 per cent
of the world's total in spite of the fact that the land surface
is about 23 per cent of the world's total. We may compare these
figures with the situation in South America that has a river
runoff of 25 per cent of the total runoff in the world, from a
surface that makes about 15 per cent of the total land area.

It is now an interesting feature of the water resources
assessment in Africa that precipitation data are better known
than data from stream flow. This calls for an improvement of the
hydrological network so that more complete data will be available
and also that the standard of recording is increased in order to
achieve international standard. But water resources assessment
is not only the question of quantity but quality is always closely related to quantity. Quantity of water, without giving reference to its quality or vice versa, may be considered to be mean-
ingless terms in most situations. However, water quality assessment as well as the assessment of suspended matter, bacteriolo-
gical, chemical and physical properties may require establishment
of water laboratories.

In Africa, as in many other parts of the world, data collection and processing as well as publishing of data is not so well
established. This must be the responsibility of some relevant
ministry. It is also necessary to establish a hydrological forecasting system which also will make it possible to develop planning strategies to avoid losses from for instance droughts and
floods. It is reported in the ECA publication just mentioned
that the water resources available will amount to about 2500
billion cubic meters. However, this is a very questionable figure
as may be seen from a comparison with data given earlier in this
paper. According to Lvovich the stable water part for Africa is
some 20 per cent lower. This fact would be the motive for an adequate study of what quantities are economically exploitable.
Moreover, one has to find out what quantities are available for
various purposes, where a principle of division could be consumptive and non-consumptive use.

Assessment of available water resources must of course include not only surface resources but also groundwater resources.
It is a fundamental hydrological fact that these two resources
are closely related which is important to bear in mind in regions
with scarcity of water. The amount of groundwater in a hydrologically well defined catchment may be adequately predetermined

through a water budget calculation, Sokolov and Chapman (21).
However, from a practical point of view much more must be known
about groundwater reservoirs before they are used for practical
purposes. Besides the available amount of water one needs information about safe yields, permissible and potential use rates,
depths, drawdowns, etc. Thus, what is needed is again a regional
as well as a national network coordinated into a general hydrological and hydrometeorological system.

We may now ask what methods are available in the assessment
and enhancement of water resources. One may introduce this subject
by saying that in industrialised countries the water assessment
techniques have successively been acquired. This has been a long
process that today includes very modern methods. As in many other
situations the transfer of knowledge to developing countries should
not necessarily imply a repetition of this sequence. The most effective way should of course consist of transfering modern methods,
the possible disadvantage being associated costs.

Among techniques to be used may be mentioned remote sensing
and the use of satellites for data transmission. Remote sensing,
which should be regarded as a complement to traditional methods
of inventory, implies the use of remote sensors that may be ground
based, airborne or satelliteborne. In hydrological studies common
photography, near infra-red photography, far infra-red scanning,
radar scanning and ultra-violet scanning have been used. In many
countries in Africa remote sensing is already a technique that is
applied not only in detecting water resources but also for other
tasks as, for example, in the study of desertification, especially
in the Sahelian zone.

The use of nuclear techniques may be another means that can
be used especially in the evaluation of ground water resources.
It may provide information about origin and mechanisms of recharge,
dynamics of ground water movement, interconnexion between aquifers
and possible interrelations between surface and ground water.
There are other methods to be mentioned too as for instance the
use of seismic techniques for ground water assessment and drilling
techniques including a wide range of well-logging devices. However, as has been pointed out by several authors, politicians
and decision makers are faced with a series of conflicting objectives such as those given by WMO (22):
- meeting water needs within a reasonable time span
- maximizing use of local personnel and equipment
- minimizing investment
- minimizing the costs of exploitation
- conserving the quantity and quality of water.

Another problem of importance is the use of improved techniques in forecasting. Closely related to this is the decision problem with regard to the construction of reservoirs for flood protection. But also the real-time forecast is an important problem that requires a set of observation stations and computers for rapid and immediate processing of the data observed. Models of various kinds and the use of computers may facilitate many hydrological studies. The International Hydrological Programme of Unesco has emphasized research and development of models intended for application in developing countries with the specific hydrological conditions prevailing there, Unesco (23).

OBSTACLES TO EFFICIENT WATER USE IN IRRIGATED AGRICULTURE

Earlier in this paper it has been pointed out that one crucial point in trying to assess the water demand for food supply had to do with the overestimation of water needs because of inefficiency of water use in agriculture. So, as far as food production is dependent on water use in agriculture, one has to consider how to improve water use efficiency.

Let us start by emphasizing again that food production by farming not only depends on water management but also on land management. A rational development must consider the joint water and land management. This is also quite obvious from a hydrological point of view. In a natural state a hydrologically well defined watershed is in a hydrological equilibrium. However, every human interference with land implying changes in land use will automatically have its effects on the water balance. A very well known example of this is the sealing off of land that occurs in urban areas which results in a large amount of surface runoff. A second effect is subsidence because the ground water storage will not be replenished. Other examples are deforestation and expansion of intensive cultivation.

Such changes may have negative or positive effects on the hydrological regime. There may for instance be a redistribution between surface and ground water resources that may have feedback effects on agriculture. This may lead to compensating adjustments in order to reestablish the premise for agriculture. Not only the quantity but also the quality of water may be affected. However, very often it is difficult to anticipate what effects are to be expected. Alterations in land management may also have other effects. For example irrigation of vast areas may result in changes of the local climate and thus affect the atmospheric part.

The combined use of land and water for food production puts great demands upon the planner of an irrigation project as well as on the farmers who in the end are responsible for the proper management of the water resources. Food production of course means much more than land and water. It is the question of selection of the appropriate crops that has to coincide with probable rains because the water holding capacity of the soil is one factor of importance in selecting the crops. As pointed out by Stewart (24) this capacity determines the storage limits in offseasons as well as in growing seasons with rainfall or irrigation.

There are also several degradation effects on land that have their repercussions on the efficiency. Such effects are very often enforced by human activities. There are several reasons for the degradation of the soil. One is very closely related to the surface runoff whether due to natural rainfall or to irrigation practices. One may, according to FAO (25), distinguish between three basic factors that may influence water erosion, namely
  - climate and the timing, amount and intensity of rainfall
  - characteristic soil properties as well as land slope and distance affected by water
  - land management that comprises possible cultivation states.

Another problem closely related to erosion is sedimentation. Material eroded from a certain land surface may very well be transported to rivers, open channels for water distributions etc. The sediment may also damage hydraulic structures like pumps etc.

Two very severe problems in irrigated agriculture are salinisation and water logging. These problems have been extensively documented in the literature. Very briefly, salinisation has to do with the fact that in most arid regions the soil contains a certain amount of salt. The irrigation process is an activity that disturbs the natural salt distribution and when evaporation from the irrigated area is high it may happen that the salt is drawn up from lower strata and enriched in the root zone and the upper layer of the soil. In ordinary irrigation techniques extra water for leaching is needed. However, the leaching procedure, applied when salinisation is at hand, very often involves a series of sometimes insurmountable difficulties. There is always the risk of causing water logging which means that too much water is retained in the root zone. Moreover, it is necessary to keep the ground water surface below the critical level for active salinity of the soil profile, Elgabaly (26). In fact, due to salinisation and water logging at least 200 to 300 thousand hectares of irrigated land is lost every year, Biswas (27). More-

over current estimates indicate that 20 to 25 million hectares
of land that are saline at present once were productive and fertile land. This corresponds to about 10 per cent of irrigated
land today.

There are some water problems associated with evaporation
and seepage. Thus reservoirs and canals, especially in arid lands,
are subject to heavy losses. It is a notable fact that the evaporation effect is not always recognized, NAS (28). It is an equally remarkable fact that the evaporation from small reservoirs and
farm ponds with a large open area (compared to the water stored
in the reservoirs) is of the same order as the water quantity used
for the agricultural production. So, drastically expressed it may
be the question of a full or an empty reservoir.

MANAGEMENT AND OPERATION OF IRRIGATION SCHEMES IN DEVELOPING
COUNTRIES

Since irrigation is an important means of increasing crop
yield management and operation of irrigation schemes may be considered to be of utmost importance. It is of course very discouraging to start with the statement by Bottrall (29) that most
irrigation schemes in developing countries have been very disappointing.

One may also start by drawing attention to the fact that
there is a distinction between large and small irrigation projects. Large irrigation projects, ranging from 5000 to 250000
hectares, are often operated by the government. One apparent problem with large projects has to do with scale. Communication and
information difficulties have to do with big distances. Sometimes
these problems will be aggravated because large systems often are
controlled by a single reservoir. Allocation of water to users
will be a special problem in such a system. Large systems also
are less flexible than small systems in spite of the fact that
they are designed to meet various demands. Anyhow medium-scale
or small-scale projects show definite advantages. Especially they
provide more flexibility for local involvement and there is in
fact more potential for improvement and experimentation in the
initial phase of the project.

Planning of large-scale projects often gives rise to conflicts during early stages. There is very often little cooperation
between agricultural and engineering agencies. Such controversy

may have its feed-back effects when the irrigation project is operating and an apparent weakness often appears when it comes to
the final allocation of water. Large irrigation projects are mechanized aiming at a better output. Better production techniques
should thus compensate for shortage in production.

However, it is difficult to transfer good results from research and pilot studies to large farming systems. Modern research planners often claim that the inefficiency is due to the
lack of skill and interest from the farmer and that they do not
accept modern ideas that aim at a better output from the system,
Gibbon (30). But modern technologies are developed at research
institutes often without consideration of psychological and other
factors that determine the level of acceptance in introducing
new ideas. From all the papers written about the advantages of
'appropriate technology' one may learn that the bottom up approach
has to be tried, that is to start from the farmer's level,
Baquer et al.(31). This means that agricultural output should not
be based only on profit goals but instead on the optimal solution
that takes into account economic as well as social aspects.

That socio-cultural aspects are extremely important has been
documented many times, Ehlers (32). Thus, if science and technology are to succeed in contributing to a potential development of
the farming system one has to start from an analysis of the prevailing political, physical, biological, social and economic situation. Conclusions from such a study have to be combined with
studies of existing systems and production processes. It may be
well worthwhile to quote three essential ideas that have been
proposed by Gibbon (loc cit), namely

-   develop a range of alternative technologies from which
    the farmer can select those most appropriate to him
-   convince planners and policy makers that science and
    technology are not enough to solve problems in agricul-
    ture
-   consider socio-political changes in conjunction with
    technological and scientific development.

Such ideas are supported by many practitioners who have
worked for many years in developing countries, see for instance
Widstrand (33), and de Schutter and Bemer (34). The advantage
of starting from the farmer's view of the irrigation project is
also made clear by the World Bank. A project in Pakistan showed
that investment in improving management at the farmer's level
was certainly a much better investment than trying to build new

reservoirs and a network of channels, Lowdermilk, Freeman and Early (35).

　　We are here faced with a problem of immense complexity. Food supply depends on agriculture, rainfed or irrigated. To make it possible to feed a rising world population we know that food production must be raised considerably during the coming years. However, agriculture is almost everywhere at a relatively low level as compared with its possible full potential. For that reason an industrial or quasi-industrial type of agriculture has emerged in developing as well as in developed countries. We already know much of the negative effects of this technology at the farmer level a fact that has to do with alienation. This technology influences the farmer's attitude towards irrigation and his potential for cooperation. Such industrial agricultural systems also imply a risk to the environment, especially to water courses and ground water because chemicals of different kinds may be discharged into these waters.

　　Finally we should mention the problems caused by water borne diseases that seriously affect the farmer as an irrigator and also his family in providing the rural household with water for the daily use. Thus the problem of diseases is a severe constraint in discussing food production. Only to mention one of these diseases, we know that, on a global scale, Schistosomiasis (bilharzia) affects 200 million people every year.

CONCLUSIONS

　　A considerable space has been used to give an account of the global water situation as a background to the great need of water for agricultural food production. Available global water resources are of constant quantity but the quality may be assumed to deteriorate continously if current practice in water use will persist. Because of the unequal global distribution of water Lindh (36) there will be a strong competition between different needs of water (domestic, agricultural, industrial etc) especially where water resources are scarce. Conflicts may be very difficult to solve, Widstrand et al. (37). If one adds to this the constraints due to low water use efficiency, the severe reduction in capacity of work caused by water-borne diseases, the negative attitudes of the farmers toward the mechanisation of agriculture (and the failure of "appropriate technology") as well as the increased need of food for an increasing world population, etc the complexity of the problems to be faced will be quite apparent. Any attempt to solve these problems conjunctively must be based

on the assumption of an operating series of networks. Such networks may for instance be social, administrative, legal, communication or distribution networks. If one or more of these networks does not work this may lead to unprecedented effects.

Finally it should once again be stressed that the importance of water is not only due to its material metabolistic functions
in different societal contexts. Its importance in fulfilling im-
material goals and values - also strongly emphasized in this paper - has to be recognized in any attempt to solve water problems
in developing countries. Therefore a proper understanding of the
complex role of water in man's life is a necessary prerequisite
for any action taken to intervene with the total environment.

REFERENCES

1. Radhakrishna, S. (ed): 1980, *Science, Technology and Global
    Problems*. Pergamon Press.

2. U.N: 1978, Journal of Development Planning, No 13,
    pp. 123-198.

3. Vlachos, E. and Flack, J.E.: 1975, *The General Socio-Econo-
    mic-Political Context in Urban Runoff; Trends and Prospects*.
    Proc. of a Research Conference on Urban Runoff. Amer. Soc.
    of Civil Eng., N.Y. pp. 36-51.

4. Vlachos, E.: 1980, *Social Impacts of Droughts. - Nato Advan-
    ced Study Institute*. Laboratorio Nacional de Engenharia
    Civil, Lisbon, Portugal.

5. Lindh, G.: 1979, *Socio-Economic Aspects of Urban Hydrology.
    - Studies and Reports in Hydrology*, No 27, Unesco, Paris.

6. Lvovich,, M.I.: 1973, *The world's water*. - Mir Publishers,
    Moscow.

7. Lvovich,, M.I.: 1977, Ambio, Vol VI, No 1, pp. 13-21.

8. Kalinin, G.P. and Shiklomanov, I.A.: 1974, *World Water Ba-
    lance and Water Resources of the Earth. - Studies and Re-
    ports in Hydrology*, No 25, Unesco, Paris.

9. Baumgartner, A. and Reichel, E.: 1975, *The World's Water
    Balance*. - R Oldenburg Verlag, München.

10. OECD: 1979, Observer, No 98, May, pp. 18-19.

11. Nikitopoulos, B: 1962, *The Influence of Water on the Distribution of the Future Earth's Population*. Athen's Center of Ekistics, Rr-ACE: 125 (COF), October 7.

12. Falkenmark, M. and Lindh, G: 1974, Ambio, Vol 3, No 3-4, pp. 113-122.

13. Falkenmark, M. and Lindh, G.: 1977, *Water for a Starving World*. - Westview Press, Boulder, Colorado.

14. de Maré, L.: 1976, *Resources, Needs and Problems*. - Dept. of Water Resources Eng., Univ. of Lund, Report No 2.

15. Balcerski: 1973, in Szesztay, K.: *The Hydrosphere and the Human Environments*. Wellington Symposium. - IASH, Paris.

16. Bassler, F.: 1975, *Water Resources in the European Community*, ENV/418/75-E.

17. FAO: 1976, *Water for Agriculture*. - U.N. Economic and Social Council. E/C.7/L.54.

18. FAO: 1979, *Land Evaluation Criteria for Irrigation*. FAO World Soil Resources Report, No 50.

19. Aziz, S.: 1977, Economic Impact, No 18.

20. ECA: 1976, *Problems of Water Resources Development in Africa*. - Economic Commission for Africa, U.N. Water Conference, Document E/CN.14/NRD/WR/1/ Rev. 2.

21. Sokolov, A.A. and Chapman, T.G.: 1974, *Methods for Water Balance Computations*. - Studies and Reports in Hydrology, Nr 17, Unesco, Paris.

22. WMO: 1976, *Technologies for the Assessment and Enhancement of Water Resources. - U.N. Water Conference*, Document E/Conf. 70/ CBP.2/ Add. 1.

23. Unesco: 1980, *International Hydrological Programme. Third Session of the Intergovernmental Council, Final Report*, IHP/JC/111, Unesco, Paris.

24. Stewart, J.I.: 1980, *Planning and Managing Irrigation Projects for Optimal Water Use Efficiency in Johl, S.S. (ed): Irrigation and Agricultural Development*, pp. 147-160, Pergamon Press.

25. FAO: 1976, *Improving the Efficiency of Water Use in Agriculture*. - U.N. Water Conference, Document E/Conf. 70/ CBP.2/ Add. 1, pp. 32.

26. Elgabaly, M.M.: 1980, *Problems of Soils and Salinity in Biswas, A.K. (ed): Water Management for Arid Lands in Developing Countries*, pp. 47-58, Pergamon Press.

27. Biswas, A.K. (ed): 1978, *United Nations Water Conference. Summary and Main Documents*, Pergamon Press.

28. NAS: 1974, *More Water for Arid Lands.* - National Academy of Sciences, Washington.

29. Bottrall, A.: 1978, Water Supply and Management, Vol 2, No 4, pp. 309-332.

30. Gibbon, O: 1980, *Rainfed and Irrigation Systems in Johl, S.S. (ed): Irrigation and Agricultural Development*, Pergamon Press.

31. Baquer, A; Nandy, A; Uberoi, J.P.S.; Ram, M., and Reynolds, N.: 1979, Mazingira, No 10, pp. 69-74.

32. Ehlers, E: 1977, *Social and Economic Consequences of Large-scale Irrigation Developments in Worthington, E.B. (ed): Arid Land Irrigation in Developing Countries.* Pergamon Press.

33. Widstrand, C.: 1978, *Water and Society, Conflicts in Development, Part 1,* Pergamon Press.

34. de Schutter, J. and Bemer, G. (eds): 1980, *Fundamental Aspects of Appropriate Technology.* - Delft University Press.

35. Lowdermilk, M.K.; Freeman, D.M. and Early, A.C.: 1978, *Social and Organizational Factors for Farm Irrigation Improvements: A Case Study in Grigg, N. (ed): Water Knowledge Transfer*, pp. 985-999. Water Resources Publications, Fort Collins, Colorado.

36. Lindh, G: 1980, *Water and Food Production in Biswas, M. and A. (eds): Food, Climate and Man.* John Wiley.

37. Widstrand, C. et al: 1980, *Water and Society. Conflicts in Development, Part 2.* Pergamon Press.

SOIL MANAGEMENT AND THE FOOD SUPPLY

E. G. Hallsworth*

Science Policy Research Unit, University of Sussex,
Brighton, U.K.

ABSTRACT. The effects of variations in the climate on the food
supply are related almost entirely to variations in the temperature regime and in the rainfall. The appropriate management techniques, by adjusting the nature of the surface cover can alter
slightly the albedo of a region. Management cannot alter the quantity of rainfall received. However, by altering the nature of the
surface cover, the organic matter input, the structure and composition of the sub-surface horizons, and the system of husbandry
practised, management can make considerable improvements in the
volume of soil explored by the plants' roots, and so increase the
quantity of water available to the crop.

In this paper the potentialities of these modifications are
discussed, mainly in the light of experiences in Australia.

INTRODUCTION

The aim of most of the attempts to manage the soil is to
enhance food production, and to reduce the variation in production
that is associated with vagaries in the climate. Rather than
attempting to review all the 16,000 or more papers on the subject
published every year, I will concentrate mainly on Australian
studies. My justification is that in terms of wheat production,
Australia has a highly variable climate, and few countries show
a greater variation in wheat yield (1).

* Formerly Chief of the Division of Soils and Chairman of the
  Land Resources Laboratories, C.S.I.R.O., Australia.

261

W. Bach, J. Pankrath, and S. H. Schneider (eds.), Food-Climate Interactions, 261–284.
Copyright © 1981 by D. Reidel Publishing Company.

TABLE 1   Adjusted coefficients of variation of wheat
yield over the 25 year period 1954-1978.*

Country	Mean Wheat Yield over period kg/ha	Adjusted Coefficient of variation** %
U.S.A.	1810	8.0
Mexico	2400	9.8
France	3299	10.2
Pakistan	991	10.3
India	1013	10.3
Spain	1211	11.1
China	1066	12.1
Argentina	1398	13.3
Turkey	1233	13.5
Canada	1552	16.0
Australia	1217	18.9
Algeria	668	22.2
Morocco	843	24.0

* Source of data, FAO.
**Calculated as (Standard Deviation/Mean) x 100. SD
  estimated after removal of linear trend effects.

The food produced in any area, whether consumed either as
plant products directly or as animal products, all comes initially
from the plant, and in any one area the quantity of plant material
produced is the result of an interplay of several different
factors.

These are:

1. Climatic   (a) Temperature - the mean annual temperature,
                   the magnitude of the annual and diurnal
                   cycles, and the frequency and timing of
                   phytotoxic events such as frosts or
                   extremely high temperatures.

                (b) The moisture regime - the annual total,
                    its distribution through the year, and
                    the extent of variation from year to year.

2. the nature of the soil - its ability to

   (a) supply nutrients

   (b) retain water and supply it to the plants

   (c) allow roots to develop freely

   (d) develop a surface structure that is
       permeable to water and air, able to
       withstand traffic, but also sufficiently
       well aggregated to resist soil erosion.

3. The topography, for the soil always tends to slide down-
   hill, giving rise to shallow soils on the ridges and
   deeper soils in the valleys. Water also flows downslope,
   carrying nutrient and toxic elements alike with it -
   usually leading to the base of a slope being moister
   than all parts upslope.

4. The organisms present. These fall into two categories -
   the synthesizer species which fix CO_2 from the air into
   more complex forms, using only mineral nutrients from
   the soil and solar energy, mediated by chlorophyll, and
   the decomposer species, which include all others than the
   green plants, and which obtain the energy required for
   growth by breaking down the organic matter produced by
   the synthesizer species.

From the foregoing it can be seen that the fertility of the
soil, i.e. its ability to produce, has three components:

(a) Its chemical fertility - or ability to supply nutrients.

(b) Its physical fertility - its ability to supply water to
    the plant, provide a stable surface and a medium suitable
    for root growth.

(c) Its biological fertility - the resultant of the inter-
    actions of the various constituents of its biota.

None of these are independent, and marked interactions occur
between them - and it is by the manipulation of these interactions
that man is able to manage the soil in such a way as to overcome
some of the adverse effects of climate, and the limitations that
the very nature of the soil imposes on plant productivity.

CHEMICAL FERTILITY

Water is the major factor limiting plant growth over perhaps two-thirds of the land surface outside the arctic and antarctic circles. It is fluctuations in the quantity and distribution of the annual rainfall that are responsible for most of the effects of climatic change on crop production. One of the earliest, and still one of the major, ways of smoothing the impact of variations in rainfall has been to irrigate – by storing water in times and in regions when there is an excess for use at times and in places where there is a deficiency. This practice is so well known as to need little further mention, except to point out that over much of the world the irrigation practises are a long way from giving optimum use of the available water, and a very serious proportion of the land that has been irrigated has been damaged by saliniz- ation.

In those areas where natural or artificial supplies of water are adequate for crop growth tremendous increases in crop yield are now being obtained by the use of fertilisers. Two of the major nutrients, nitrogen and phosphorus show some contrasts in behaviour which are of great importance in those areas where water supply is limiting.

Nitrogen is a mobile nutrient, absorbed mainly as nitrate by mass flow in the transpiration stream moving through the soil to the plants' roots. Assuming a transpiration ratio of 500 and a plant containing 2%N, the required mean concentration of nitrogen in the transpiration stream of maize is, according to Nye and Tinker (2), of the order of 40 p.p.m. Values of this magnitude do occur in fertile soils, and less fertile soils can easily be supplemented to this figure by the use of nitrogenous fertiliser.

Phosphorus, on the other hand, is largely immobile. The required concentration in the transpiration stream of maize, according to Barber et al. (3), is of the order of 4 p.p.m. if the plants' requirements are to be met by mass flow only. This level is nearly two orders of magnitude above the P levels observed in the soil solution. Barber et al. found a modal value of 0.05 p.p.m. P in the topsoils of 145 soils in the mid-west of U.S.A. Hence the main mechanism of phosphorus uptake is by diffusion to the root, and the ability of a plant to obtain phosphorus from the soil depends on the quantity and distribution of its roots.

Since potential maximum yield at any site varies with water availability, the difference in behaviour of these two nutrients influences the extent to which they may be used to counteract climatic changes. Theoretical curves for response to nitrogen are shown in Figure 1 (1). As the maximum yield increases, the ability of the plant to obtain more N from the soil does not increase, and

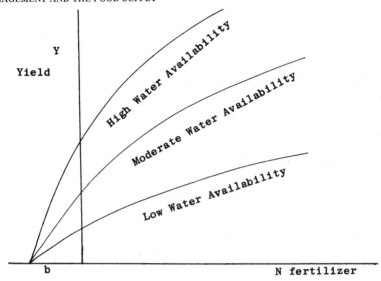

Fig. 1   Effect of increasing water availability on maximum crop
         yield and on yield response to applied nitrogen

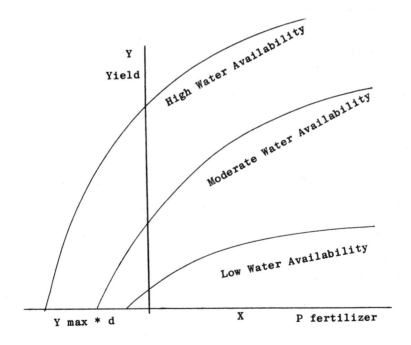

Fig. 2   Effect of increasing water availability on maximum crop
         yield and on yield response to applied phosphorus. The in-
         creasing contribution of phosphorus from soil sources as
         maximum yield (Y max) is Y max * d where d is a constant.

the response curve becomes steeper. Theoretical yield response
curves for phosphorus on a soil low in phosphorus are given in
Figure 2. As potential maximum yield increases, the ability of
the plant to obtain more phosphorus also increases, due to the
greater root growth and consequently the greater volume of soil
explored.

A consequence of this is that yield response to phosphorus
fertiliser under dry conditions and low maximum yield can be proportionately high, whereas responses to nitrogen fertilisers are
proportionately low under dry conditions.

It has been shown by Wark and Cook (4) that the N-response
of stubble-sown wheat in South Australia is positively related
to the May-October rainfall, and the effect is well illustrated
in Figure 3 by Russel's data for four research stations over the
six years 1956-1961 (5). The yield response in 1959 was very low,
by contrast in 1956 the yield response was substantial.

From this and related studies it is clear that however valuable fertiliser nitrogen is in making the best use of water in
areas of higher rainfall, it cannot be used to counteract the adverse effects of a climatic moisture deficit.

Can they be used in areas of variable rainfall to take advantage of the higher rainfalls that occur in some years? The theoretical curves are not encouraging (Figure 1). The response curve is
steeper the higher the potential yield, i.e. the higher the available rainfall, but a farmer who attempted to maximise his ten-
year production by always applying sufficient N-fertiliser to take
advantage of the good rainfall years would be seriously out of
pocket if the 'good' rains did not eventuate. To apply N to the
growing crop, when better than average rains have fallen is a
possibility, but not one without increased cost. Were it possible
to adopt a system of husbandry which, by increasing the level of
soil organic matter, enhanced the level and availability of soil
nitrogen, a better N supply to the crop in years of higher rainfall
would automatically be provided. In general, to increase or maintain a high level of soil nitrogen, a leguminous crop must be present in the rotation, otherwise the nitrogen level declines (6).
The reduction occurs rapidly at higher temperatures (Figure 4).
Using data from long-duration experiments, Russel (7) has developed a model which relates cropping practice to soil nitrogen level.
The levels predicted by this model, for various rotations continued
for 100 years are given in Table 2.

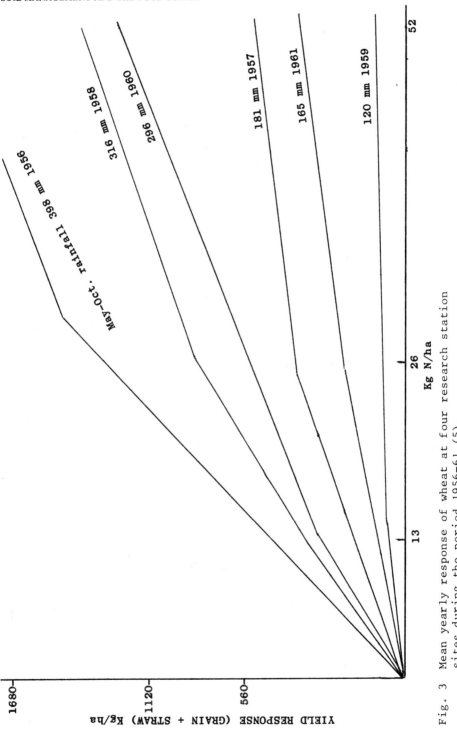

Fig. 3   Mean yearly response of wheat at four research station
sites during the period 1956-61 (5).

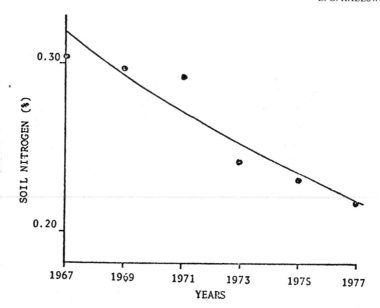

Fig. 4   Changes in the soil nitrogen content of continuous sorghum
         plots at Narayen Research Station, Queensland, Australia
         and the estimated changes using the fitted model.

TABLE 2   Ranking of projected soil organic nitrogen values
          beginning with an initial soil nitrogen content
          of 0.100% and running the model for 100 years

Crop Sequence	Projected Soil Nitrogen Percentage after 100 years
PP	0.217
WPPP	0.146
WPP	0.129
WWPP	0.118
WP	0.111
WWP	0.101
WWWP	0.091
FWPP	0.085
FWP	0.062
WW	0.059
FW	0.040

The absorption of phosphorus by the plant is related to the ability of the plant's roots to explore the soil. This is affected by all the soil factors which affect root growth, some of which can be manipulated by management. Some of the factors limiting root growth are chemical, some are physical and some are biological. The chemical factors are those related to acidity and to salinity. The latter affect the water balance long before any toxic effect of salt on the plant manifests itself. The factors of acidity include the actual pH level of the soil solution and the presence of manganese and aluminium ions. Many plants show a restricted root growth at pH 4.0 or lower, even in the absence of ions of Mn and Al, but the presence of the latter two cause restriction to root growth at appreciably higher pH values that 4.0. Aluminium goes into solution at levels below pH 5.6, whilst manganese goes into solution mainly under reducing conditions. Aluminium produces a particularly characteristic stunting of root growth, with the root initials failing to develop, and in addition restricts the rate of entry of phosphate into the cell sap. Its effect is not mediated appreciably by increasing the calcium ion concentrations of the solution, but only by decreasing the acidity. Manganese toxicity, however, is directly controllable by competition from calcium ions, even under reducing conditions. The toxic effects of both aluminium and manganese are mediated by climatic conditions (8) which favour root growth, temperatures below 20°C or a reduction in light energy received being accompanied, within limits, by a reduction in root growth. Physical conditions in the soil, such as resistance to penetration and periodic waterlogging both increase the adverse effects of soil acidity.

The management to combat the adverse effects of too much rain is firstly to ensure that waterlogging does not occur during the growth of the crop, and secondly to raise the pH by adding lime or limestone. In many parts of the world where liming materials are in short supply, this may be too costly, and for such areas the only feasible alternative is to use plant species or cultivars that are tolerant to the effects of acidity, as has been done for wheat (9) or the legumes (10).

PHYSICAL FERTILITY

The ability of the soil to supply water to the plant depends on the volume of the soil that can be explored by the plant roots, by the permeability of the soil to water, particularly that of the soil surface; and by its ability to store water. Clay soils, which have the greatest water-holding capacity are often intractable and difficult for plant roots to penetrate, whilst the presence of impervious layers, either of clay or of sesquiosides will also deny the roots access to the lower horizons of the soil. All these are aspects of soil structure. The aim of soil management directed to

overcoming the adverse effects of climatic variations - particularly as related to rainfall - is to develop a structure soft
enough for easy root and shoot penetration but sufficiently porous
to allow rapid penetration of water arriving at the surface into
the deeper parts of the profile.

SOIL VOLUME

The first step is to maximise soil volume. This is of little
consequence in deep, well-structured soils, but most of the soils
of the world do not fall into that category. Many have only a
shallow layer of soil over a much heavier subsoil and on all soils
the growth of weeds can seriously compete for water. Until
recently the standard method for weed control has been to cultivate the soil with a tined machine, to cut off the weeds at their
roots. This treatment also destroys the surface roots of the
crop. This is less important with crops such as wheat and barley
which receive little cultivation but much more important in tree
crops. Thus Cockcroft and Wallbank (11) showed that with peaches
on the Shepparton Fine Sandy Loam in northern Victoria, no roots
grew in the top 75mm of soil kept free from weeds by cultivation,
compared with $2 cm/cm^3$ where weeds were controlled by weedicides.

SOIL COMPACTION

Soil compaction involves a rearrangement of the solid
particles of the soil which brings them closer together, with an
increase in density and mechanical strength. Compaction of the
surface soil, either by traffic or by heavy rains makes it more
difficult for the rainwater to enter the soil, or for seedlings
to break out. Compaction within the body of the soil makes it
more difficult to cultivate and for roots to penetrate. A compaction that will resist a pressure of 300 p.s.i. $(21 kg/cm^2)$
will prevent root growth. Actually roots can apply a pressure of
only 120 p.s.i. $(8.4 kg/cm^2)$, but by expanding cylindrically
when their forward movement is halted, they weaken the strength
of the soil ahead of the root tip. The less dense the soil, the
lower is its resistance to penetration, the more readily can the
roots penetrate and the greater is the volume of soil explored -
and the greater the volume of water or phosphate available. The
effect of compaction, as measured by a penetrometer, on root growth
is illustrated by Figure 5 (12). Note that as soil temperature
increases - at least up to 20°C, root penetration increases at all
levels of compaction. A climatic change which leads to a lowering
of soil temperature during the growing period will consequently
lead to a reduction in root elongation, resulting in a decline in

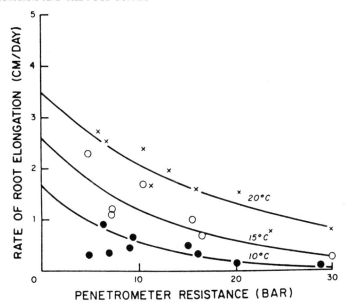

Fig. 5 Rate of root elongation for wheat as a function of resistance of soil to penetration and soil temperature.

the volume of soil explored by the root system and a consequent
reduction in the plant's ability to obtain phosphate and water.
A first step towards controlling the adverse effects of lower temperatures is to use techniques that will enhance root growth, i.e.
by adding phosphate, by increasing the depth of rooting by sub-
soiling, or by drainage to avoid water-logging, by reducing the
numbers of root destroying fungi, eelworm or other root pathogens,
by using clear plastic mulching to increase the soil temperature,
by liming to reduce acidity and exchangeable aluminium, or by
maintaining a good soil structure.

Numerous studies throughout the world have emphasised the
importance of maintaining the level of organic matter for the
maintenance of soil structure. In connection with N-supply it has
been shown (Table 2) how this will decline under some cropping
systems, due to the activities of the soil biota. The Shepparton
Fine Sandy Loam, for example, respires at 0.75×10^{-9}g carbon per
second per c.c. of topsoil. It consequently requires approximately
11 tonnes/hectare of an organic material such as straw if it is to
maintain a stable structure and an optimum infiltration-rate (13).
The problem of maintaining an adequate nitrogen supply to a crop
in an area where rainfall is periodically limiting consequently
can be met at least in part by the same management technique as
is required to make best use of the water that falls - provided
the manager pays due attention to the nitrogen content of the
organic matter added.

THE SOIL WATER BUDGET

The major factor affecting plant growth in the field, after
temperature, is water supply. Reduction in the supply of water to
the plant's roots leads to a reduction in leaf water potential and
hence to reductions in rate of photosynthesis and rate of growth
(Figure 6).

For the crop plant to grow satisfactorily it needs to have
water held in the soil at tensions at which the plant roots can
extract it. The total of water available can be determined by two
methods - the laboratory method, where each layer of soil is
sampled and the water held at field capacity (F.C.) and at the
wilting point (W.P.) is measured. In this case

Available water content = water held at F.C.
                          - water held at W.P.
     or the field method, where
Plant available water = difference between water content of
                        the wettest profile experienced
                        during the year and the water content
                        of the driest.

The water content in this case is being measured by the neutron
probe. In a well structured soil there is little difference
between the figures obtained by the two methods, but in any soil
containing a horizon which impedes root growth, the difference
can be large as shown in Figure 7 (14).

In considering the water budget in areas where water is
limiting, and the prospects for improving the quantity of plant
available water, there are two situations:

(a) Where the soil is not wetted beyond rooting depth, and

(b) Where the soil is wetted beyond rooting depth, but not
    all the water is extracted. This is due to either limit-
    ed rooting depth or limited rooting density. In the
    field the depth to which roots penetrate can be found by
    measuring the depth to which water is being extracted,
    whilst the distribution and efficiency of the roots is
    shown by the quantity of water extracted over a given
    period.

To overcome or to reduce the effect of limited water supply
where the soil is not normally wetted beyond rooting depth, management has several options. These are

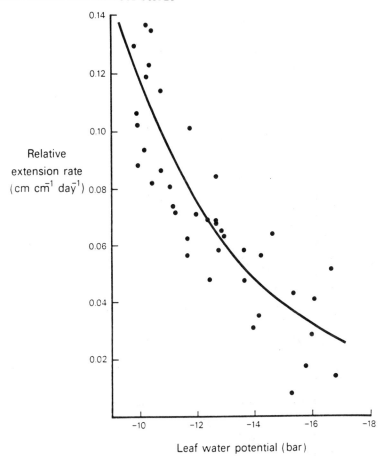

Fig. 6 Relative extension rate as a function of measured leaf
water potential. Critical LWP to give a 50% reduction
in RER is -12.5 bar.

(a) to increase the infiltration rate

(b) to reduce the evaporative loss from the soil surface

(c) to increase the quantity of water infiltration into
the soil.

These three options are not independent, and they all relate to
the management of the manner in which organic matter is returned
to the soil.

The proportion of the rain falling in an area that can be stored in the soil and be available during the growing period of the crop can also be manipulated by the appropriate systems of management.   These are

(a) by increasing the effective area of catchment for the crop

(b) by increasing the time water lies on the surface by reducing the rate of run-off by reducing the effective slope of the land

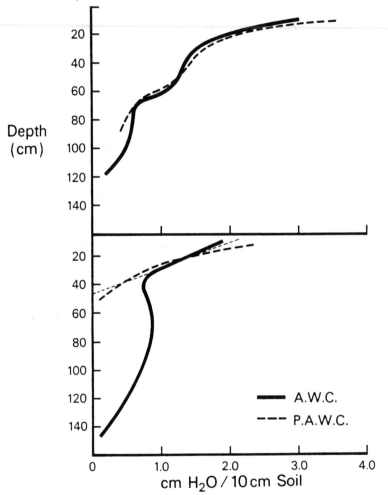

Fig. 7   (a) The Available Water Capacity (AWC) and the maximum root water extraction profiles for a black earth in the Burdekin Irrigation Area (14).
(b) The Available Water Capacity and the maximum root extraction profiles for a solodic-solodized solonetz in the Burdekin Irrigation Area (14).

(c) by increasing time available for accumulation of water
for one crop by cropping the land only one year in two.

Control of Infiltration Rate and Evaporative Loss

In the first case, where the quantity of water in the soil
is less that it need be, some water has run off because the
infiltration rate is too low due to the formation of a surface
seal. In soils of finer texture than sands this is due to a
breakdown of the soil aggregates at the surface, due to bombard-
ment with raindrops or to surface traffic. This can be overcome
by increasing the organic matter content, and this procedure may
be enough for soils on gentle slopes in areas where storms are of
short duration or the rainfall of low intensity, as in much of
the cool temperate and winter rainfall areas of the world.

In much of the tropics and sub-tropics, however, the rainfall
intensity in some storms is very high, and the surface crumbs are
quickly broken down. In this case, as well as increasing the
organic matter in the surface soil, it is necessary to leave a
layer of plant material on the surface to provide protection for
the soil crumbs. In cropland, the system is known as minimum
tillage, the essential feature of which is killing the weeds and
other plants present by killing them with herbicides rather than
by ploughing them in. This system has been shown to be cost-
effective in several parts of the world for wheat, e.g. in England
and Western Australia (15,16) and to be particularly effective in
wet tropical conditions (17,18). There are several related
effects - thus when 4 tonnes/h. of mulch was applied, the water
available for plant growth increased from 13% to 14.5%, the
temperature two weeks after planting dropped from 34.5°C to 31.5°C
in the first season and from 30°C to 28°C in the second season.
There was a marked change in root density (Table 3).

Table 3.   Effect of mulching at 4 tonnes/hectares on rooting
           density of some crops at Ibadan, Nigeria

Crop	No till	Conventionally tilled
(in mg x 10^{-2} cm^{-3} at 15 cm from the row and 0-10cm deep)		
Maize	103.7	38.8
Maize-cowpea	42.7	41.8
Cowpea	32.7	24.0
Soybean	65.0	13.6

The minimum tillage plot also gave a much lower soil loss
by erosion than the conventionally tilled plot. This latter
effect is due to the protection against the destructive effect
of raindrops that is afforded by the layer of dead vegetation.
The practice of leaving a cover of dead vegetation on the soil
surface has been developed in the United States under the term
'stubble mulching' and is being used very effectively on the
soils derived from loess in eastern Washington.

An alternative to this provision of a protection by the use
of a tree or shrub cover, the crops being grown in between the
trees. This old practice is now being advocated again in the
tropics under the term "agro-forestry". Its effectiveness depends
on finding a tree or shrub crop which grows under the climatic
restraints of the area and can be fitted into the economics of the
system.

For pastures under semi-arid conditions, i.e. rangeland, it
has been shown in central New South Wales that soil organic matter
content, total soil water, and depth of water penetration all
increase as one moves from the open ground to the bole of the tree
or shrub (19).

At the other extreme of the range of soil texture, on the
sandy soils that are almost devoid of clay, the condition of non-
wettability may develop (20). The problem here is that the decomposition of the organic matter of the roots of some plants gives
rise to a hydrophobic compound which coats the surface of the
mineral grains so that they do not wet. As a result the water
landing on the surface from light showers stays on the surface,
or in the surface pores and is readily evaporated. Under pastoral
conditions management can do nothing about this. Under arable
conditions, however, if the seed is planted at the base of a
furrow, the non-wetting character can be turned to advantage and
the water arriving at the surface from light rains is concentrated
in the area of the seeds, resulting in more water penetrating
around the seeds and hence in earlier germination.

Control of Water Available from Plant Growth by the Control of
Time and Space Relations

Under natural conditions the density of plant cover is
adjusted to the available water supply, and in low rainfall
regions the plants are more widely scattered, giving a larger
potential rooting area to each plant. A system of crop management has been introduced in Isreal (21) that makes use of the
same principle - the run off/run on technique. In this, the rain
falling is caused to run-off that part of the land on which it
falls, and run-on to strips of land occupied by a crop.

The effect is to double, treble, etc., the quantity of water
arriving on the surface of the cultivated strip, and is accomplished by treating the surface of the land in between the cultivated
strips in such a way as to make it impervious to water. This
procedure has been used so far only with fruit trees but could be
extended to row-crop production, by driving the tractor wheels
along the same strips of soil every year.

The second procedure is to increase the time water lies on
the surface by decreasing the velocity of run-off. There are
several variations, ranging from the use of stubble-mulching or
minimum tillage, through the use of contour furrows with check
bars at intervals, to the alteration of the original slope of the
land to the level terraces used by Shansi and Kiangsu in western
China.

The third alternative is to increase the time for water
infiltration relative to the time of water loss by transpiration,
by reducing the proportion of time that the land is occupied by
plants. This is the system of dry land farming, perhaps best
developed in the western United States and Canada. In this system
the crop is planted only every other year, and during the period
without crop the surface soil is kept free from weeds so that,
theoretically, the rain that falls will enter the soil and be
stored for use during the subsequent cropping period. The problem
with this approach is that to prevent weed growth and so minimise
water losses it has been necessary to have frequent cultivations.
This has led to breakdown of the surface crumb structures, development of a surface seal, with effects contrary to the purpose of
the operation, and to enhanced erosion. The incorporation of
stubble mulching into this practice, with the maintenance of a
heavy mulch cover during the non-crop period, and the development
of machinery capable of sowing the crop through the mulch has
proved very successful on the loess-derived soils of Washington
State. In Australia, in the main, the yields attained by taking
a crop every year are considerably greater that half of the yield
obtained by taking one crop every two years and appears to be
associated with a restriction of root growth (21).

FACTORS RESTRICTING ROOT GROWTH

The restriction of rooting depth and density in soil horizons
where the water content is above -15 bar appears to be due to

1. high mechanical impedence, due to cemented
   or compacted horizons

2. poor aeration associated with water-logging

3. excessive acidity

4. salinity

5. lack of nutrients

6. damage to roots by root attacking fungi, eelworms,
   etc., i.e. biologically caused infertility

The effects of high mechanical impedence on root penetration
has been considered earlier. The management technique to overcome
this problem is deep ripping and the incorporation of organic
matter and gypsum into the ripped layers. This procedure leads
to better infiltration of water and to an increased leaching of
salt, where salt is present (Figure 8).

Poor aeration associated with water-logging in clay subsoils
has long been realised to be associated with poor root growth of
many commercial crops, and the improvement in yield of many crops
that follows the installation of a drainage system is too well-
known to require much further discussion. I would make two points.
There are parts of the world, particularly in the winter rainfall
areas, where the plants suffer from water-logging in winter - and
hence limited root growth - and from drought in summer. Some form
of drainage in such conditions would lead to an increase in plant
available water at the time of the year when otherwise the plants
would suffer water stress, by increasing the volume of soil that
the roots could explore. At the other extreme, in parts of the
world where rainfall is excessive, as on the heavy clays of the
Guyana front-lands, where in their natural state the soils would
remain water-logged to the surface for most of the year. Here
the old practice of heaping the soil into broad banks with surface
drains between, with consequent loss to production of part of the
surface, can be replaced by the use of mole drains, drawn freshly
for each sugar crop.

Excess acidity in the subsoil is often associated in tropical
areas with the acid sulphate soils - the so-called cat clays. Here
the acidity is produced by the oxidation of sulphides and sulphur-
containing compounds, associated with buried organic matter, to
sulphuric acid. The acidity increases in this case when the soil
is drained - a classic case occurring in Trinidad when the newly
independent government obtained a grant to drain the Oropouche
swamp - and rendered the land almost totally unproductive.

Salinity is always likely to be a cause of trouble in low
rainfall areas, and gives rise to a patchy appearance to growth
even where the concentrations are not high enough to be toxic.
Where salinity problems have developed, as in many of the older

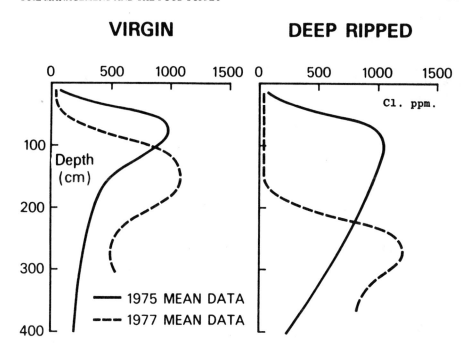

Fig. 8   Changes in the salinity profile of a normally cultivated
         and a deep ripped strongly sodic clay after two 'wet'
         seasons of ponding (14).

cultivated lands of the semi-arid world, the only solution is
leaching the salt out by provision of drainage and applying
sufficient water, but this can only be done in irrigated areas,
and will only be done for high value crops. The effects of salinity will be made worse by any climatic change that causes a
reduction in water supply – by fall in rainfall – or increases the
evaporative demand by a rise in temperature. The deleterious
effects of such changes are almost impossible to correct. The
best approach in the face of possibly adverse climatic change is
one that must be taken at the planning stage in the development
of new areas, namely the exclusion from cultivation of those parts
of the land that at present contain high levels of salt, or which
will – following cultivation – cause salinization of soils lower
down the watershed.

    For high value crops an increase in salinity in the water
used to compensate for the rainfall deficit can be met by changing
the irrigation system practised from furrow or sprinkler systems
to "drip" or "trickle" irrigation. All other systems of irrigation
supply water intermittently to the plant roots, and as a result in
between each irrigation the osmotic pressure of the soil water
rises as its volume is reduced and the salinity consequently increases. Trickle systems, which supply water to the plant roots

continuously, avoid this fluctuation and consequently allow the
use of more saline water.

Another factor preventing the plants' roots from making full
use of the water held in the soil is the level of nutrients in the
lower horizons in which the unused water is stored. This factor,
which appears to be largely responsible for the failure of growing
crops in alternative years - is illustrated simply in Figure 9.
It can be seen that as the soil dries out the roots present in
the relatively phosphate-rich upper layer of the soil die away.
The plant ends its growth cycle whilst temperatures are still
favourable and moisture is still present in the lower horizons,
because presumably there is inadequate phosphate present in these
lower horizons to sustain root growth. The soils of eastern
Washington, and of those parts of South Australia and Victoria
where the alternate wheat-fallow system has been successful all
contain higher levels of available P down the profile than is
available in those soils where it has failed.

BIOLOGICAL FERTILITY

The last soil factor affecting root growth is the presence
and numbers of root disease organisms in the soil - either soil
fungi, nematodes or other soil fauna. Their effect is simply
to reduce the effective length of the rooting system. As a
result the plants' uptake of both water and phosphate is reduced.
Management to control this effect has at present three options.
The first is simply to sterilize the soil with chemicals. In the
early work in Adelaide we used methyl bromide and chloropicrin.
At present level of prices this procedure is uneconomic. Very
considerable increases in yield were obtained - wheat grown on
16 in. of rain rose from 24 bushels/acre (1296 kg/h) to 64 bushels/
acre (3000 kg/h), which was associated with the leaves remaining
green for about four weeks longer into the summer on the treated
plot.

The second is to utilise plants that can be innoculated with
mycorrhiza, which has the effect of extending the root system and
allowing growth to continue in much drier soil than would otherwise be possible. Great advances in knowledge of the endogenous
mycorrhiza have been made in the last five years.

The third option is to manipulate the soil organisms. The
work pioneered by Baker in California (23) and exemplified by the
work of Rovira and Cook (24) in South Australia, where development
of large numbers of Pseudomonads appeared to reduce infection of
wheat by Gaeumannomyces graminis.

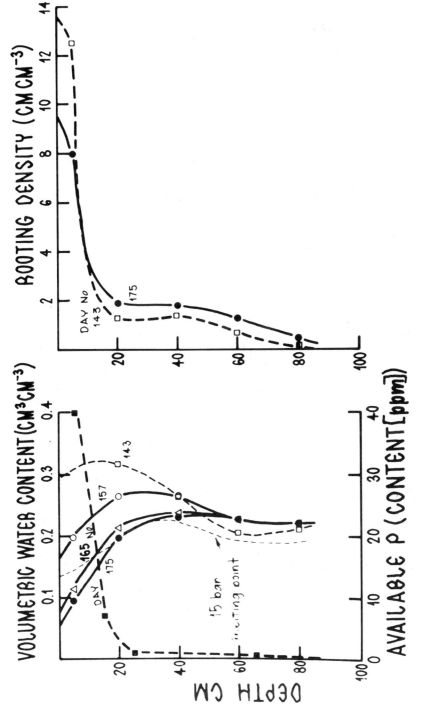

Fig. 9   Root growth in a solonised brown soil in South Australia, with available phosphate concentrated near the surface and varying water contents at different times of the year (22).

Table 4. Effect of resistant variety on yield of
          subsequent crop

(a) Numbers of cysts on Condor and Festiguay wheat
    at flowering

Site	Variety	No. cysts/plant
Bute	Condor	60
	Festiguay	6
Calomba	Condor	84
	Festiguay	5

(b) Effect of a 1975 Festiguay crop on CCN numbers
    and grain yield of Halberd wheat in 1978

Variety in 1975	No. cysts/plants	Yield (t/ha)
Sabre	58 + 7.4 S.E.	2.05 + 0.06 S.E.
Festiguay	2 + 0.5	3.13 + 0.21

The fourth option is to manipulate the soil organisms by
manipulating the cultivar grown. Thus Rovira has shown (Table 4)
that in areas where the effective root size of wheat has been
reduced by infection with cereal cyst nematode, a marked improvement in yield can be obtained by periodic introduction into the
cropping sequence of a resistant strain of wheat which will
drastically reduce the level of infection in a subsequent crop
of a non-resistant cultivar (25).

ACKNOWLEDGEMENT

I should like to thank all my colleagues in C.S.I.R.O. for
keeping me supplied with accounts of the progress of their work
in the fields concerned here, and in particular to E.L. Greacen,
A. Rovira, J. Russell, A.J. Peck, G.H. Harrington, J. Harris,
also to P. Ungar of the U.S.D.A., Nazeer Ahmad of the University
of the West Indies, Trinidad, and to N. Collis George of the
University of Sydney, N.S.W., Australia.

REFERENCES

(1) Russell, J.S., 'Nitrogen in dryland agriculture', Proc. of Int. Conf. on Dryland Agriculture, Adelaide, August 1980 (in press).

(2) Nye, P.H., Tinker, P.B. (1977), 'Solute movement in the soil-root system', Studies in Ecology, Vol. 4, Blackwell, Oxford.

(3) Barber, S.A., Walker, J.J. and Vasey, E.H. (1962), Trans. Joint Meeting Comm. IV and V, Int. Soc. Soil Sci., New Zealand, 121-4.

(4) Wark, D.C. and Cook, L.J. (1939), S. Australian Dept., Agric., Bull. No.340.

(5) Russell, J.S. (1967), Aust. J. Exprl. Agric. Anim. Husb., 7, 453-62.

(6) Hallsworth, E.G., Lemerle, T.M., Gibbons, F.R. (1954), Aust. J. Agric. Res., 5, 422-447.

(7) Russell, J.S. (1980), 'Models of long-term soil organic matter change' in "Simulation of nitrogen behaviour in soil-plant systems" (eds.) Frissel & Veen. Pudoc. Wageningen.

(8) Rorison, I.H., Sutton, C.D., Hallsworth, E.G. (1964) in "Experimental Pedology", Hallsworth, E.G. and Crawford, D.V. Butterworth, London.

(9) Foy, C.D., Arminger, W.H., Briggle, L.W., Reid, D.A., (1965), Agronomy J., 57, 413-417.

(10) Anon. (1979) Ann. Report C.S.I.R.O. Div. of Tropical Crops and Pastures, St. Lucia, Queensland, Australia.

(11) Cockcroft, B., Wallbank, J.C. (1966), Aust. J. Agric. Res., 17, 49-54.

(12) Cockcroft, B., Barley, K.R., and Greacen, E.L. (1969), Aust. J. Soil Res., 333-8.

(13) Cockcroft, B., Tisdall, J.M. (1978) "Modification of soil structure", ed. W.W. Emerson, R.D. Bond, A.R. Dexter, pp.387-391, John Wiley & Sons, London.

(14) Gardner, E.A. (1980), Queensland Dept. of Primary Industry Agric. Chem. Branch Rept.

(15) Stonebridge, W.C., Fletcher, I.C., Lefray, D.B. (1973), Outlook on Agriculture, 7, 4, pp.155-161.

(16) Finney, J.R. and Knight, B.A.G., (1973), J. Agric. Sci. Camb., 50, 435-442.

(17) R. Lal, (1977), in Soil Conservation and Management in the Humid Tropics, Eds. D.J. Greenland, R.Lal.

(18) R. Lal (ed.) (1979), Soil Tillage and Crop Production.

(19) Harrington, G.H. (1979). Private communication.

(20) Bond, R.D. (1972), Rural Research in C.S.I.R.O., 77, 2-5.

(21) Grierson, I.T., French, R.J. (1975) in Tillage Practices of the Wheat Crop , C.S.I.R.O., Land Resources Laboratories Discussion Paper No. 1.

(22) Hallsworth, E.G. (1977), Search, 8, 58-67.

(23) Baker, K.F. and Snyder, W.C. (eds.) 1965. Ecology of the Soil borne plant pathogens, U. Calif. Press, Berkeley.

(24) Cook, R.J., Rovira, A.D. (1976), Soil Biol. Biochem. 269-273.

(25) Rovira, A.D. (1979), Aust. Field Crops Newsletter, p.90.

FURTHER READING

Russell, J.S. and Greacen, E.L. (1977), 'Soil Factors in Crop Production in a Semi-arid Environment', Queensland University Press, St. Lucia, Queensland, Australia.

Greenland, D.J. (1981), 'Soil Management and Soil Degradation', British Journal of Soil Science, 32 (in press).

Unger, Paul M. (1981), 'Management of crops on clay soils in the tropics', Tropical Agriculture, 58 (in press).

Handbook of Australian Soils (1968), Stace, H.T.C., Hubble, G.D., Brewer, R., Northcote, K.H., Sleeman, J.R., Mulcahy, M.J., and Hallsworth, E.G., Rellim Press, Adelaide.

INTERRELATIONS BETWEEN PESTS AND CLIMATIC FACTORS

F. Klingauf

Biological Control Institute, Fed. Biol.
Res. Centre for Agriculture and Forestry
Darmstadt, Germany

ABSTRACT   The role of plant pests and pathogens is decisively
influenced by climatic factors, mainly temperature and humidity. Moreover, climatic factors may directly stress the
crops. The geographical distribution patterns of damaging
agents and the fluctuation of injuries and damage are essential indicators of weather and climate effects. Influences of
climatic factors on the development of pest populations, their
dispersion and activity, and on the effectiveness of pest
control are exemplified. Modern plant protection is directed
towards integrated pest management involving all compatible
methods in controlling injuries and damage. Attention is drawn
to climate and weather as a resource in an improved pest management, outlined in some modern trends such as selection of growing areas, prognosis, and biological control.

INTRODUCTION

     Plant pests and pathogens as well as their host plants are
living organisms and as such they are in themselves as well as
in their mutual relations affected by all climatic factors.
Moreover, crops may be directly impaired by stresses of climate
and weather. There are dramatic examples of pests devastating
a crop in connection with special and outstanding weather conditions. Thus, in Ireland the wet seasons of 1845 and 1846
favoured for some weeks an epidemic dispersal of the potato
attacking parasitic fungus Phytophthora infestans which led to
a total loss of crop with disastrous consequences for man and
country. Similar extensive damages caused by weather in a direct
way are the steady yield losses by water shortage being common
in dry areas, or damages by frost and tempest.

W. Bach, J. Pankrath, and S. H. Schneider (eds.), Food-Climate Interactions, 285–301.
Copyright © 1981 by D. Reidel Publishing Company.

The role of climate and weather in agricultural production and its amplifying or attenuating effect on pests and diseases is an important field of study in phytomedicine - the science dealing with diseases and pests of plants, their prevention and control - particularly within its branches of population dynamics and epidemiology. This paper refers to some principles and examples of the influences of climate and weather on the significance of both pests and diseases; in addition, methods of modern pest management in crop production considering those influences are discussed.

Man is involved in many ways in the relationships between crops and pathogens or pests in their environmental dependence. The scheme in Fig. 1 indicates the complexity of these interrelations. An example of the activity of man is the breeding of plant varieties with different properties of susceptibility or resistance to damaging agents; another measure is the growth of plants in different agricultural areas with diverse types of climate. Moreover, man influences the system by various methods of management such as ploughing, special planting, irrigation, and pest control.

CLIMATE AND GEOGRAPHICAL DISTRIBUTION OF DAMAGING AGENTS

Pests and diseases are not uniformly distributed over the area where their host plants are growing. Furthermore, there are usually significant differences in attack frequency within their distribution area. One of the best investigated examples is the distribution and disease frequency of a group of pathogenic fungi belonging to the family of powdery mildew, the Erisyphaceae. Based on the data by Hirata (9), Weltzien (25) compiled a global map with the numbers of host species reported from different regions of the world. As shown on this map, Erisyphaceae are not known in large parts of Africa and Asia, whereas Central Europe, California and some other regions have an accumulation of more than 500 reported host plants (Fig. 2). Of course, world maps like these have to be supplemented with the progress of knowledge. Nevertheless, they give some understanding of the main distribution zones of pests and pathogens. More detailed information on the distribution of some pests is available for selected smaller areas where more data has been gathered on, e.g., the number of pest organisms per sample unit or damaged plant, the frequency of lesions, symptoms etc.

The available maps allow a classification of some typical distribution patterns. Cosmopolitan agents occurring in most parts of the world are, e.g., the green peach aphid Myzus persicae or the above mentioned Phytophthora infestans. Some pests and diseases show a more or less endemic distribution as, for instance, the cereal aphid Diuraphis noxia or the fungus Venturia

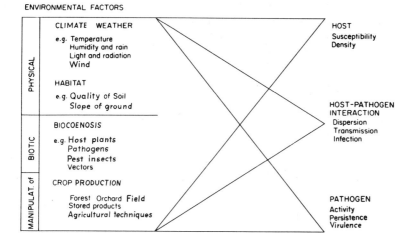

Fig. 1   Scheme of principal environmental factors acting upon
          host-pathogen (pest) interrelations, after Franz (5).

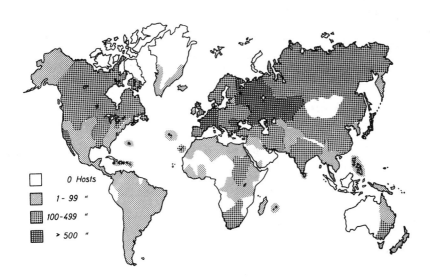

Fig. 2   World map of host frequency of powdery mildews, after
          Weltzien (25), based on data by Hirata (9).

cerasi on cherry. In other cases two closely related pests occur
in different but neighbouring areas on nearly the same host plants
with an overlapping zone in between the distribution areas, e.g.,
the Egyptian cotton leafworm Spodoptera littoralis in southern
Europe, Africa and the Near and Middle East and its closely related species, the tabacco caterpillar Spodoptera (Prodenia) litura
in the Far East. Weltzien (26) distinguishes another type of
distribution pattern according to the continuity of distribution.
Taphrina deformans, the cause of peach and almond curl, covers a
rather continuous area; discontinuous or disjunct areas are known
for Phytophthora phaseoli. A special case of distribution is
represented by such pests and diseases which, after leaving their
endemic sites, have spread over new infestation areas, as often
happens by the carelessness of man. Examples of this kind are
the expansion of the Colorado beetle Leptinotarsa decemlineata
and of the grape disease caused by Plasmopara viticola.

Based on the principles of geobotany, Reichert (12,13) and Reichert
and Palti (14,15) developed the concept of pathogeography, and
Weltzien (22,23,26) developed the fundamentals of geophytopath-
ology, which is defined as the science dealing with the "distribution patterns of plant diseases, the causal understanding of
these patterns, and the geographic aspects of disease control"
(26). Taking into consideration the medical aspects in activities
to achieve plant health, Weltzien (24) arrived at the term
"geophytomedicine".

It is evident that the distribution of different pests and diseases depends on the distribution of their host plants. On the other
hand, the host is not the only limiting factor as demonstrated
by the Spodeoptera species or the original occurrence and new
colonizations of quarantine pests. Geographical barriers, interference of two species and especially climatic conditions of the
environment and climatic preferences of the species are important
factors for different distributions. This is clearly demonstrated by the abundance of pests and pathogens in relation to
climatic conditions in different regions or by the fluctuation
of injuries and damages within a given region in successive time
intervals due to the variability of weather or climate.

As demonstrated in the pestograph (Fig. 3) of the cowpea aphid,
Aphis craccivora, temperature and humidity are the main factors
for favouring outbreaks of the pest. The importance of both
factors in the relationship between pests and host plants is
shown in Fig. 4 which displays the effective and potential
areas of parasitic attack.

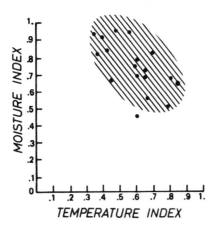

Fig. 3   Pestograph of cowpea aphid, Aphis craccivora, after
           Gutierrez et al. (6).

LEVEL OF INFESTATION AND CLIMATIC CONDITIONS

In organisms without self-regulation of temperature the activity
is directly correlated to weather conditions. Climate and micro-
climate thus influence the infestation level. Yearly reports
of the regional frequency of important pests and diseases are
published e.g., by the Federal Biological Research Centre for
Agriculture and Forestry in Germany.

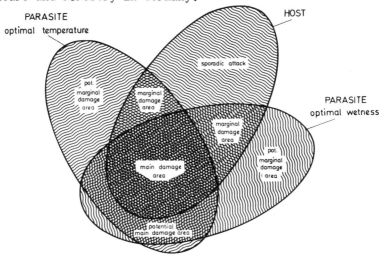

Fig. 4   Model of the effective and potential occurrence of a
           plant pest or pathogen, controlled by temperature and
           wetness, after Weltzien (26).

Development of pest populations

The development of pest populations depends on climatic factors
in a variety of ways. Thus, many fungal pathogens need high
humidity for germination and growth, especially those forming
zoospores like Oomycetes. In wet years or periods such parasites
may find good conditions for development. The newly hatched
maggots of the beet fly, Pegomyia betae, a common leaf miner on
sugar beets, survive and pierce the leaf only when the surface
of the leaf is wet. Insects with a thin cuticula such as aphids
prefer warm and humid weather. Cool and humid weather promotes
the development of P. infestans; at lower temperatures the
sporangia form zoospores, whereas at higher temperatures they
develop only a single germinating hypha which cannot pierce the
leaf to build a new sporangium.

Usually the number of generations is higher in warm climates or
vegetation periods. Table 1 shows the number of generations of
the codling moth, Laspeyresia pomonella, in different regions of
the Iran as influenced by climate. The humid-warm climate in
the Caspian area stimulates the highest number of generations
which results in high infestation levels up to 100% in unsprayed
orchards.

Table 1   Number of generations of the codling moth in some
           Iranian districts, after Bayat-Asadi (1).

location	Teheran		Esfahan	Azer-baidshan	Gorgan
	Damavand	Karadj			
altitude (m)	1900	1320	1590	1710	100
number of gen.	2	3	3	2	4

Dispersion and activity of pests

The dispersion of pests, and thus infestation and damage, is
directly affected by climatic conditions. Rainfall, water and
wind play an important part in the transport of small organisms
like spores of fungi, bacteria, nematodes, and insects. Nematodes in irrigated fields are often spread by water in the
direction of the flow. This results in a particular distribution of the injured plants (namely in rows) in the field which
can be used as an indicator of the nature of the pest. Small
insects and spores drift passively in the wind and they
can be caught in traps for investigation. Often the part
of a field which faces the main wind direction shows the
highest infestation level. Forecasting and effective control
can profit from this phenomenon.

Insects which orientate themselves to odour show a characteristic flight behaviour with respect to the wind. This behaviour
can be used to study the olfactory potential of a species as
has been done for the sugarbeet fly (16). Figure 5 shows a
cage of adult flies placed in a leeward or windward position
with respect to the sugarbeet field.

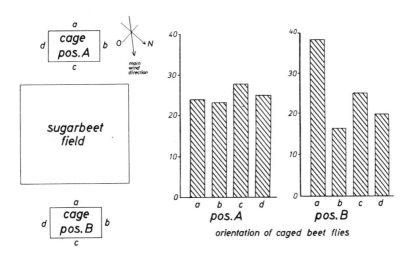

Fig. 5. Wind, host plant odour and olfactory orientation of
         the sugarbeet fly, after Röttger (16).

Temperature, besides other environmental factors, also influences
the flight activity and mating behaviour of the adult insects.
Traps baited with female pheromon attract more male codling moths
when the temperature is higher during the flight activity at dusk
and early night. Low temperatures and rain stop the flight
activity. The efficacy of the trap changes according to temperature conditions in early spring as demonstrated in Fig. 6.

A strange effect of climatic factors of so far unknown significance has been demonstrated for the ingestion activity of aphids
as measured by the number of faeces droplets (10): About one or
two days before a change in air pressure the aphids showed an
unusual activity, irrespective of whether the atmospheric pressure
changed to a high or to a low (Fig. 7). Some hours before a
thunderstorm they showed a decreased activity. It is suggested
that air ions are the causative agent (10). It has already been
demonstrated that reproduction and moulting rhythms of insects
may depend on air ionisation (7).

INTERACTION BETWEEN PEST CONTROL AND CLIMATIC FACTORS

Pesticide activity

Depending upon climatic factors pesticides may show different
degrees of effectiveness. They are much more degraded in strong
sunlight (photochemcial effects, especially by the shortwave
light) and at high temperatures. The activity of granulate
formation depends on moisture, as it must be dissolved to become
effective. In large parts of Germany the dry summer of 1976 led
to high losses in beet crops by beet yellowing. This was due to
the reduced effectiveness of granulated insecticides in controlling the aphid vectors.

Many pesticides are more active and spread more easily over the
plant at higher temperatures. In contrast to this, DDT shows
better effectiveness at lower temperatures. Insects ingest
different dosages of pesticides which vary with their feeding
activities at different temperatures. Sublethal dosages ingested
at low temperature may increase pesticide resistance which is
promoted under medium selection pressure.

Another problem arises with the development of insect pathogenic
viruses for control of insect pests. Virus particles are inact-
ivated by ultraviolet sunlight which is more intensive at low
latitudes. In many studies of this kind UV-protective ingredients
for virus preparations have been recommended. Brassel and Benz
(4) and Krieg et al. (11) studied the influence of ultraviolet

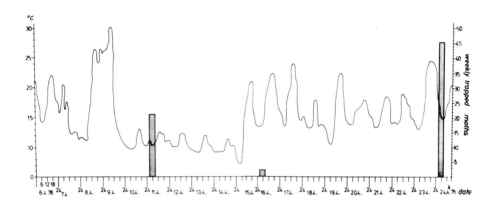

Fig. 6   Male codling moth, weekly baited by pheromon traps, in
          relation to temperature in early spring in the Gorgon
          district (near to Caspian Sea), Iran, after Bayat-Asadi
          (1).

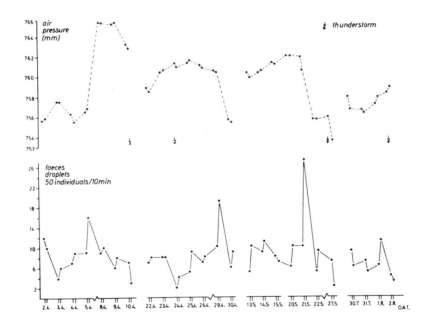

Fig. 7   Black bean aphid, Aphis fabae. Increased excretion
          activity one to two days before change of air pressure
          and decreased number of faeces droplets some hours before
          a thunderstorm, after Klingauf (10).

radiation on a granulosis virus to be used in codling moth control.
A comprehensive review of the influence of different environmental
factors on microbial control agents is given by Franz (5).

Pesticide resistance

Special problems in plant protection arise indirectly from the
close correlation of pest populations and climate. With high
generation numbers in warmer climates the chance for development
of pesticide resistance increases since continuous pesticide
applications favour the selection of a resistant strain. Moreover, a lot of pests and disease agents multiply asexually or
parthenogenetically under suitable weather conditions so that
their genes are not mixed. Thus, the properties of resistance
are directly transferred to the following generations and may
multiply in numbers. Only the sexual generation occurring in
autumn (fungi, aphids) offers a chance of mixing resistant and
susceptible individuals which may reduce or eliminate the
acquired degree of resistance.

Agricultural techniques

For many reasons, modern agricultural methods may promote pest
and parasite infestations, i.e., offering suitable microconditions.
High plant density in intensive cereal cultures, for example,
favour the reproduction of aphids and some fungal pathogens.

Not only pests but also defence (resistance) mechanisms of crops
are affected by agriculture techniques and climatic conditions.
In certain cases, a fast development of the crops may save them
from pest attack since older plants are often more resistant,
e.g., by the rigidity of their tissues. On the other hand, weak
tissues grown in high humidity and in the shade may more easily
fall prey to pests and pathogens. In other cases early
planting may decrease, or even avoid, infestation. This
method is practised by German farmers for broad beans to
prevent the black bean aphid attack (Aphis fabae).

The next chapter shows how known interrelations between pests,
pesticides and climatic factors can be used to good advantage
in crop production.

CLIMATE AND WEATHER AS A RESOURCE IN PLANT PROTECTION

Selection of growing areas

A classical method of prophylaxis in plant protection is to

grow plants in noninfested areas, this favours crops but not
their main pests. Since such regions are only available on a
very limited scale, this is feasible only in special cases, such
as the production of healthy seed or planting material which
lacks effective pest control methods. An example is the seed
potato production in the Netherlands where the danger of virus
infection is lower. At present virus diseases cannot be controlled directly. Rather the control is directed against the virus
vectors. The holocyclic green peach aphid, the most important
transmitter of potato viruses, is overwintering on peach trees
which are almost nonexistent in the Netherlands. Anholocyclic
races which may survive mild winters on different crops, especially cabbage, are rare in most parts of the Netherlands. Thus,
the aphid arrives in the seed potato area during its summer
migration only late in the season, and the virus infection may
be hindered by the destruction of the herbaceous parts or by
early harvesting depending upon the time of plant infestation.
A reliable forecast of the aphid flight is a prerequisite for
the exact timing of the control measures. Another example is
the relocation of crops to regions with less pest problems in
order to avoid or reduce pesticide contamination. This is the
case, for instance, with carrots grown for baby food in some
areas of southern Germany, where the beet fly, Psila rosae, does
not occur.

Forecasting and pest control

Many forecasting methods in plant protection are based on the
interaction of pest attack and weather conditions. Routinely
the plant protection service refers to temperature and humidity
in forecasting possible damages. For example, it is known that
mild winters can reduce the winter mortality of insects. Thus,
after moderate winters one of the first warnings for gardeners
and nurseries in the Rhineland refers to the picea-damaging
aphid Liosomaphis abietina which is easy to control if chemical
treatment is done in time.

Another example may demonstrate the importance of weather information in forecasting diseases. The infestation of wheat by
fungus Cercosporella herpotrichoides mainly depends on humidity
and temperature in autumn. The preferred range of temperature
for disease development is between 3 and 14°C. Temperatures of
3 and 4°C promote the formation of conidia, higher temperatures
with an optimum at 8 to 9°C favour the infestation. Thus, the
fungus finds best conditions of development if the temperatures
range between both optima for a certain period of time (for
infestation not less than 15 hours). The relative humidity -
measured 2m above ground - has to be 80% at a minimum. In
infested plants the fungus develops best at 10°C. Higher
temperatures, e.g., in spring and early summer, stop the develop-

ment. Since the incubation time takes 4 to 5 weeks and an
effective chemical control is possible only until approximately
40 days after infection, a weekly forecast based on temperatures
and humidity in springtime offers a good chance for timing the
control measures (21).

A simple method of forecasting the occurrence of insect pests in
springtime is to measure and to add up daily the mean temperatures which exceed the minimum temperature required for the development of a species. Investigations by Bayat-Asadi and Klingauf
(2) on codling moth may serve as an example. To calculate the
mean effective temperature the following formula after Wildbolz
(27) was used

$$\Sigma temp. = \frac{^{0}C \text{ at } 6 + 12 + 18 + 24 \text{ hours}}{4} - 10^{0}C \qquad (1)$$

($10^{0}C$ = minimum development temperature).

Table 2 gives the results for the main occurrence of the adult
moths after hatching from overwintering and the peaks of flight
of all generations within four years. The values varied more or
less from year to year within a certain range but showed distinct
differences between the generations. The lowest variance was
obtained for the activity peak of the first generation. Here
the flight maximum was nearly equal to the temperature sum of
about 200 which may be easily calculated from daily records of
spring temperatures.

Table 2   The relationship between peaks of male codling moths
           from pheromon traps (Gorgan, Iran) and the cumulative
           temperature obtained from equation 1 (2).

year	overwint.gen. date	$\Sigma^{0}C$	1st gen. date	$\Sigma^{0}C$	2nd gen. date	$\Sigma^{0}C$	3rd gen. date	$\Sigma^{0}C$	4th gen. date	$\Sigma^{0}C$
1976	7.5.	201	22.5.	349	15.7.	1065	25.8.	1712	10.9.	1965
1977	22.5.	198	7.5.	318	15.7.	1216	25.8.	1859	10.9.	2089
1978	24.4.	207	22.5.	422	16.7.	1119	6.8.	1490	10.9.	2478
1979	24.4.	219	29.5.	520	16.7.	904	7.8.	1244	10.9.	1927
$\bar{x}$		206		402		1076		1576		2115

Well established in agricultural practice is the forecast of
Phytophthora infestans. At the time when the fungus is a
severe threat the agro-meteorological service issues on a weekly
basis two critical data: 1. the date before the economic injury
level is reached, due to existing weather conditions, and when
fields have to be supervised to initiate control measures in the
case of overwhelming symptoms, and 2. the date when pesticide
applications become necessary since weather favours a quick dispersion of the disease. The Phytprog-service is based on a
method of Schrödter and Ullrich (17,18,20). For the calculation
of the data the following values are measured: (a) with respect
to infection or germination of the spores all wet periods of more
than 4 hours duration at temperatures of 10 to 12°C (humidity
above 90% or 0,1mm or more precipitation); (b) with respect to
formation of sporangia or fructification wet periods of more
than 10 hours duration at temperatures between 20 to 22°C;
(c) with respect to mycelium growth the hours with temperatures
of 15 to 20°C ignoring the wetness, and finally (d) with respect
to the inhibition of development by dry periods (which limit the
survival of the sporangia) the hours with humidity below 70%.
This complicated method provides an effective control mechanism
which enables the farmer to spray against the disease as late as
possible and at the right time, thus saving money for pesticides
and application procedures.

A similar concept was developed by Blaeser and Weltzien (3) for
controlling Plasmopara viticola (grape vine downy mildew).
Besides incubation periods the following factors play an important role in parasite development: 1. Infection may occur when
the combined effects of temperature (°C) and duration of wetness
(h) reach at least 50; 2. sporulation needs darkness, at least
13°C and 98% relative humidity in the canopy; and 3. viability
of sporangia (max. 6 h above 30°C) and dispersal of sporangia by
windblown rain. With a temperature-leaf-wetness-recorder the
climatic conditions can be recorded. Whereas usually the farmer
controlls the disease by routine spraying (5-8 applications at
the present time), this technique allows a proper pesticide
application resulting in a reduction of the number of treatments
(Fig. 8).

Towards an integrated management

Climatic factors also play an important role when beneficial
organisms are used in pest management systems. For instance,
the entomogenous nematode Neoaplectana carpocapsae strongly
depends on temperature and humidity. A temperature and humidity
forecast promotes nematode growth. Therefore, a good chance
exists for the use of the nematode to control pest insects, e.g.
in the favourable climate of the Nile Delta (Fig. 9).

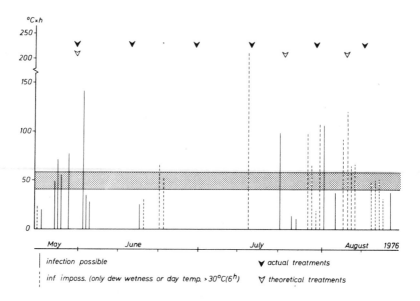

Fig. 8   Actual and theoretical treatments against the grapevine
           downy mildew, Plasmopora viticola, in relation to temp-
           erature and wetness (3).

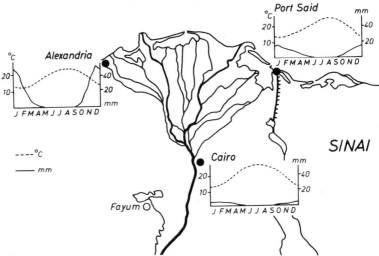

Fig. 9   Egyptian agricultural production regions and average
           monthly temperature and precipitation levels. Winter
           season in Northern Egypt offers favourable conditions
           for cotton leafworm control with the entomogenous
           nematode Neoaplectana carpocapsae (19).

Furthermore the following example may illustrate the importance of ecological factors in biological control. One of the most important pests in several greenhouse crops is the white fly Trialeurodes vaporariorum. Since chemical treatment has to be stopped some time prior to harvesting to avoid residues of pesticides, the insect is difficult to control. The life cycle of this species lasts 63 days at $16°C$ and only 23 days at $24°C$.

An effective parasitoid, Encarsia formosa, shows a life cycle of 90 days at $13°C$, 30 days at $18°C$, and only 17 days at $24°C$. At $18°C$ the period of development of both host and parasitoid is nearly the same but the fecundity of the white fly (oviposition rate) exceeds that of its parasitoid. At about $26°C$, however, reproduction of the parasitoid is twice as high as that of the host. Thus, a successful application of the parasitoids includes the management of a suitable temperature regime (8).

Modern plant protection is directed toward an integrated pest management involving all compatible methods of controlling pests and diseases. Besides the proper use of pesticides which avoids unnecessary applications and minimizes side-effects on man and ecosystems, the integrated concept propagates especially the use of all non-chemical control methods including improved agricultural techniques, plant resistance, and biological control. A better understanding of the role of climatic factors in pest attack may considerably promote the realization of an integrated pest management.

ACKNOWLEDGMENT. I am grateful to Dr. A.M. Huger for his help in preparing the manuscript.

References

1 Bayat-Asadi, H.: 1980, Biologie, Populationsdynamik und
  Bekämpfung des Apfelwicklers, Laspeyresia pomonella L., im
  Apfelanbaugebiet der Provinz Gorgan/Iran, Diss. Agricultural
  Faculty Univ. Bonn, pp 1-108.

2 Bayat-Asadi, H., and Klingauf, F., unpublished.

3 Blaeser, Marlene, and Weltzien, H.C.: 1979, Z. Pfl.krankh.
  u. Pfl.schutz 86, pp. 489-498.

4 Brassel, J., and Benz, G.: 1979, J. Invertebrate Path. 33,
  pp. 358-363.

5 Franz, J.M.: 1971, in Microbial Control of Insects and Mites,
  Burges, H.D., and Hussey, N.W., Eds., London New York, Academic Press, pp. 407-457.

6 Gutierrez et al.: 1970, in Climate and Food: Climate fluctuation and U.S. agricultural production, The Nat. Acad. Sci.,
  Washington, D.C.

7 Haine, E.: 1961, Nehmen luftelektrische Faktoren Einfluss
  auf den Aktivitätswechsel kleiner Insekten, insbesondere auf
  die Häutungs- und Reproduktionszahlen von Blattläusen?, For-
  schungsberichte des Landes Nordrhein-Westfalen, pp. 1-80.

8 Hassan, S.A., and Meyer E.: 1980, Auswertungs- und Informat-
  ionsdienst für Ernährung, Landwirtschaft und Forsten (AID),
  Bonn, Konstantinstr.124, no. 30.

9 Hirata, K.: 1966, Host Range and Geographical Distribution
  of the Powdery Mildews, Niigata Univ., Japan.

10 Klingauf, F.: 1976, Z. ang. Ent. 82, pp. 200-209.

11 Krieg, A., Gröner, A., Huber, J., and Matter, M.: 1980,
   Nachrichtenbl.Deut.Pflanzenschutzd., Braunschweig 32, pp.
   100-106.

12 Reichert, I.: 1953, 7th Proc. Int. Bot. Congr., pp. 730-731.

13 Reichert, I.: 1958, Trans. N.Y. Acad. Sci. (2) 20, pp. 333-339.

14 Reichert, I., and Palti, J.: 1966, 1st Proc. Congr. Mediterr.
   Phytopathol. Union, pp. 273-280.

15 Reichert, I., and Palti, J.: 1967, Mycopathol. Mycol. Appl.
   32, pp. 337-355.

16 Röttger, U.: 1977, Untersuchungen zur Wirtswahl der Rübenfliege, Pegomya betae Curt. (Diptera Muscidae), Diss. Agricultural Faculty Univ. Bonn, pp. 1-153.

17 Schrödter, H., and Ullrich, J.: 1965, Phytopathol. Zschr. 54, pp. 87-103.

18 Schrödter, H., and Ullrich, J.: 1966, Phytopathol. Zschr. 56, pp. 265-278.

19 Sikora, R.A., Salem, I.E.M., and Klingauf, F.: 1979, Med.Fac. Landbouww.Rijksuniv. Gent 44/1, pp. 309-321.

20 Ullrich, J., und Schrödter, H.: 1966, Nachr.Bl.Dt. Pflanzenschutzdienst, Braunschweig 18, pp. 33-40.

21 Unruh, M.: 1974, Anleitung für die Beobachter des Pflanzenschutz-Warndienstes, Schriftenreihe der Landwirtschaftskammer Rheinland, Bonn, Rhein. Landwirtschafts-Verlag G.m.b.H., pp. 1-213.

22 Weltzien, H.C.: 1967, Z. Pflanzenkr. (Pflanzenpathol.) Pflanzenschutz 74, pp. 175-189.

23 Weltzien, H.C.: 1972, Annu. Rev. Phytopathol. 10, pp. 277-298.

24 Weltzien, H.C.: 1973, Geogr. Z., Beih., Fortschr. Geomed. Forsch., pp. 110-114.

25 Weltzien, H.C.: 1978, in The Powdery Mildews, Spencer, D.M., Ed., London New York, Academic Press, pp. 39-49.

26 Weltzien, H.C.: 1978, in Plant disease, advanced treatise, 2, Horsfall, J.G., and Cowling, E.B., Eds., New York, San Francisco London, Academic Press, pp. 339-360.

27 Wildbolz, Th.: 1965, Schweiz. Zeitschrift für Obst- und Weinbau 101, pp. 572-579.

FOOD, ENERGY, AND CLIMATE CHANGE

David Pimentel

College of Agriculture and Life Sciences
Cornell University, Ithaca, New York   14853

ABSTRACT.   America and Europe are currently using about 17% of their
total energy for their food systems.   About 6% is used directly
for agricultural production.   In developing countries, the amount
of energy used in the food system ranges from 30 to 60%.   More
fossil energy will be expended in the future as we try to feed the
expanding world population.   As most of the arable cropland in the
world is already in production, the only way to increase crop yields
will be to use more fossil fuel in intensive management systems.
Climatic changes in the future will probably alter the relationship
of energy and land use in world food production.   The increased
burning of fossil fuels together with the trend to remove forests
and other vegetation for fuel, agriculture, housing and highways
will be responsible for the rising level of atmospheric CO_2.   The
projection is that the CO_2 level in the atmosphere will about double
by the middle of the next century.   For North America estimates are
that the temperature will rise about $2°$ C and rainfall in certain
regions will decline 10%.   These climatic changes will reduce wheat
and corn yields from 10-25%.   Forest growth should also be reduced.
Contributing to reduced yields will be more severe pest problems,
particularly insect pests caused by the warmer temperatures.   By
fertilization the increased CO_2 level should increase the growth
rate of plants.   Unfortunately, this fertilization probably will be
more than offset by the reduced plant growth rate due to high
temperatures, low moisture, and increased pest damage.   As a result,
the increased levels of CO_2 will probably reduce food and forestry
production in North America.

W. Bach, J. Pankrath, and S. H. Schneider (eds.), Food-Climate Interactions, 303–323.

INTRODUCTION

The explosive growth of the human population and the stress
it imposes on the earth's resources means that the earth is losing
its capacity to provide adequate supplies of food and the other
resources essential for man's survival and a quality life (1).

One can not help speculating how large the world supply of
food will have to be to feed the world population in the years
2000 and 2100. Can mankind provide his numbers with a nutritionally satisfactory diet if, as projected, the world population
escalates to about 6.5 billion (10^9) by 2000 and 10-12 billion by
the year 2100? We know that food supplies are influenced by such
factors as arable land, water, climate, fertilizer, and fossil
energy (2,3). Food supplies are also affected by the status of
public health, losses due to pests, availability of labor, the
extent of environmental pollution, and the lifestyle desired by
the people.

Although these factors are important and are interdependent
variables in the food system, it is nearly impossible to consider
the impact of all factors as they function together in one worldwide system. Therefore, only the interactions of energy, food,
land, water, and climate will be discussed.

ENERGY USE IN AGRICULTURE

Mankind's ability to use fossil energy to manipulate and
manage ecosystems is undoubtedly one of the most important reasons
for rapid population growth in the world (Figure 1). Energy has
been used extensively to raise food and protect man from deadly
diseases. At present, each American and European requires about
1,500 liters of oil to produce, process, distribute, and prepare
his food each year. In the United States the amount of energy used
to supply food represents about 17% of the total energy expended
(1). Of this about 6% is used in agricultural production, while
food processing and packaging, transport, storage, and preparation
combined use about 11%.

Thus, in an industrialized nation like the United States,
fossil energy has become as vital a resource for crop and livestock production as the availability of fertile land and ample
supplies of water. The dominant uses of energy in the U.S. agricultural system are fuel to run farm machinery and to produce
fertilizers and pesticides (Table 1). Nitrogen fertilizer is made
primarily from natural gas while pesticides are made primarily
from petroleum.

As a result, approximately 600 liters of gasoline equiva-

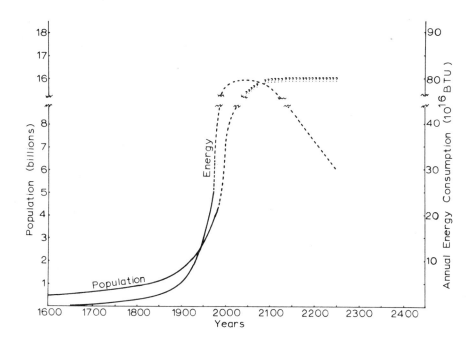

Figure 1.   Estimated world population numbers ·(———)· from 1600 to
            projected numbers (————) (????) to the year 2250.
            Estimated fossil fuel consumption (———) from 1650 to
            1975 and projected (————) to the year 2250 (4,5).

lents are required to raise a hectare of a typical crop like corn.
This amounts to an expenditure of about 1 liter of gasoline per 9
kg of corn produced or about 3 calories of corn for each calorie
of fossil energy input (Table 1).   On average grains produced in
the United States yield from 2 to 3 calories of grain per fossil
energy calorie expended (Table 2).

     Producing other kinds of food products, however, is not as
energy efficient as grain production.   For example, in apple and
orange production, about 2 calories of fossil energy are expended
per 1 calorie of fruit produced (Table 2).   Culturing vegetables
varies from 1 to 5 calories input per calorie produced.

TABLE 1. ENERGY INPUTS PER HECTARE IN U.S. CORN PRODUCTION IN
          1975 (1).

	QUANTITY/HA	KCAL/HA
INPUTS		
Labor	12 hrs	5580
Machinery	31 kg	558,000
Diesel	112 litres	1,278,368
Nitrogen	128 kg	1,881,600
Phosphorus	72 kg	216,000
Potassium	80 kg	128,000
Limestone	100 kg	31,500
Seeds	21 kg	525,000
Irrigation	780,000 kcal	780,000
Insecticides	1 kg	86,910
Herbicides	2 kg	199,820
Drying	426,341 kcal	426,341
Electricity	380,000 kcal	380,000
Transportation	136 kg	34,952
Total		6,532,071
OUTPUTS		
Corn yield	5394 kg	19,148,700
kcal output/kcal input		2.93
Protein yield	485 kg	

Although fruits and vegetables require larger energy inputs
per food calorie than grain, neither are as energy-expensive as
producing calories in the form of animal protein. From 10 to 90
calories of fossil energy are required to produce 1 calorie of
animal protein. The major reason that animal protein products are
significantly more energy-expensive than plant protein foods is that
first the forage and grains have to be grown and then consumed by
the animal, which is eventually used as food. In addition, the
forage and feed that maintain the breeding herd are additional
energy costs. For example, about 1.3 head of breeding cattle have
to be maintained to produce one feeder calf per year. Of great
concern is that many of the grains fed animals are entirely suitable
for human consumption. Indeed, in the United States today 90% of
the grain produced is cycled through livestock to produce milk,
eggs, and meat (6).

TABLE 2. ENERGY INPUTS AND RETURNS FOR VARIOUS FOOD AND FEED CROPS PRODUCED PER HECTARE IN THE UNITED STATES (1).

Crop	Crop Yield(kg)	Yield in Protein (kg)	Crop Yield in Food Energy (10^6 kcal)	Fossil Energy Input for Production (10^6 kcal)	Kcal Food/Feed Output/ kcal Fossil Energy Input	Labor Input Manhours
Corn	5,400	485	19.1	6.5	2.9	12
Wheat	2,060	247	6.8	2.8	2.4	7
Oats	1,730	242	6.7	2.2	3.1	6
Rice	6,160	462	22.4	14.4	1.6	17
Sorghum	3,030	344	10.5	5.4	2.0	12
Soybean	1,880	640	7.6	1.8	4.2	10
Beans, dry	1,460	325	5.0	2.7	1.8	10
Peanuts	3,720	320	15.3	10.9	1.4	19
Apples	17,920	36	9.6	18.0	0.5	175
Oranges	19,040	193	6.8	18.3	0.4	173
Potato	34,380	722	19.7	16.0	1.2	35
Spinach	11,200	358	2.9	12.8	0.2	56
Tomato	49,620	496	9.9	16.6	0.6	165
Brussels Sprouts	12,320	604	5.5	8.1	0.7	60
Alfalfa	6,830 1/	1,127	15.4	2.5	6.2	13
Tame Hay	5,000 1/	200	8.6	1.7	5.0	16
Corn Silage	31,020	393	25.3	6.3	4.0	15

1/ dry.

TABLE 3. ENERGY INPUTS AND RETURNS PER HECTARE FOR VARIOUS LIVESTOCK PRODUCTION SYSTEMS IN THE UNITED STATES (7).

Livestock	Animal Product Yield(kg)	Yield in Protein(kg)	Protein as Kcal(10^3)	Fossil Energy input for production (10^6 kcal)	Kcal Fossil Energy input/ kcal protein output	Labor input manhours
Broilers	2008	186	744	7.3	9.8	7
Eggs	910	104	416	7.4	17.8	19
Pork	490	35	140	6.0	42.9	11
Sheep (grass-fed)	7	0.2	0.8	0.07	87.5	0.2
Dairy	3270	114	457	5.4	11.8	51
Beef	60	6	24	0.6	25.0	2
Dairy (grass-fed)	3260	114	457	3.3	7.2	50
Beef (grass-fed)	54	5	20	0.5	25.0	2
Catfish	2783	384	1536	52.5	34.2	55

If soybean protein production per hectare is compared to that of pork, about 20 times more protein is produced raising soybeans than producing pork (Tables 2 and 3). Note also that potatoes yield relatively large quantities of protein per unit area. Overall, animal protein products require significantly larger inputs of land and energy than plant proteins.

THE EFFECT OF CO_2 LEVELS ON CLIMATE AND THE ENVIRONMENT

Assessing the impact of increased CO_2 levels is difficult because we do not fully understand the quantitative aspects of the "sources and sinks of carbon and the transfers of carbon between the atmospheric, biospheric, and oceanic reservoirs" (8).

Nonetheless, we know the burning of fossil fuels and organic matter increases the levels of atmospheric CO_2. Projections are that levels of CO_2 may rise from 300 to 400 ppm by 2000, and reach 600 ppm by 2500 (9).

The management of the CO_2 problem is probably impossible even if all the sources and sinks were understood. This is particularly true if the major contributions of CO_2 to the atmosphere are caused by burning fossil fuels and a reduction in the forests of the earth (10, 11). The rapidly growing human population and extensive poverty in most of the world undoubtedly will necessitate the greater use of fossil fuels to feed the people and improve the quality of life, especially of the poor. The clearing of forest areas for agricultural land and for fuel wood will exacerbate the problem.

In addition to identifying the causes of CO_2 buildup and estimating the extent of the increase, the probable effect this will have on such environmental factors as climate and CO_2 fertilization must be anticipated. These will have profound influences on agricultural productivity.

CARBON CYCLING IN U.S. AGRICULTURE AND FORESTRY

The amount of fossil energy expended in the United States annually is 18×10^{15} kcal (Table 4). This releases an estimated 1.3 billion tonnes of carbon annually into the atmosphere.

The 1.3 billion tonnes of carbon released from fossil fuel consumption is nearly equal to the amount of carbon that is cycled annually in the biotic community of the United States (Table 5). This assumes that the amount of carbon fixed in plant productivity equals the amount released by the biotic community.

Table 4.   Consumption of fossil energy in the U.S. and the World (12).

World				U.S.			
Fuel	Amount	Kcal x 10^{15}	Carbon t x 10^6	Fuel	Amount	Kcal x 10^{15}	Carbon t x 10^6
Coal	2.5 x 10^9 tonnes	17	1263	Coal	0.5 x 10^9 tonnes	3.5	246
Oil	21.7 x 10^9 bbls	30	2228	Oil	6.8 x 10^9 bbls	9.5	706
Gas	47.8 x 10^{12} ft^3	15	1114	Gas	18.8 x 10^{12} ft^3	5.0	372
TOTAL		62	4605	TOTAL		18.0	1318

The situation, however, is not that simple. For instance, the annual productivity of the plant community may be increasing because of the increased CO_2 level in the atmosphere (13). When CO_2 levels are increased, the growth of crops and other plants is significantly greater than normal. Balanced against this is the fact that in the United States farm land is being decreased at an alarming rate. Just from 1945 to 1975 about 30 million hectares, an area about equal to the size of Ohio and Nebraska combined, has been covered with roads and housing (14). Eventually this trend can be expected to result in a reduced plant community. Which of these two alternatives will have a greater impact on food supplies remains to be documented.

In the United States about 85% of the carbon that is fixed as biomass in plants is found in cultivated crops and forests (Table 5). Of this the bulk is in our standing forests, indicating the vital role of forests. Little carbon is associated with the biomass of other vegetation or with the standing biomass of micro-organisms and animals (Table 5).

The atmosphere and oceans appear to be the sinks for most of the CO_2 being released from the burning of fossil fuels used in agriculture and the burning of biomass. Clearly the United States significantly contributes to the atmospheric buildup of CO_2 in the world, as do other highly industrial nations, because of the high level of fossil energy consumption in these nations.

CARBON CYCLING IN WORLD AGRICULTURE AND FORESTS

Annually an estimated 62×10^{15} kcal of fossil energy is used in the world (Table 4). This releases nearly 5 billion tonnes of carbon into the atmosphere. This represents nearly 1/5th the total amount of carbon taken up annually in plant biomass productivity in the world (Tables 4 and 6). The world situation differs from that in the United States where at present the amount of carbon released by burning fossil fuels about equals the amount of carbon used annually by plant biomass productivity (Tables 4 and 5).

About 2/3rds of the carbon used by the world plants is used by agricultural and forest plants (Table 6). As in the United States, the largest amount of carbon is in the forests. The amount of carbon held in the standing biomass of other organisms including microorganisms is extremely small or less than 1% (Table 6).

Previous estimates of annual plant productivity and carbon cycling appear to be high. For example, it is reported that forest productivity of biomass in the temperate zone averages about 10 t/ha per year (15,16). Foresters that were consulted give estimated the total world biomass production was calculated to be 63×10^9 t per year (Table 6). This is substantially lower than the previous estimate of about 177×10^9 t.

The amount of carbon held in humus in the world is estimated to be 2.2×10^{12} tonnes (Table 7). This is nearly 6 times the amount of carbon present in the total standing biomass of the world (Tables 6 and 7). Thus, the total carbon stored in the biosphere is calculated to be 2.5×10^{12} tonnes.

If, as suggested by several workers (10,11,19,20), the release of carbon from the biosphere (including humus) about equals the carbon released by consumption of fossil fuels, then each year only 0.2% of the carbon would have to be released from the biosphere to equal that released from fossil fuels. This appears to be a reasonable figure because land continually is being cleared for highways, housing, and agriculture (2) and forests are being cut for lumber and fuel wood (2,11,21).

TEMPERATURE AND RAINFALL CHANGES

Temperature and rainfall trends on the eastern U.S. seacoast from 1738 to 1975 do not suggest any major departures from the mean (Figure 2). However, temperature conditions were warm for the late 18th century; cool for the 19th century; and warm for the first half and for the past 7 years of the 20th century. The warming trend thus far in this century is about $2\,^\circ C$.

TABLE 5. ESTIMATED ANNUAL CARBON CYCLED IN PLANT BIOMASS AND CARBON IN THE STANDING PLANT BIOMASS AND OTHER ORGANISM BIOMASS IN THE UNITED STATES.

Terrestrial	Area in Hectares 10^6	Annual Plant Biomass Productivity			Plants t/ha	Total Carbon in Plants t x 10^6	Total Carbon in Plants t x 10^6	Standing Biomass		
		Biomass t/ha	Total Biomass t x 10^6	Total Carbon t x 10^6				Other Organisms t/ha	Total Other Organisms t x 10^6	Total Carbon in Other Organisms t x 10^6
Farmland										
Cropland	135	6	810	365	2	270	121.5	0.6	81.0	36.5
Cropland, idle	21	4	84	38	2	41	18.5	0.6	12.6	5.7
Cropland in pasture	36	4	144	65	6	216	97.2	0.4	14.4	6.5
Grassland in pasture	183	3	549	247	6	1098	494.1	0.4	73.2	32.9
Forest & Woodland	45	3	135	61	110	4950	2227.5	0.4	18.0	8.1
Farmsteads, roads	11	0.1	1	>0	2	22	9.9	0.4	4.4	2.0
Other										
Grazingland	117	2	234	106	4	468	210.5	0.3	35.1	15.8
Forestland	202	3	606	273	110	22,220	9999.0	0.4	80.8	36.4
Other land (urban, marshes desert, etc.)	167	0.1	17	8	5	835	375.8	0.1	16.7	7.5
Total	917	---	2,580	1,163						
Aquatic										
Lakes and Rivers	132	3	396	178	0.1	13.2	5.9	0.3	39.6	17.8
Grand Total	1,049	---	2,976	1,341		30,134.2	13,559.7		375.8	169.2

TABLE 6. ESTIMATED ANNUAL CARBON CYCLING AND CARBON IN THE STANDING PLANT BIOMASS AND OTHER ORGANISMS BIOMASS IN THE WORLD.

	Area in Hectares 10^9	Annual Plant Biomass Productivity			Plants t/ha	Total Plants t x 10^9	Total Carbon in Plants t x 10^9	Standing Biomass		
		Biomass t/ha	Total Biomass t x 10^9	Total Carbon t x 10^9				Other Organisms t/ha	Total Other Organisms t x 10^9	Total Carbon in Other Organisms t x 10^9
Cropland (temperate)	0.75	6	4.5	2.03	1	0.75	0.34	0.6	0.45	0.02
Cropland (tropical)	0.75	6	4.5	2.03	1	0.75	0.34	0.5	0.39	0.18
Grassland Pastures (temperate)	0.90	3	2.7	1.22	6	5.4	0.24	0.3	0.27	0.12
Grassland Pastures (tropical)	1.5	3.5	5.25	2.36	7	10.5	4.73	0.1	0.15	0.07
Rangeland (scrub)	2.6	2	5.2	2.34	15	39	17.55	0.1	0.26	0.12
Forests (temperate)	3.2	3	9.6	4.32	110	352	158.4	0.4	1.28	0.58
Forests (tropical)	2.4	4	9.6	4.32	160	384	172.8	0.2	0.48	0.22
Other	2.4	0.5	1.2	0.54	2	1	0.45	> 0	> 0	> 0
Freshwater										
Lake and Stream	0.2	3	0.6	0.27	0.1	0.02	0.009	0.3	0.06	0.03
Swamp and Marsh	0.2	5	1.4	0.45	40	0.8	0.36	0.5	0.10	0.05
Continent Total	14.9		44.15	19.88		794.22	355.219		3.44	1.39
Ocean	35.9	0.5	17.95	8.08	0.01	0.36	0.16	0.05	1.80	0.81
Estuaries	0.2	4	0.8	0.36	2	0.4	0.18	0.40	0.08	0.04
Marine Total	36.1		18.75	8.44		0.76	0.34		1.88	0.85
Grand Total	51.0		62.90	28.28		794.98	355.559		5.32	2.24

TABLE 7. ESTIMATED CARBON IN HUMUS MASS IN THE WORLD[1].

	Area in Hectares 10^9	Humus t/ha	Total Humus t x 10^9	Total Carbon t x 10^9
Cropland (temperate)	0.75	260	195	88
Cropland (tropical)	0.75	140	105	47
Grassland Pastures (temperate)	0.90	400	360	162
Grassland Pastures (tropical)	1.5	80	120	54
Rangeland (scrub)	2.6	30	78	35
Forests (temperate)	3.2	240	768	346
Forests (tropical)	2.4	200	480	216
Tundra	0.8	440	352	158
Other	1.6	100	160	72
Freshwater				
Lake and Stream	0.2	400	80	36
Swamp and Marsh	0.2	1,400	280	126
Continent Total	14.9		2,978	1,340
Ocean	35.9	50	1,795	808
Estuaries	0.2	200	40	18
Marine Total	36.1		1,835	826
Grand Total	51.0		4,813	2,166

1/ Most of the humus estimates for the terrestrial habitats
were based on a survey by Schlesinger (18).

In general higher rainfall is associated with cool weather
(Figure 2)(24). Precipitation was generally high from 1940 to 1960.
Since then the trend in rainfall appears to be downward, with
estimates for a 15% decrease (25). Note the extreme fluctuations
in rainfall during the last 10 years, when both the lowest and highest annual rainfall ever recorded occurred (Figure 2).

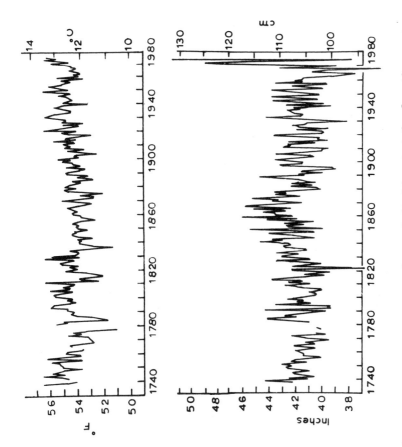

Figure 2.   Annual average temperatures and precipitation totals for the eastern seaboard of the United States for the period 1738 to 1975. The data are from a representative, reconstructed series centered on Philadelphia from Landsberg (22) for 1738 to 1967 and from 1967 to 1975 for the Philadelphia Station (23).

No one knows what the forced climatic changes will be from
the increase in CO_2 in the atmosphere. If there is substantially
less rainfall and warmer temperatures, then agricultural and forestry production will be adversely affected. Most agricultural
crop plants and forest trees need large quantities of water for
growth. An acre of corn, for example, requires about 500,000 gallons (1 gal = 3.79 liters) of water just for transpiration and an
acre of rice requires 1.5 million gallons per growing season (26).

Each year, of all the water that reaches the streams in the
United States (1260 billion gallons per day (bgd)), one-fourth, or
420 bgd is withdrawn. And of this only about 95 bgd is consumed
(27). Agriculture accounts for the use of 83% of the consumed
water whereas industry and urban areas together consume less than
17%.

Although industrial and urban areas return most of the used
water to streams and lakes, much is polluted when it is returned.
Agriculture on the other hand uses most of the water and does not
return it to streams and lakes. Thus it is obvious that a significant decrease in available water has many far-reaching effects.

For example, in 17 western states water is already in short
supply and about 85% of the water consumed in those arid regions
is for irrigation. If water supplies decrease, then conflicting
needs for the available supplies among agriculture, urban and
industrial use as well as fossil energy mining are inevitable.
One of the strongest competitors for water in certain parts of the
western United States will be coal and synfuel production (28,29).
Among the four competing groups, evidence suggests that when in
short supply the proportion of water allotted to agriculture will
decline (30). This is because the economic yields from agriculture at present are far less than the yields from industry, mining,
and some recreation.

POTENTIAL IMPACT OF INCREASED ATMOSPHERIC CO_2 ON CLIMATIC PATTERNS

The projected doubling of the CO_2 content of the atmosphere
by the middle of the next century is expected to influence both
temperature and rainfall patterns but exactly how this forced atmospheric change influences climate is not known. If temperatures
in North America, as projected rise at least 2^0 C (31), then this
warming trend might affect rainfall patterns and result in drier
conditions.

The combined effect of warm temperatures and reduced rainfall
could result in low water levels in streams and lakes. In addition,
the warm temperatures will increase the evapotranspiration rates
in plants. Thus, the warm temperature and reduced rainfall should
have a detrimental effect on agricultural and forest production.

CROP PRODUCTION SUSCEPTIBLE TO CLIMATIC CHANGE

Let us suppose that a 2° C rise in temperature occurs by the middle of the next century and also that rainfall declines about 10% (other possible changes include an altered yearly rainfall distribution pattern and/or a rise in evaporation rate that might result in moisture stress). Temperature and moisture changes of the type mentioned might be expected to have an adverse impact upon food production.

With corn, for example, estimates are that a 2° C rise and 10% drop in rainfall would result in a 25% reduction in corn yield in the major corn belt states of Indiana, Iowa, Missouri, and Illinois (32). Certainly, reducing corn yields from about 6,272 kg/ha (100 bu/A) to about 4,704 kg/ha (75 bu/A) would be significant and have a major impact on the U.S. food supply.

Such a change in temperature and rainfall patterns could be expected to reduce wheat yields by about 12% in the four major producing states of North Dakota, South Dakota, Kansas, and Oklahoma (33). This amounts to a reduction of wheat from about 1,610 kg/ha to about 1,410 kg/ha and because wheat is one of the major crops of the United States, this decreased yield would have an immense impact on the nation's food system.

The incidence of pest loss would also be affected by an increase in temperature and decline in rainfall. At present annual losses in agriculture and forestry caused by pests are significant: about 37% for agriculture and 25% for forestry (34,35). Losses of this magnitude are occurring in spite of all chemical and nonchemical pesticidal controls now in use.

If temperatures in the United States possibly increase an average of 2° C and rainfall declines 10%, the pattern and extent of pest outbreaks can be expected to change and these in turn could affect the agricultural and forest production.

In general, insect pest populations increase with a rise in temperature. Some insect pests, for example, produce 500 offspring per female and go through a generation in 2 weeks. With a warmer and longer growing season these pests may pass through an additional 1 to 3 generations. The exponential increase of some insect pest populations under the more favorable environment could drastically increase insect numbers and make their control increasingly difficult and more expensive.

Another factor contributing to insect outbreaks would be changes in winter temperatures and snow cover. If overwintering temperatures were warmer, then insect mortality during the winter would be less and the initial spring populations would be larger.

This would result in larger summer populations on the crop, because the initial base populations starting in the spring would be larger than those exposed to the more usual, severe winters.

On the other hand, reduced snowfall over winter would have an opposite effect on the insect populations. Reduced snow cover could result in the exposure of larger numbers of overwintering insects to freezing conditions and thereby reduce surviving populations.

Warm/dry conditions in combination with increased CO_2 fertilization of crops and forests may alter the nutrient makeup of the plants. If altered, some plants may become more attractive to certain insects with the result that the rate of increase in the pest populations would be increased. Thus, the warm/dry conditions plus increased CO_2 fertilization from atmospheric supply may increase insect pest outbreaks.

In addition, the higher temperatures may also reduce the effectiveness of insecticides against insect pests. At high temperatures some insects are better able to detoxify insecticides and thus are able to escape control.

Increasing the temperature while reducing moisture levels will favor certain species of weeds whereas others will decline (36). Although a change in weed species should not necessarily increase crop losses to weeds, the climatic change mentioned (less water) in combination with new weed pests might result in greater crop losses. The warm/dry conditions, for example, will probably increase the intensity of competition between weeds and crops for the limited available moisture.

Most important perhaps would be the reduced effectiveness of herbicides on weeds under the warm/dry conditions. Under dry conditions not only is the uptake of herbicides reduced but the toxic effects of herbicides are also reduced because the physiological activity in weeds is suppressed.

If this happened, weed control would depend more heavily upon mechanical cultivation than at present and would require added machinery and labor. The result would be an increase in fossil fuel burned and an increase in crop production costs.

In general under warm/dry conditions, plant pathogens should be less of a problem. Therefore, losses from plant diseases should be reduced with the projected climatic change.

Currently, about 13% of U.S. agriculture is irrigated (37) and if rainfall did decline 10%, it is logical to expect that more irrigation would be necessary to culture arid lands. The deciding factors are priorities of other segments of the economy as the energy costs and benefits are weighed (1).

The energy supply and how it relates to irrigation is well illustrated with the U.S. corn crop. About 14 million liters or 14 metric tonnes of water are needed to produce about 5000 kg of corn per hectare under arid conditions (38). The energy cost to pump this water from a depth of about 100 meters is about 21 million kcal. If we use irrigation, this increases the energy use in crop production about 400% (Table 1).

Clearly, a 400% increase in energy use in agricultural production will adversely affect the cropping system in the United States. Already in parts of the United States some low value crops such as alfalfa can no longer be irrigated (39). The reason for the problem with low value crops is the large amount of water that is required for irrigation and the current high price of energy. This situation is expected to worsen as fuel prices rise. Hence, irrigation as a means of offsetting reduced rainfall does not appear to be an encouraging alternative.

CO_2 FERTILIZATION OF CROP PLANTS

Under optimal temperature and moisture conditions, increasing the CO_2 content of the atmosphere increases plant growth. Based on data assembled by Wittwer (13,40), leaf photosynthesis increases from 40 to 135% with elevated CO_2 levels. Other studies that have included some natural environmental stress have projected much lower growth responses in crop plants (41).

The major question to answer is: will the rise in CO_2 level increase crop yields sufficiently to offset the projected decline in crop yields caused by the temperature rise and rainfall decline? It is doubtful whether the CO_2 fertilization will be sufficient to offset the reduced growth and production of crops due to changes in average temperatures and rainfall.

CONCLUSION

The American and European societies are currently utilizing about 17% of their total energy for their food system. About 6% is used directly for agricultural production. In developing countries, the amount of energy used in the food system ranges from 30 to 60%.

Because the world population continues growing rapidly and even now one-half billion humans are malnourished, more fossil energy will be expended as we endeavor to feed the expanding world population. This is especially true since most of the arable cropland in the world is already in production. The only way to increase yields on the limited land is to manage it more intensely and this means a greater use of fossil fuels.

Climatic changes in the future will probably alter the relationship of energy and land use in world food production.

The increased burning of fossil fuels together with the trend to burn forests and other vegetation for fuel, agriculture, housing, and highways, will be responsible for the rising level of atmospheric CO_2. The projection is that the CO_2 level in the atmosphere will about double by the middle of the next century.

If temperature in North America rose about 2° C and the associated rainfall declined by an estimated 10%, these climatic changes might reduce crop yields anywhere from 10 to 25% as measured by wheat and corn crops. Forest growth might also be reduced. Contributing to the reduced crop and forest yields might be more severe pest problems, particularly insect pests caused by the warmer temperatures.

The increased CO_2 level might possibly, through fertilization, raise the growth rate of plants. This fertilization probably will not offset the reduced plant growth rate due to high temperatures, low moisture, and increased pest damage. As a result, the increased levels of CO_2 might reduce food and forestry production in North America.

REFERENCES

1. Pimentel, D. and Pimentel, M.: 1979, *Food, Energy and Society*, Edward Arnold, London.

2. Council on Environmental Quality: 1980, *The Global 2000 Report to the President*, Vol. II, The Technical Report, U.S. Government Printing Office, Washington, D.C.

3. Kellogg, W.W., Schware, R., and Friedman, E.: 1980, *The Earth's Climate*, The Futurist, October, pp. 50-55.

4. Environmental Fund: 1979, *World Population Estimates*, Washington, D.C.

5. Linden, H.R.: 1980, *Importance of Natural Gas in the World Energy Picture*, paper presented at International Institute of Applied Systems Analysis Conference on Conventional and Unconventional World Natural Gas Resources, Laxenburg, Austria, June 30–July 4.

6. Pimentel, D., Oltenacu, P.A., Nesheim, M.C., Krummel, J., Allen, J.C., and Chick, S.: 1980, *Grass-fed Livestock Potential: Energy and Land Constraints*, Science 207, pp. 843–848.

7. Pimentel, D., ed.: 1980, *Handbook of Energy Utilization in Agriculture*, CRC Press, Boca Raton, Florida.

8. Williams, J., ed.: 1978, *Carbon Dioxide, Climate and Society*, Pergamon, New York.

9. American Association for the Advancement of Science: 1980, *Workshop on Environmental and Societal Consequences of a Possible CO_2-Induced Climate Change*, Annapolis, Maryland, April 2–6, 1979, USDOE Contract No. AS01-79EV10019.

10. Woodwell, G.M.: 1978, *The Carbon Dioxide Question*, Sci. Am. 238, pp. 34–43.

11. Tolba, M.K.: 1980, *The State of the World Environment 1980*, The 1980 Report of the Executive Director of the United Nations Environment Programme.

12. United States Bureau of the Census: 1979, *Statistical Abstract of the United States 1979*, U.S. Government Printing Office, Washington, D.C.

13. Wittwer, S.H., 1980: "Carbon Dioxide Fertilization of Crop Plants," in *Workshop on Environmental and Societal Consequences of a Possible CO_2-Induced Climate Change*, Annapolis, Maryland, April 2–6, 1979.

14. Pimentel, D., Terhune, E.C., Dyson-Hudson, R., Rochereau, S., Samis, R., Smith, E., Denman, D., Reifschneider, D., and Shepard, M.: 1976, *Land Degradation: Effects on Food and Energy Resources*, Science 194, pp. 149–155.

15. Lieth, H.: 1975, "Primary Production of the Major Vegetation Units of the World" in H. Lieth and R.H. Whittaker, eds., *Primary Productivity of the Biosphere*, Springer-Verlag, New York, pp. 203–215.

16. Lieth, H.: 1978, "Vegetation and CO_2 Changes," in J. Williams, ed., *Carbon Dioxide, Climate and Society*, Pergamon, Oxford, pp. 103–109.

17. L. Tombaugh, Michigan State University; R. Morrow, Cornell University, R.E. Phares, U.S. Forest Service: 1979, personal communications.

18. Schlesinger, W.H.: 1977, *Carbon Balance in Terrestrial Detritus*, Ann. Rev. Ecol. Syst. 8, pp. 51-81.

19. Wong, C.S.: 1978, *Atmospheric Input of Carbon Dioxide from Burning Wood*, Science 200, pp. 197-200.

20. Adams, J.A.S., Mantovani, M.S.M., and Lundell, L.L.: 1977, *Wood Versus Fossil Fuel as a Source of Excess Carbon Dioxide in the Atmosphere: A Preliminary Report*, Science 196, pp. 54-56.

21. Eckholm, E.P.: 1976, *Losing Ground*, Norton, New York.

22. Landsberg, H.E.: 1970, *Man-Made Climatic Changes*, Science 170, pp. 1265-1274.

23. National Oceanic and Atmospheric Administration: 1968-75, *Climatological Data*, Vols. 19-26, Environmental Data and Information Service, National Climatic Center, Asheville, N.C.

24. Katz, R.W.: 1977, *Assessing the Impact of Climatic Change on Food Production*, Climatic Change 1(1), pp 85-96.

25. Thompson, L.M.: 1975, *Weather Variability, Climatic Change and Grain Production*, Science 188, pp. 535-541.

26. Penman, H.R.: 1970, *The Water Cycle*, Sci. Am. 223(3), pp. 99-108.

27. Murray, C.R., and Reeves, E.B.: 1977, *Estimated Use of Water in the United States in 1975*, U.S. Geological Survey Circular 765.

28. McCaull, J: 1974, *Wringing out the West*, Environment 16(7), pp. 10-16.

29. Office of Technology Assessment: 1980, *Energy from Biological Processes*, OTA-E-124, Washington, D.C.

30. Gertel, K. and Wollman, N.: 1960, *Rural-Urban Competition for Water: Price and Assessment Guides to Western Water Allocation*, J. Farm Econ. 42(5), pp. 1332-44.

31. U.S. Environmental Protection Agency: 1978, *Research Outlook 1978*, Research and Development, EPA 600/9, 78-001, June.

32. Benci, J.F., Runge, E.C.A., Dale, R.F., Duncan, W.G., Curry,
     R.B., and Schaal, L.A.: 1975, "Effects of Hypothetical Climat-
     ic Change on Production and Yield of Corn:, in *Impacts of
     Climatic Change on the Biosphere*, CIAP Monograph 5, Part 2,
     Climatic Effects, U.S. Department of Transportation, Washing-
     ton, D.C., pp. 4-3 to 4-36.

33. Ramirez, J.M. and Sakamoto, C.M.: 1975, "Wheat", in *Impacts
     of Climatic Change on the Biosphere*, CIAP Monograph 5, Part 2,
     Climatic Effects, U.S. Department of Transportation, Washing-
     ton, D.C., pp. 4-37 to 4-90.

34. Dasmann, R.F.: 1972, *Environmental Conservation*, 3rd Ed.,
     John Wiley, New York.

35. Pimentel, D., ed: 1980, *CRC Handbook of Pest Management*, CRC
     Handbook Series in Agriculture, CRC Press, Boca Raton, Florida,
     in press.

36. Waggoner, P.: 1979, Connecticut Agricultural Experimental
     Station, personal communication.

37. U.S. Department of Agriculture: 1977, *Agricultural Statistics
     1977*, U.S. Government Printing Office, Washington, D.C.

38. Addison, H.: 1961, *Land, Water and Food*, Chapman and Hall Ltd.,
     London.

39. Larson, D.L. and Fangmeier, D.D.: 1977, "Energy Requirements
     for Irrigated Crop Production," in R.A. Fazzolare and C.B.
     Smith, eds. *Energy Use Management*, Pergamon, New York, pp.
     745-750.

40. Wittwer, S.H.: 1980, *Carbon Dioxide and Climatic Change: An
     Agricultural Perspective*, J. Soil Water Conserv. 35, pp. 116-
     120.

41. Lemon, E.: 1977, "The Land's Response to More Carbon Dioxide",
     in N.R. Andersen and A. Malahoff, eds., *The Fate of Fossil
     Fuel CO_2 in the Oceans*, Plenum, New York, pp. 97-130.

TOWARDS A CONSERVATION STRATEGY TO RETAIN WORLD FOOD AND
BIOSPHERE OPTIONS

Gerrit P. Hekstra

Ministry of Public Health and Environmental Protection,
Leidschendam, Netherlands.

ABSTRACT. This paper challenges widely held views (World Bank,
UNDP, FAO), that even a growing world population could be fed
many times over (Cornucopia) by increasing capital and technology investments in agriculture (Green Revolution). The fact is
that fertile soils are lost more rapidly than new arable land
is being gained. Monopolization of the seed trade by chemical
industries furthers both the erosion of genetic resources and
unsustainable energy-intensive farming. Loss of self-reliance,
ecological instability and societal disruption is greatly increased, in particular in the Third World. A strategy that combines conservation with development is needed to preserve the
world's genetic diversity in order to meet future climatic and
substrate opportunities.

1.   INTRODUCTION

     In retrospect the nineteen seventies display a remarkable
shift from optimism to pessimism with regard to world food perspectives. In the mid-sixties agriculture fell under the enchant-
ment of the Green Revolution, but since the mid-seventies undeniably several alarming facts on the depletion of gene-pools
and ecosystems dominate. We find ourselves in a spell of great
uncertainty, with widely varying views as to the direction agricultural policies should take. All the elements of hope and fear
about hunger in the world are well presented in an article on
Food in Scientific American of September 1980. On the one hand
it fosters the belief that western capital and technology are
needed to boost agriculture in the Third World. On the other
hand it makes clear that most of the farmers in developing coun-

W. Bach, J. Pankrath, and S. H. Schneider (eds.), Food-Climate Interactions, 325–359.

tries are "poor but efficient" and that their types of agricultural development deserve more attention as a strategy survival. In its ambiguity, the article is representative of enlightened western circles with true sympathy for developing countries, but they lack a consistent socio-ecological perspective on sustainable development, and they are afflicted with the bias of western value judgements and lifestyles(1).

On the other side there are angry Third World spokesmen like G.A. Semlini (2), who sharply condemn western encroachment when saying:"Where agriculture is forced to adopt a policy of producing raw materials for export to other countries rather than food for feeding the indigenous people, that policy is both perverse and suicidal... In return for the exported raw materials, the poor nations import food and manufactured goods at ever higher prices." The clash of views is clear. Though much of the last decades' aspiration for a Horn of Plenty - the World of Cornucopia - has withered, actual western encroachment on the Third World agriculture is greater than ever before. The process is fueled by such beliefs as expressed in the subheadings of the above mentioned article:
"The task of feeding everyone adequately calls for an investment of more than one hundred billion dollars in the agriculture of developing countries", and
"Without a fairer distribution of income, many will still go hungry".

The latter statement suggests that a little more western morality and organization might indeed help relief much of the misery in many of these "backward" countries. The first statement seems to reinforce the system of western encroachment, channeled through the World Bank, UNDP and multinational corporations, with the support of bilateral and multilateral aid agencies and peace-corps volunteers. Of course, most or all of them are consciously or unwittingly operating with the entrenched belief in western values and supremacy. Thus, the social disruption and ecological ravages in the Third World - rather than being recognized as inherent in forceful western encroachment - are most often imputed to corruption or incompetence of the Third World leaders, which could be overcome through the adoption of more western morality and organization.

What is attempted here, is to analyse the contrast between aspirations and ecological realities. As an undercurrent there is discussion on power structures and conflicts between short-term profit and long-term sustainability. Therefore, an outline is given on how global life-support systems and processes (soil fertility, genetic variability and ecological diversity) are systematically eroding, thus rendering the biosphere increasingly less capable of meeting future climatic challenges - opportunities

as well as hazards - and of adapting to changes in the substrate
(soil, water, nutrients).

Of particular concern is the preservation of gene-pools.
This refers to husbanded species (crops, pharmaceuticals, livestock) as well as the larger number of wild species, which are
faced with extinction even before their cursory investigation
of potential usefulness. The World Conservation Strategy, launched
in 1980, should not be regarded as just another cry of some un-
practical alarmists and idealistic nature-lovers, but rather
as a serious strategic instrument for a stepwise integration
of conservation with development.

2.   CORNUCOPIA VS. ECOLOGICAL AND POWER REALITIES

2.1 Maximum food projections

Overriding in the world food debate is still the notion
that the world can feed its present population many times over.
The idea stems from a MOIRA-study(Model of International Relations and Agriculture)(3).

Taking into account the possibilities of irrigation and
the limitations in crop production caused by local climate and
substrate conditions (soil, water, nutrients), the absolute maximum production expressed in grain equivalents of a standard cereal was computed as 49.830 million tons per year, that is almost
40 times the present cereal crop production. The maximum production of the area cultivated with cereal crops (in 1965 already
65 percent of all the land cultivated) could potentially be increased 30 times. South America and Africa south of the Sahara
would be the most promising areas, Australia the least. The potential agricultural land of the world was suggested to be 3419
million hectares, that is 25 percent of the land area of the
world, of which 470 million hectares could be irrigated. Maximum
use of surface run-off and ground water and the transfer of water
from the world's major rivers to fertile desert soils is assumed.

A subsequent MOIRA-study presents even more optimistic figures: 3687 million hectares potentially suitable agricultural
land with a maximum of 58.058×10^8 kg consumable protein production (1965: 1.882×10^8 kg). Yet, according to FAO figures
of 1978, only about 11 percent of the world's land area (excluding Antarctica) offers no serious limitation to agriculture;
the rest suffers from drought, mineral stress (nutritional deficiencies or toxicities), shallow depth, excess water, or perma-
frost(5). FAO still believes that the world's cropland could
at least be double the present 1400 million hectares. Such estimates foster false hopes for the starving since reality has shown

that land losses over the last decades are greater than the number of hectares brought under cultivation - and "high yielding
varieties" are not that promising.

2.2 The world of reality

In a Swedish Government Report Curry-Lindahl concludes
that during the last 20 years through ill-designed land use projects, development assistance has contributed considerably to
the tremendous decline in productivity of ecosystems(6). The
objective was to increase the arable acreage and food production
as well as to ameliorate the standard of living for the people.
But in fact over the last twenty years:

. Every year more arable land is lost than gained.
. Nearly fifty developing countries formerly self-supporting
   in food have become food importers, and their number increases every year.
. Desertification is continously increasing at an accelerating
   rate.
. The severely detrimental consequences of erosion are increasing in both lowlands and highlands and through sedimentation in rivers, lakes and the sea.
. Water resources are decreasing, ground water tables are
   sinking, previously permanently-flowing rivers fall dry,
   marshes and lakes are drained, with no long-term benefit.
. Vegetation disappears, fertile grasslands are transformed
   to dry thornbush before turning into desert, gallery forests
   are cut down, lowland forests are eliminiated, mountain
   rain forests are destroyed, causing flood catastrophes in
   the lowlands.
. Wild animals, important protein resources and ecological
   stabilizers, have vanished from many areas where they were
   abundant, many species are in danger of extinction.
. Toxic chemicals are dispersed in ecosystems in larger quantities every year making pest species more resistant.

All these alterations are the aftermaths of detrimental
land-use methods and technical assistance programmes have accelerated them. Conservation organizations have tried for two
decades to draw the attention of the UNDP, the FAO and the World
Bank to the danger of neglecting ecological realities in their
planning and implementation of development projects. Hardly a
month passes without news about these kinds of projects which·
non-governmental conservationists try to stop or modify. But
usually decisions have already been made. Both the UNDP and the
World Bank proclaim consciousness about the environment, but
give priority to economic growth and measure their results in
figures related to growth not to environmental quality and ecological resources. Of course, it is easy to increase economic

growth in terms of GNP if consumption of the capital (of natural
resources) remains invisible in the balance sheets. "Through
the projects they initiated, encouraged and financed, as well
as their philosophy, the UNDP and the World Bank constitute the
two most serious obstacles to conservation measures"(7).

2.3 Earlier ecological warnings

What was said about careless land-use development projects
is far from being new. Ecologists have worried for several decades, but their anxieties have always been ignored by the widely held optimism of public leaders, who believed that agriculture
was adapting rapidly to scientific methods leading to an increase
both in yields (through the Green Revolution's "high yielding
varieties") and in ecological balance (through the so called
"integrated pest control"). Neither of the two - HYV and IPC -
has alleviated starvation in developing countries, nor have they
convinced ecologists. On the contrary: "The theoretical calculations of the world food potential are eagerly accepted as reality and used by naive idealists as well as less naive but short-
sighted politicians and smart salesmen full of gumption... The
Wageningen School of Agriculture is to be blamed for their global calculations of potential food production that do not take
into account ecological constraints and effects and that evoke
an image of cornucopia, detached from reality. They are utter-
ly dangerous because they distract attention from ecological
risks, foster phantasies with a political background, and thwart
the battle of the ecologists(both domestic and abroad) against
commercial short-term profits and the shortsightedness of political leaders."(8).

Warnings on what can go ecologically wrong have been given
by the economist John Stuart Mill in the middle of the last century shortly after the monumental work of G.P. Marsh (1864) "Man
and Nature" appeared(9). Since then, warnings of ecologists have
been heard in regular intervals... and ignored. The last two
decades have seen an increasing amount of literature about the
degradation of ecosystems by ill conceived modern agriculture
and mismanaged land, of which Rachel Carson (10), Jean Dorst
(11), M.T. Farvar & J.P. Milton (12), Erik Eckholm (13) are
some wellknown examples. The most broadly conceived warning and
recommendation for action is that of the World Conservation Strategy (1980) (14), to be discussed in more detail in section 5.5.

It seems, that in spite of all the ecological evidence to
the contrary, land use development and agricultural innovation
in the Third World are continually fueled by the basic beliefs/
myths:

. that the population explosion requires Draconian development strategies such as the Green Revolution, and
. that the First World has the technological answers to increased food security with a high-energy-input type of production.

2.4 The myth of power-sharing paving the way to survival

Almost concurrently with the World Conservation Strategy
appeared the Brandt Report of the Independent Commission on International Development Issues (15). From its recommendations
and the introductory plea for "Peace, Justice and Jobs" it is
quite clear that the Commission's main concern is to remedy within the next two decades what is considered today as the world's
gravest problem: the imbalance of power and development between
the North and the South. This report fits well the United Nations'
debate on a New International Order (NIO).

The Brandt-Commission must be commended for paying attention
to the societal factor in the whole process of development. But
the proposed steps fall still within Western interests and charity. Thus the Commission recommends to "increase food aid" (a
disputable proposal) while it should rather be "linked to the pro-
motion of employment". What the Commission forgot to recommend was
a) to reduce exploitation of the best agricultural land in the
Third World for export products to the First World, b) to make
better use of that land for sustainable domestic food production
in harmony with the environment, and c) to strive for amelioration of the regional climate, ground water management and the
conservation of genetic resources. Thus, it seems that the proposed New International Order is far from being an ecological order.

2.5 The other side of western supremacy

As Dammann (16) has clearly shown, present poverty in Africa,
Asia and Latin America is far from being indigenous, but the result of Western expansion. In the latter part of the Middle Ages
(about 1500 AD) the Sudanese culture from Abyssinia to Senegal
compared well with contemporary European civilization. About 1325
the kingdom of Mali - the same size as Europe - impressed the entire Islamic world by its magnificence and opulence, and the Lord
of Mali was included among the most important rulers of the Islam.
About 1500, Timbuctoo was a celebrated place of learning, several
hundred years old, as well as a centre of religion, commerce, and
literacy south of the Sahara.

In the course of the next 400 years, when most of the European
growth in trade and economy was based on buying and selling human
beings, the fall of Africa was gradual but definitive. Between fif-

ty and a hundred million of the most able-bodied men and women
were exported as slaves or died on transfer. But even more detrimental than physical losses of the strongest and healthiest
young adults was the complete demoralising effect: turning the
value of neighbours and neighbourhood into merchandise and profit.

Besides, for four hundred years all profit went into one direction. "The slave trade did not contribute, like other trade,
to any new economic activity during the four hundred years it lasted, but rather acted as a means of suppressing all cultural
growth and activity during this period. It established the ra-
cial arrogance of the Europeans as being an acceptable attitude
and at the same time eroded African culture and self-confidence
bit by bit" (16).

Nowadays many educated Africans strive to live like Europeans. They are the best intermediates to introduce Western interests and supremacy to Africa. There can be no doubt, according
to Dammann, that the "slave trade and what it entailed, laid the
foundation to the unique economic growth which gave Europe, and
subsequently the United States, the world record in material su-
perfluity."

"Europe had still not completed her efforts aimed at rendering the African incapable of managing his own affairs, at depriv-
ing him of the rights over his natural resources, to own his land,
and at crushing even the last vestiges of self-respect and independence. This finally happened in 1884 as a result of the last
Colonization Conference in Berlin, where the big powers divided
Africa among themselves, and drew arbitrary lines across the
lands of black people without consideration of language or eth-
nic groups. Each square would now be a 'state' in which a European
nation could help itself freely to the labour and natural riches."

In two subsequent chapters Dammann demonstrates that the
role of Europeans in Asia and America was no less abject, although, due to the longer distance, pressures were comparatively
less severe and the effects less detrimental. He concludes that
"Africa is perhaps the most sensitive part of European history."

2.6 Should population be controlled before giving equal access
     to resources?

The correlation between poverty, level of education and population growth, witnessed today in many developing countries,
has striking similarities with Europe between 1700 and 1900.
Over those two centuries the population increased by 295 percent
despite the fact that more than 50 million emigrated to other
parts of the world during that time. "Europe solved its problem
by outmigration and exploitation of regions belonging to other
races" (16). But Europe (and North America, Australia and New

Zealand) do not allow them to do the same today, or to get equal
access to resources. The developing countries as a whole are by
no means densely populated. Only India with 170 inhabitants per
km^2 reaches European numbers (200 - 400 per km^2). Even China is
still below 100 people per km^2. The success of how Europe sur-
vives with these high population densities and still maintains
a high standard of living lies in the use of over a hundred million hectares of arable land in the Third World. The poor countries are unable to use our natural resources in the same way
(16). It is too easy to accuse corruption or even the power-hun-
gry and self-interested dictatorships in most of the developing
countries of showing too little social concern of the misery of
the poor in their countries. Despite almost universal endorsement
of human rights and democratic rule, the majority of humankind
is currently governed in a repressive militarized manner(17).
Authoritarian rule, military suppression and neo-fascism tend
to increase with greater resource constraints around the world
(18).

The Brandt Report states very clearly that the world's dilemma - destruction or development - is very much the result of
overdevelopment or rather maldevelopment in the North. It is,
however, realistic enough to see that the North will not give
up much of its privileged position voluntarily, and therefore
stresses the idea of mutual interest between North and South,
but is vague about how this has to be worked out. The reasoning
goes as follows: "Whoever wants a bigger slice of the cake cannot
seriously want the entire cake to become smaller. Therefore developing countries cannot ignore the need for economic growth in
industrialized countries, as on that depends their willingness
to participate in a more constructive transfer of resources".
Thus economic growth of industrial nations should go on. But
the fallacy is that this growth can only be attained by a greater exploitation of the Third World's resources. Mutual interests
then become an euphemism for securing First World interests first.

What the Brandt Report is not saying - for obvious political
reasons - is that the entire cake cannot and will not grow much
more, as it is limited by ecological constraints.

3.   DESTITUTION OF ESSENTIAL LIFE-SUPPORT SYSTEMS AND PROCESSES

3.1 Land loss

As was mentioned above, most of the best land in the world
is already being farmed (5). Unfortunately large areas of prime
quality land are taken permanently out of production for build-up.
One example from Great Britain: the 850 square miles (2200 km^2)
of previous farmland at the Thames Estuary that have been built
on were examined as to what happened to the land. It appeared

that for every unit of land used for providing homes and shops, six have been used for factories, sixteen for roads, fifteen for lawns, gardens and play-grounds ("tended space"), nine have become derelict and 61 have been turned into wasteland. It was pointed out that the whole of the housing programme of that region could have been built on pre-existing derelict land or many times over on pre-existing wasteland(19). By 1974 California had lost about 0.7 million hectares of prime agricultural land by urbanization and highways and another 0.4 million hectares were zoned for urban development by 1985. The loss of arable land nation-wide for the USA has been estimated at 5.1 million hectares between 1958 and 1974(20). From an OECD report (21) it appears that in developed countries at least 3000 km^2 of prime agricultural land is lost each year to urban sprawl: between 1960 and 1970 Japan lost 7.3 % of its agricultural land to buildings and roads, and Europe lost from 1.5 % (Norway) to 4.3 % (Netherlands). Such losses are by no means limited to industrialized countries. In Egypt, despite the efforts to open new lands to agriculture, the total area of irrigated farmland has remained almost unchanged over the past two decades. As fast as additional hectares are irrigated with water from the Aswan High Dam, old producing lands on the Nile are converted to urban areas and roads(22). According to a UN paper (23), close to one third of the world's arable land will be destroyed in the next 20 years if current rates of land degradation continue.

Rather than restructuring built environments in order to accommodate housing, building and traffic requirements or increasing the efficiency of the urban land use, most governments resort to urban expansion and urbanization of the countryside as the politically easier solution, since farmers, when defending their rights to the land, almost invariably are in the weaker position.

3.2 Erosion

Soil erosion is a natural process, but in undisturbed ecosystems with a protective cover of vegetation the soil is regenerated about the same rate it is removed. Silt loads of rivers reflect the amount of protection measures taken in the catchment areas. Under natural conditions it would take one to four centuries to generate one centimeter of topsoil, so generally speaking one can say that once the soil has gone it has gone for ever.

More than half of India suffers from soil degradation. Out of her total of 3.3 million km^2 land surface, 1.4 million km^2 are subject to increased soil loss (erosion of topsoil), and 0.27 million km^2 are being degraded by floods, salinization and alkalization(24). An estimated 6000 million tons of soil are lost every year from 0.8 million km^2 alone; with them go more than 6 million tons of nutrients - more than the amount annually applied in the form of fertilizers(25).

Much erosion is caused by neglect of watershed management, clearing of mountain forests, logging and cutting for fuel and subsequent overgrazing, and - at last but not least - irresponsible road construction. The results can be extremely expensive. It costs Argentina over ten million dollars a year to dredge silt from the estuary of the Rio de la Plata to keep Buenos Aires open to shipping. Of the hundred million tons of sediment, 80% comes from only 4% of the drainage basin - the heavily overgrazed and eroded catchment area of the Rio Bermejo, 1800 km upstream(26). As 10 % of the world's population lives in mountainous areas and another 40% in the adjacent plains, the lives and livelihoods of half of the world population depend directly on the way in which watershed ecosystems are being maintained(27).

Erosion by wind is another form of severe loss of topsoil. John Steinbeck's classic "The Grapes of Wrath" describes so graphically the dust bowl of the American Great Plains in the 1930s. Lack of proper soil conservation has already led to an average loss of one third of the topsoil of the United States in less than one century(28), and in one third of the USA the loss of topsoil is more than 75%(29). Wind erosion is far from being mastered. During the dry spring and summer of 1976 about 2 to 3 cm of topsoil was blown away from plowed agricultural land in The Netherlands and Germany, respectively(30). Saharan dust is probably the best investigated form of wind erosion. It is found as far west as the coast of Brazil and the Caribbean Islands(31).

3.3 Desertification

It is estimated that the spreading of desert-like conditions in drier regions and heavy erosion in more humid regions present global losses of around 6 million hectares a year of arable land, 3.2 million hectares of rangeland, 2.5 million hectares of rainfed cropland and 0.125 million hectares of irrigated farmland. At presently estimated rates of desertification, the world's desert areas(now 800 million hectares) would expand almost 20 percent by the year 2000. An area about two and a half times the size now classified as desert is endangered by desertification(32).

The outlook for reducing erosion and desertification in the industrialized countries and elsewhere is not good. Food and forestry projections imply increasing pressures on soils throughout the world, especially in North and Central Africa, the humid and high-altitude portions of Latin America and much of South Asia. In addition the increased burning of dung and crop wastes for domestic fuel will deprive the soil of nutrients whereas the loss of carbon (humus) will degrade the soil's ability to hold moisture.

The estimated 113 million tons of nutrients that are potentially available to developing countries from human and livestock
wastes should be used as far as possible to fertilize the land,
while crop residues make a useful cover to protect the soil
against the sun and erosion.

3.4 Irrigation

As of 1975, 230 million hectares - 15 percent of the world's
arable land - were being irrigated; an additional 50 million
hectares are expected to be irrigated by 1990. But already half
of the world's irrigated land has been damaged to some degree
by salinity, alkalinity and water-logging(22).

What is perhaps the classic example of misplaced development aid, is the construction of the Aswan High Dam by the Soviet Union, which was completed in 1970. As predicted, the conversion of some 800 km of the Nile floodplain from a one crop
system of irrigation to a four crop rotation system by perennial
rather than intermittant irrigation, created ideal conditions
for the spread of a snail, an intermediate host to the parasitic
blood flukes, causing the serious disease bilharzia or schistosomiasis. The average incidence of the disease along the Nile
between Aswan and Cairo was about 5% in 1937, but has increased
to 35% in 1972. This figure includes about half of the Egyptian
population. In addition, changing the flow of the Nile has strongly decreased the fertility of the soil in the delta because most
of the nutrients are now deposited in Lake Nasser behind the
dam. This causes Egypt to import more fertilizers with atten-
dant ecological, let alone financial, complications. If an effort equal to only half of that involved in the Aswan Dam had
been channelled into rural development(including agriculture
and population control) it could have benefited Egypt far beyond
the rewards of the Dam, which will soon be rendered useless because of siltation(33).

Irrigation systems need not be based on grand-scale constructions. Most of the Far East countries(Indonesia, the Philippines) make use of a traditional method to irrigate their
rice fields called sawahs that has proven useful over thousands
of years. The introduction, however, of farm machinery is now
largely destroying this ecologically balanced system, and is
a major cause of erosion, deforestation and ultimately landslides,
as has recently occurred in Indonesia(34). Irrigation can have
climatic effects as well. About 1970 it was estimated that the
total continental runoff of the world had decreased 5%, whereas
total continental evaporation had increased by about 2%. Evaporation of the irrigated areas alone can increase by 100 to
1000%, but the areas covered by artificial reservoirs are only
0.2% of the land surface(35).

3.5 Shifting cultivation and agro-forestry

In areas under shifting cultivation (milpa), forests act
to restore soil fertility. More than 200 million people occupying about 30 million km^2 of tropical forest could continue this
stable and productive practice if the population itself were stable and if there were no other external pressures on the forest
ecosystems, e.g. logging operations, that force the shifting
cultivators to shorten fallow periods or to move on to the upland forest on steep slopes where the resulting erosion will be
severe. Along the Ivory Coast, shifting cultivation reduced the
forest cover by 30 % between 1956 and 1966, and today only one
third of the forests existing at the beginning of this century
are left(5). Similarly, shifting cultivation clears about 3.500
km^2 each year in the Philippines - in Mindanao alone it cleared
10.000 km^2 between 1960 and 1971(36).

If land is eroding so rapidly that it must be abandoned,
there is seldom any alternative for the farmer. As is being indicated in chapter 5 of the World Conservation Strategy, this situation can be avoided by promoting the integration of some modern technology with the traditional system of resource management, e.g. by shortening the fallow period gradually through
mixed cropping practices, limited use of some inorganic fertilizers, recycling of organic waste materials, replacing of natural
cover during fallow periods by cover crops that can be used economically - such as pasture for livestock in mixed agri-pastoral
systems, or tree crops in mixed agri-silvicultural systems(14).

Large-scale agriculture may indeed not be possible in the
humid tropics. Agro-forestry borrows heavily from milpa traditions in such practices as planting many different crops together, as many as 20 different species in a small clearing. These
small, family-run, labour-intensive farms are invariably more
efficient and ecologically stable than highly mechanized agriculture in the same area. Trees are planted with shade-loving
crops, resulting also in reduced pest problems. An important
principle is that the soil must always be covered to avoid sun-
baking (lateritic soil), leaching of nutrients and erosion.

3.6 Wetlands cropping

Many freshwater wetlands, seagrass beds and coral reefs,
though of great importance for fisheries, are being destroyed
all over the world with severe effects on the economies that
depend upon them. Virtually all relevant studies point to an
increasing destruction of these ecosystems, a resource on which
60 - 80 % of commercially valuable marine fish species depend
for habitat at some point in their life cycle. The cost of damage to US marine fisheries caused by degradation of coastal
wetlands has been estimated to almost $ 86 million a year (37,38).

In Sri Lanka repeated removal of corals for the production of
lime is so extensive that local fishery has collapsed; mangroves,
small lagoons and coconut groves have disappeared and local wells
have been contaminated with salt(39).

The losses to African floodplain fisheries due to the building of dams and reservoirs are many times greater than the gains
in fisheries within the reservoirs(40).

Another worry about the rapid loss of wetlands around the
world is the "erosion" of ecological diversity and genetic resources. These ecosystems, both below and above the water level,
carry the greatest number of species per km^2 and compare well
with rain forests in species diversity and complexity of food
chains. They are well-known for their high productivity (41),
an aspect that is usually valued higher (in terms of quick agricultural profits) than their long-term ecological value of being
a genetic reservoir.

3.7 Impact of air pollution on productivity and diversity

Air quality is likely to worsen worldwide as increased
amounts of fossil fuels, especially coal, are burned. Emissions
of SO_2 and NO_x are particularly troublesome because they combine with water vapour to form acid rain. In large areas of Scandinavia, Eastern Europe, Canada and the Eastern United States
the pH of rainfall has dropped below 4.5 which is very acidic
(42), and in some lakes the accumulation has even led to a pH
value below 3 (43). Freshwater fishes are heavily impaired in
waters with a pH below 5 and many lakes no longer contain fish
(44). Damage is also done to the photosynthetic activities of
forests and crops (45); pine forests are particularly sensitive
(45), and an impairment of growth of up to 10 % is possible.

The OECD has given great attention to monitor and study
long-range air pollution(46, 47). A survey of many relevant studies is given in 48, 49, 50 and 51. Recently the United Nations
Economic Commission for Europe (ECE Geneva) has established -
with the support of UNEP's Global Environmental Monitoring System - an action programme on transboundary air pollution, involving an all-European evaluation and monitoring programme of air
pollution (Eastern and Western Europe and North America). It
also encourages the setting of national ceilings to maximum annual emissions of SO_2 (52). Only the Netherlands and Sweden
have, so far, set officially such ceilings - 500 and 300 million
kg of SO_2, respectively. Other countries have indicated their
willingness to do so. There is, however, neither a guarantee
that the sum of these annual national emission ceilings will
in the long run be ecologically acceptable, nor is it an assurance that the limits will be kept(53).

A reduction of photosynthesis on the land as well as in
the oceans (phytoplankton) must theoretically be expected also
from increasing radiation of ultraviolet-B on ground level when
the stratospheric ozone-layer becomes affected by air pollutants,
such as chlorofluorocarbons (used in cooling liquids, propellant
gases etc.), N_2O (mainly from bacterial breakdown of surplus nitrogen fertilizer in soils), other NO_x (mainly from combustion
and supersonic flights) and several other trace gases. They can
dissipate into the stratosphere where they can react with O_3
and thus affect the ozone layer's capacity to filter out much
of the UV-B radiation from the sun. It is not yet clear whether
the decrease in photosynthesis of UV-B-sensitive plants will
be compensated by increased photosynthesis in non-sensitive plants
and phytoplankton. It is likely that most plant species will be
affected, as most of them evolved under conditions of low UV-B
radiation.

The UNEP-Coordinating Committee on the Ozone Layer recently
reviewed the evidence (Bilthoven, November 1980) and found that
the production and use of CFC's has not declined since 1974.
Neither is N_2O decreasing. With present trends continuing, the
reduction of the ozone layer might theoretically amount to 10 %,
but actual measurements show variations falling still within de-
tection limits. The uncertainty, which cannot at present be quantified, demands a close surveillance over the coming decades(54).

3.8 Impact of water and soil pollution on productivity

Pollution of rivers, lakes and groundwater around the world,
and in particular in developing countries by excessive use of
pesticides, is becoming a major threat to human health, the environment and productivity. Farmers in some parts of Asia are reluctant to stock the wet rice fields and ponds, because fish are
being killed by pesticides. The lack of fish not only means a
serious loss of high quality protein for the diets of rural families, but also an increased risk in malaria incidence, as these
fishes would have consumed large amounts of Anopheles larvae in
the shallow waters. Downstream from intensive irrigation projects
the water is rendered useless because of high salinity and high
pesticide content(22). The Japanese experiences with fish being
rendered unsuitable for consumption by heavy metal pollution
(Minamata and itai-itai diseases) are all too painfully known.
But also in the North Sea, due to pollution, the number of fish
with apparent cancerous abnormalities is rapidly increasing(55,
56). The Common Seal has almost completely disappeared from the
Wadden Sea, due to a number of polluting factors, notably the
PCB's(57). Most of the PCB's and the other pollutants come down
the River Rhine, which still is the main source of drinking water for about 20 million people. The high salinity of the Rhine
water, due to emissions from potassium-mining in France, rendered
the water no longer suitable for some sensitive types of irriga-

tion in The Netherlands(58, 59).

Affection of soil fertility by pollution has only recently got major attention after some much publicized experiences in highly industrialized countries (Love Canal, USA, and Lekker-kerk and Waterland in The Netherlands). Though publicity is main-ly focused on the health impact for people living at or near these chemical dumps, the long term impact on soil fertility and ecological potential of the wider area has only started to be investigated. Dumps of chemical wastes are like a time-bomb whose timing mechanism no one knows(60). A very quick assess-ment in The Netherlands has revealed at least 300 such dumps, that have to be wholly or partially excavated over the next de-cade at the cost of several billion guilders, and the number of other dumps to be carefully checked might well be up to 3000 (61). Though The Netherlands is a small country with a relative-ly high density of chemical industry, there is every reason to believe that similar situations occur in other industrialised regions of the world. Moreover, as standards of control become stricter in most western countries, the tendency of industries - in particular multinational corporations - to transfer their most polluting products to developing countries is already ob-vious.

4.   GREEN REVOLUTION AND THE SEEDS OF THE EARTH

To sustain a food system that would meet world needs within world and regional climatic variations, regional and local envi-ronmental as well as social and economic constraints, full advan-tage should be taken of all genetic resources of the earth as they have developed during evolution in response to climate and substrate (soil, water, minerals). If climate and substrate op-portunities are to be fully employed, the basic resource of the world's genetic stock must be available in abundance and variety to meet these and hitherto unknown but thinkable opportunities. If better standards of living in the Third World are to be achieved, it makes no sense to apply a few high yielding varie-ties of the so-called Green Revolution.

4.1 The Green Revolution's addiction to agrichemicals

The two general components of the Green Revolution - in-creased use of HYV's (especially wheat, rice and maize) and "in-puts" (fertilizers, irrigation water and pesticides) - are al-ready an integral part of the agriculture in industrialized coun-tries; and, since the 1960s, became increasingly imposed upon do-mestic acreage of the developing nations. By 1975 already 43 million hectares (i.e. 32 million hectares between 1965 and 1975) in the developing countries of Africa, Asia and Latin America were planted with HYV's of wheat and rice alone(62).

The first rice variety developed by the International Rice
Research Institute in the Philippines - IR-8 - can produce several times the harvest of traditional rice from a given area,
if handled correctly. Some new strains perform even better, but
all strains are extremely responsive to and dependent upon fertilizers, water and pesticides. HYV's, however, outperform the
traditional varieties by 10 to 25 % only when given equal amounts
of fertilizer and water (63), and the traditional varieties require definitely less pesticides. The high dependence on fertilizer made fertilizer shortages and price inflation a serious
problem for agriculture in the developing countries (64) - in
particular after the 1973 energy crisis. As an optimum they require 70 - 90 kg per hectare but in most developing countries
applications are a fraction of that (14 - 33 kg/ha) (65); in Europe often double or triple this amount is applied.

Modern high yield agriculture can be described as a system
that turns fossil fuel into calories of food while replacing
man-power. But Japan and Taiwan attain yields similar to those
of American farmers with much lower fossil energy inputs; and
China in particular has rapidly and successfully increased its
food production in the past two decades with only a fraction
of the energy costs typical of Europe and the USA by much greater use of human labour and by relying less on heavy machinery
(66). China, however, is now at the verge of introducing more
western types of development which raises doubts that they can
stay free of the inherent problems of increased pollution, ecological disturbance and social disruption (notably unemployment).

The developing countries' use of fertilizer - notably nitrogen is constantly far behind planning, since production costs
go up with energy prices. The two ways of becoming less dependent on fertilizer production would be to increase the use of
natural nitrogen fixation, and synthetic production using solar
energy. The UNEP-UNESCO-ICRO Panel on Microbiology is mainly
concerned with the first. In their advisory capacity they can
develop and promote an international network for the preservation and exchange of cultures of micro-organisms, their use as
natural resources, and the training of people to utilise them.
Microbiological research centres concentrating on nitrogen fixation have been established at several places in the world, and
several courses are being organized to improve Rhizobium technology and grain legumes cropping - usually intercropping with
cereals. The World Data Centre on Microorganisms has been established at the University of Australia at Brisbane, Queensland.
It maintains an important culture collection and houses the ma-
ster copy of the World Directory of Collections of Microorganisms;
441 collections are registered from 65 countries. Since 1 Jan.
1977 approximately 14.900 forms have been entered in the computer programme(67).

A major drawback to the HYV's is that they perform well only in a narrow band of weather conditions - if rainfall or temperatures are too low or too high, yields may decline substantially. While producing higher absolute yields, HYV's tend to give more variability in absolute yield(68).

Since the mid-1970s, average yields from HYV's in most developing countries are constantly dropping. Besides, they are increasingly vulnerable to pests. There is a saying among angry farmers in Pakistan that the miracle rice has led to miracle locusts(69). By 1978 the Wall Street Journal was writing a front page feature saying: "there isn't anything left in the Green Revolution's bag of tricks - the Revolution, in fact, has turned against itself"(70). Nowadays the HYV's often do not produce as well as the old known species. In many situations, because seed supplies have been used for food, the people could not get back to their original strains. This has resulted in food shortages in many areas.

Within the span of two decades the varied agricultural systems of the Third World and their supporting social structure have been uprooted, overthrown and replaced by a new western model. Both the crops and economies of Asia, Africa and Latin America have been hauled into the western market economy under the pretext of feeding the hungry(71).

Rockefeller Foundation staff now candidly admit that their work for the Green Revolution has done little to aid the poor(72). The poor are driven from their fields into the cities where they have to buy expensive cereals grown in fields where they once harvested inexpensive legumes.

Already in 1975 it was clear in the USA that small family farmers were being bought up and that small local seed traders were becoming extinct because city based corporations do not purchase seeds locally. Farmers loose their independence to absentee corporations and the uniformity of crop productions is accelerated(73).

4.2 Preservation of the world's crop genetic variability

When assessing the seed situation of the world's most common crops, it first appears that the old centres of crop diversity are vanishing rapidly. These so-called Vavlov-centres, named after the Russian botanist N.I.Vavilov, are, through a combination of varied topography, climate and cultivation methods, the areas where almost all food crops originated(74). Prehistoric people found food in over 1500 wild plants and they used an equivalent or even greater number of wild plants for medical purposes. At least 500 were used in ancient cultivation

for food. The world's vegetable food diversity has narrowed to
200 species grown by backyard gardeners and to 80 species favoured by market gardeners. Only 20 plant species are used in
field cultivation(75). Less than 10% of 300,000 higher-order
plants have been given even the most cursory scientific examination. Less than 3000 have been studied in detail(76). Yet,
most of the ancestors of the cultivated species and a large number of unexploited species are vanishing at an appalling speed
through manmade habitat changes - mainly through modern agriculture. With every disappearing variety, ten to thirty other
plant or animal species dependent on it for survival, disappear,
too(77).

Superficially, the genetic resources of the world's major
crops appear to be well protected by a network of 60 nationally
controlled gene banks and eight international crop research
centres and seed laboratories, which are under the supervision
of the International Board of Plant Genetic Resources (IBPGR)
based at the FAO in Rome(78). Given its role as a coordinator,
the board does not anticipate large sums for itself, only $ 3
million by 1981. Donors are UNEP and only nine countries: Belgium, Canada, Germany, Netherlands, Norway, Saudi Arabia, Sweden,
UK and USA. The lion's share of the fund goes to the documentation programme in the National Seeds Storage Laboratory at Fort
Collins, Colorado, USA(79).

It would, however, seem unwise to look mainly or solely
to the USA to conserve the world's genetic resources; it is con-
jectured that the USSR has the world's most representative col-
lection(80). Apart from the government-owned collections, the
tendency is towards an increase of the major multinational agrichemical, petrochemical or pharmaceutical firms' interest in
seedbanks(80).

4.3   The seed gamble of agribusiness

Control of the world seed industry is only the second phase
of the Green Revolution: those controlling the seed are close
to controlling the entire food system, what crops will be grown,
what fertilizers and pesticide inputs will be used, and where
the products will be sold, thereby controlling the switches for
market speculation and profit for agribusiness.

In the Philippines, Esso developed a network of some hundred
"agriservice-centres" where farmers can purchase seed, pesticides
and farm implements as well as fertilizers from the Esso dealer.
The Indonesian Government went even further. "It contracted with
several international chemical companies which agreed to encourage farmers to adopt the new technology. In return the government agreed to pay the firm $ 20 for every acre of land they

succeeded in having planted with the new seeds. The results were
disastrous. The programme was riddled with corruption. Needed
farm inputs failed to arrive... the pesticides killed the fish
which provided a major portion of the protein for the people.
Finally a major famine occurred"(72). Although the basis for
the global seed industry was laid by the Green Revolution, it
is the multinational and bilateral aid programmes that back it
up. Phase II logically evolved with the turning over of the seed
programme to private enterprise, half of it subsidized by governments. In response an increasing number of developing countries
began to establish their own national seed industries. However,
the strategy frequently adopted to achieve this, envisages foreign seed companies to participate in this effort through the
establishment of joint ventures with local investors. The American Seed Trade Association is e.g. organizing a consortium of
42 US firms to expand seed sales in several Asian and Latin American countries(81).

     As stated by one of the proponents: A seed industry can
give a high return on a relatively low investment; the seed industry is probably the best catalyst for expanding the agricultural marketing system within a country and may equally be a
stimulant to foreign trade(82).

     In the last ten years or so, the large non-seed multinatio-
nal enterprises like Shell, Ciba-Geigy, Sandoz, Pfizer, Upjohn,
Monsanto, Union Carbide, ITT, Diamond Shamrock, Rhône-Poulenc,
PUK, Dutch States Mines, Exxon, Standard Oil/Chevron and Occi-
dental Petroleum have acquired several hundred seed companies
with sales of $ 5 million or more(83).

4.4 Plant breeders' rights (PBR) and similar patents

     The idea of getting control over the seed market would not
work without a system of exclusive rights and enforcement of
it by national legislation. With the advent of protection legis-
lation(in Europe, the USA and Japan), and through organizing
themselves internationally in the Union for Protection of New
Varieties of Plants (UPOV), - semi-officially linked to the UN
system through the International Intellectual Property Organization, which is the world's patents bureau - the buyers of seed
companies can expect to add royalty payments on a global scale to
their profits. More importantly, plant breeders' rights provide
dominant companies an opportunity to achieve market control in
specific crops. It appears that companies with a background of
chemicals are the most likely candidates to move into seed genetics, because of similarities in marketing and distribution characteristics, production technologies, research activities, and
patenting breeders' rights; germ plasm thus becomes a marketable
commodity. The acquisition pace may now be accelerating as global

firms anticipate a restrictive legislation in Australia, Ireland and Canada and the entrance of more countries into the UPOV-
convention(84).

With patent-equivalent protection, the longterm future of
the industry seems sufficiently secure to attract investors.
A hybrid variety cannot technically be stolen by a rival firm,
and royalty payments for non-hybrid varieties can pay the costs
of hybrid research. Only larger firms can afford the initial
investments of time and advertising; smaller firms will get out
of the market. Thus plant breeder's rights reduce the number
of competitors. They also bring another benefit. Once governments have passed such legislation, how can public breeding institutions continue to compete with the private sector offering
the same crop varieties? The new seedsmen will demand that un-
fair government competition be removed(84).

Although a few developing countries have P.B.R. legislation
- Argentina is a notable exception - the legislation scramble
in the industrialized world will ultimately affe. t genetic resources in the Third World. Bilateral and multilateral aid programmes from Europe to Africa often resulted in the distribution
of inappropriate seeds, because European advisors promoted the
best-advertised seedbrands from home(85). Therefore, in the tropical world it is time to conserve as many land species and wild
varieties as possible. The Association for the Advancement of
Agricultural Science in Africa has called for the immediate survey of African genetic resources and wants to see a conservation
strategy at national policy levels(86).

Is there a chance of a good division of labour between government and private enterprise to secure the seeds of the earth?
There is no reason for optimism: at least two of Canada's most
heavily subsidized companies - Stewart and Warwick Seeds - have
already been bought up by multinational agrichemical industries.
Most outspoken are the views of Ciba-Geigy and of KWS(Germany)
that the public sector (government subsidized laboratories)
should continue to develop basic germ plasm collections, while
the private sector exploits the final cultivar in the market
place(87). This means that government agricultural research becomes a massive subsidy to corporate breeders. This is already
commonplace in other sectors of the western agricultural system.

4.5   The bias of yield uniformity and getting farmers back to
      the company

Due to European Community legislation, Europe is losing
perhaps 400 varieties a year. Concerned European Governments
have agreed to do all in their power to collect threatened germ
plasm and have it stored at the Wellesbourne station. However,

they have not agreed on funding of the preservation of the germ plasm. Third World-oriented charity institutions are now considering funding the Wellesbourne centre because of the importance of Europe's vegetable germ plasm for developing countries (88).

In essence, the (European) Common Catalogue leaves only patented varieties on "national lists", while eliminating considerable competition for traditional varieties. The "illegal" varieties can neither be sold nor grown close to commercial "legal" varieties. For example, the fine in the UK for doing so could be as high as L 400 (88). Between 450 and 500 new cultivars are licensed each year. Most represent minor genetic advances and "fine-tuned" adjustments to changes in production, harvesting, processing and marketing procedures(89).

Government policy-makers might well ask themselves if the corporate breeders also possess biases counter-productive to the profitability of farmers and the nutritional requirements of society. The corporate breeders may be more interested in yield uniformity, processing ability and appearance, while the public breeder may be more concerned with plant hardiness and disease-resistance characteristics. Plant breeders from agrichemical industries may rely more heavily upon inputs of fertilizers and biocides, while public breeders may look for natural resistance. Corporate breeders would like to have the farmer acquire hybrid seeds, while public breeders may look for improvement of perennial varieties of apomictic strains which allow farmers to save their seeds, since they do not need hybridization (90).

Besides of being less expensive to farmers, apomictic strains are cheaper from the research standpoint and take substantially less time to bring to market (one year in contrast to four to ten years for regular hybrids). Apomictic strains in crops such as millet and sorghum have the potential of making a significant contribution to feeding the world's hungry at affordable prices. "Since multinational agrichemical companies - when looking at world markets - see little gain from this kind of research and development, it should be a main responsibility of governments in the Third World with the support of bilateral and multilateral aid programmes"(91).

4.6 The agrichemical bias

Another bias of multinational agrichemical companies is to breed varieties that need certain chemicals. It is only legitimate to suspect that chemical companies will link chemical research to plant varieties they are developing (and vice versa).

For example, University of Florida tomato breeders - encouraged by Union Carbide/Amchem financing - produced a tomato which would only ripen when sprayed(92). Given the high costs to develop a new drug or pesticide and the considerable danger that environmental agencies will withdraw the pesticide agrichemical involvement in the seed industry represents a form of "hedging" - i.e. a way of assuring that one product line needs the other.

The pesticide-connection of agrichemical firms in the seed business also looms large. Despite the great number of pesticides - the American market offers 63,000 biocides to farmers, backyard gardeners and householders(93) - crop losses have remained at about one-third of the total cereal crops ever since World War II(94). Half of all pesticides are sprayed on cotton. Crop pests and disease vectors take the rest. As long as governmental subsidies are available, sales to the Third World are promising and...poisoning! The WHO claims that over 500,000 of the world's poor become seriously ill each year due to inappropriate spraying(e.g. "dusters" fly over fields, spraying peasants and crops alike, often without warning). Besides, UNEP adds that over 300 insect species have developed resistance to one or more pesticides and are now uncontrollable by chemicals. In many countries the incidence of malaria increased (often multiplied) in the early seventies despite - or rather due to - constant sprayings(95).

Often with the help of bilateral or multilateral aid programmes - western companies were allowed to dump unregistered or deregistered (banned) biocides on Third World markets(96), among these are aldrin, endrin, dieldrin, lindane, BHC, toxaphene, strobane heptachlor, parathion, methylparathion, proxpur and leptophos. The leading producers are Shell, Bayer and a number of others(97). "Will agrichemical firms working in seeds be any more concerned about the needs of the Third World and responsible in their export practices or innovative in their research than shown so far in their chemical business?"(98).

4.7   Recommendations from "Seeds of the Earth"

The broad analysis by the International Coalition for Development Action (ICDA) of the agrichemical business' involvement in the world's seed trade concludes by saying that, if it cannot be reversed, a least it should be stopped.

The financial resources and technical expertise required to collect and conserve endangered germ plasm is well within the political reach of governments and agencies. The key to mobilize political "will" lies in the understanding of some major myths as mentioned in section 2.3 and especially in the false

hope that agrichemical industries will bring innovation and
creativity to plant breeding rather than uniformity and chemical dependence.

We can still stop it and develop a way to achieve increased
crop genetic diversity and better conservation measures, Mooney
(99)said in an interview with NCRV television on 8 Nov. 1980,
where he also explained that the business people provide less
information since "Seeds of the Earth" has been published.

His recommendations for action fall into six categories:

. Regarding germ plasm conservation emphasis is laid on a
  campaign to be launched by the International Board for
  Plant Genetic Resources for the establishment of genetic
  reserves in the Vavilov centres.
. Regarding international legal arrangements it is argued
  that plants be considered as common heritage and basic
  human rights and therefore unsuitable for any form of exclusive rights, patents, trade marks, etc. This has to be
  laid down in the Code of Conduct for Transnational Corporations and the Code of Conduct for the Transfer of Technology.
. Regarding international action, "seeds" should become a
  UN agenda item and should be incorporated in the programme
  of action for the Third Development Decade. The UN Centre
  of Transnational Corporations should undertake a study on
  the world seed business.
. Regarding Third World options it is recommended that these
  nations take immediate inventory of their endangered plant
  genetic resources and press for international cooperation
  to collect and preserve such resources.
. Regarding the role of voluntary agencies emphasis is laid
  on collective self-reliance of subsistence farmers and
  action against varietal legislation and plant breeders'
  rights.
. Regarding personal response of the individual it is recommended to plant your own garden and land with non-hybrid
  varieties and thus save seeds for future gardening.

5. PRESERVATION OF THE WORLD'S GENETIC RESOURCES

5.1 Wild gene pools of crops, pharmaceuticals and their asso-
    ciated organisms

A country's dependence on its own diminishing store of genetic diversity is likely to grow as the country develops. This
applies to food crops as well as medicinal plants. Several developing countries are setting up their own pharmaceutical industries to keep supply of essential drugs at acceptable costs.

More than 40 out of 90 species of medicinal plants in Africa,
Asia and Latin America are exclusively available from the wild
and another 20, though cultivated, can also to be taken from
the wild(100).

Since many varieties of domesticated plants and animals
are disappearing, so too are many of the associated microorganisms and animals. An estimated 25,000 higher plant species
and more than 1000 vertebrate species and subspecies are
threatened with extinction(101,102). These figures do not take
account of the inevitable losses of small animals, particularly
molluscs, insects and corals, whose habitats are being destroyed,
so that between one half to one million species of the smaller
creatures will have disappeared by the end of this century,
mostly by habitat destruction, and most of them having not even
cursorily been investigated(103). Losses are especially great
in the humid tropics due to the rapid rate of deforestation.
Only about one in six of the estimated three million kinds of
tropical organisms has been catalogued. Although the vast majority are relatively small, the lack of knowledge applies also
to the higher plants, of which between 10 and 15,000 species
in Latin America alone have not yet been described and listed.
Of the approximately 5000 kinds of fishes in the Amazon Basin
about 40% are still to be described and catalogued. Worldwide,
the number of scientists competent to undertake largescale studies of tropical ecology is only a few dozen. The total budget
for basic biological research on tropical biology is estimated
at $ 40 million only, half of which comes from US scientific
budgets(104).

The US National Academy of Sciences concluded: "After
weighing all available measures for preserving endangered species under controlled conditions, we are repeatedly forced to
conclude that the only reliable method is the natural environment"(106). Furthermore, the US National Research Council issued
a set of recommendations for "Research Priorities in Tropical
Biology" (1980), based on a careful analysis of conversion rates
of tropical moist forest, landuse, population growth and food
production requirements, present preservation practices, and
scientific and social rationale for learning more about the
tropics(107).

It is ironic to note that, while more than a quarter of
all the forests in Central America has been destroyed during
the past twenty years to produce beef for the North American
market, the average consumption of beef in these countries has
declined steadily during that time(104). In Brazil a country
of some 121 million people, of which about one third are malnourished at present, about 95% of the land that can be put into sustained agriculture is already under cultivation. With an
expected doubling of the population in the next 23 years, and

if the current plans of the Brazilian government are implemented
to alleviate their energy shortage, which is acute, by planting
sugar cane for the production of alcohol, approximately one-
fifth of the currently productive area will be taken out of
food production. This sacrifice can certainly not be sustained
without an even greater assault on the already diminishing
forests of Brazil(105).

5.2 Losses of tropical moist forests

Although covering less than 10% of the earth's land surface, tropical forests harbour about half of the earth's 5-10
million species of organisms.

According to Myers, the present status and future prospects
of tropical moist forests is alarming(108). Commercial exploitation of timber is usually followed so rapidly by the forest farmer, fuel wood gatherer and cattle rancher, that there is no
chance at all for forest regrowth. His calculation of 200,000 km^2
lost every year to timber extraction is only a "working" estimate, to which should be added 25,000 km^2 lost to fuel wood
cutting and 20,000 km^2 lost to forest ranching. Myers' total
figure of 245,000 km^2 per year is far beyond that of FAO's in
1976, which, for the sake of caution, is always kept very low
(120-170,000 km^2/y).

In hardly any area is the rate of conversion likely to decline. Virtually all lowland forests of the Philippines and
Peninsular Malaysia seem likely to become logged over by 1990
at the latest, or even earlier. Much the same applies for most
of West Africa. Almost all of Indonesian lowland forests have
been scheduled for timber exploitation by the year 2000, and
at least half of them by 1990.

Little could remain of Central America's moist forests
within another ten years or less. Extensive areas in western
Amazonia, in Columbia and in Peru are likely to be claimed for
cattle ranching and various forms of cultivator settlements by
the end of the century, and something similar holds for much
of the eastern sector of Brazilian Amazonia(109).

The twelve exceptionally endangered areas include all of
Peninsular Malaysia, Philippine's lowland forests, Indonesian
lowland forests of Sumatra and Kalimantan (Java has already lost
hers), Sri Lanka's rainforest, all of Central America's three
rainforest complexes, three areas of Brazil's lowland forests
of Amazonia, most of Ecuador's and Peru's Amazonian rainforest,
Ecuador's Pacific coast rain forest, Brazil's relict strip of
Atlantic coast forest, Madagascar's eastern rainforest, East
Africa's relict mountain forest, Ivory Coast's southwestern
lowland forest, and finally New Caledonia's rain forests. All

other forests undergo broadscale or moderate conversion but at
slower rates(110).

Profits more often than not go to industrialized countries
(directly or via the bank accounts of some privileged elites
in the developing countries themselves), leaving erosion, ill-
ness and pollution to the local populations(111). Yet forests,
without logging, have unquestioned importance for industry and
commerce. The value of the world's annual production of forest
products (logging excluded) exceeds $ 115,500 million, and in-
ternational trade is worth about $ 40,000 million a year(112).
Thirty countries, eight of them developing countries, each earn
more than $ 100 million a year from export of forest products
- and five of them earn more than $ 1000 million a year(113).
These facts deserve more attention when governments design long-
term forest policies.

5.3 Conservation of timber and fuelwood - reforestation

Logging is a wasteful use of forests as a resource. Of
Amazonia's many thousands of tree species only about 50 are
widely exploited, even though as many as 400 are known to have
commercial value. Africa exports only 35 principle species with
10 accounting for 70% of the total. In Southeast Asia loggers
concentrate on less than 100 tree species, with exports consis-
ting mainly of only a dozen or so. When a patch of forest is
exploited, only a few trees - five to twenty out of 400 per
hectare - are actually taken. Most of the remaining trees become
damaged beyond recovery, far more than would be the case in a
temperate forest(114). The logs themselves are only a fraction
of the biomass affected or destroyed. Sixty million years of
evolution is being converted into tissue and wrap paper, wood-
pulp, veneers and constructions for which there are so many
alternative products available.

Japan, Europe and the USA are the main importers of tropical timber (35,18, and 8 million cubic meters respectively in
1975). Therefore, conservation of tropical timber should start
with reducing demands in these consumer regions. Each year the
average American consumes about as much wood in the form of paper as the average resident in many developing countries burns
for cooking(115).

About half of all the wood cut in the world each year is
for fuel, mainly by one-third of humanity who still rely on it
for cooking and heating. Here tremendous progress can be made
in saving wood by improving the stoves. This will also strongly
reduce the time and effort spent in collecting firewood. In most
developing countries daily firewood is very expensive and can
cost between 10 and 30% of the daily cost of living. At the
Technical University of Eindhoven, The Netherlands, a simple

type of family cooker was developed that can save about 75% of
the fuelwood in comparison to the traditional open fire-place.
All types of solid fuels, soft and hard wood, branches, bark,
bamboo, coal, coconut shells, sawdust briquets, peanut shells,
hard grasses, etc. can be used(116). The family cooker has been
tested in various forms and sizes in Ghana by the Foundation
Mondiaal Alternatief, Zandvoort, The Netherlands(117).

With regard to reforestation it should not come as a sur-
prise that the first success is reported from China, and that
the schemes have been set up without western interference or
help. An estimated 30 to 60 million hectares of new forests -
including those planted for environmental improvement and those
meant to supply village or industrial wood needs - have been
successfully established over the last quarter-century(115).
Chinese forestry efforts have inspired a great deal of thought
elsewhere about community forestry's possible design and accomplishment, and it can only be hoped that China's present lea-
ning towards more western technology and development will not
change that pattern too much. But authoritarian communism is
not a "must" for successful reforestation. Another success is
the rapid reforestation of South Korea after the Korean war,
and especially worth mentioning is the example from Gujarat,
India(118).

Methods of re-greening barren soils and reforestation of
eroded hills should always go together with reforming agricultural practices such as slash and burn, and the indiscriminate
burning of vegetation(119,120). It is obvious that phasing out
such ways of exploitation and adoption of more ecological methods require some time, persuasion, tactics and legislation
(121,122). It certainly should go parallel with offering new
opportunities, such as the introduction of new family cookers
as described above. Further suggestions for national plans to
halt desertification and other plans for transnational cooperation for planned regenerative land use reform were drawn up
by De Weille(123). Some of his ideas were taken up by the Gha-
naian-German Development Project, and more specifically, in the
Agomeda Agriculture Project in southern Ghana. The latter project includes community development and ecological ways of gardening and field cropping, making use of traditional Ghanaian
methods harmoniously integrated with small-scale techniques,
newly adapted from old European practice. Further details can
be obtained from the Foundation Mondiaal Alternatief, P.O.box
168, 2040 AD Zandvoort, The Netherlands.

5.4 Biosphere Reserves

The UNESCO-Programme on Man and the Biosphere (MAB), started in 1971, has in the course of the 1970s developed several

international cooperation projects that are highly relevant for
the conservation of nature, in particular in the tropics, concerning forest as well as grasslands, deserts and agro-ecosystems. Project 8 regards "Conservation of natural areas and the
genetic material they contain"(124). The project is better known
under the title Biosphere Reserves, as the idea is to have a
system of reserves all over the world that contains representative areas of the major biomes. Each BR should meet high standards of conservation to ensure that all the major ecological
components and processes are sustained under as natural conditions as possible. Each country submitting an area to be classified as a BR should indicate the steps that are or will be taken
to ensure that these standards will be met. An expert panel of
MAB then decides whether the protected area deserves BR status.
By November 1979 a total of 162 sites in 40 countries have been
listed(125).

When analysing the list of BR's it is unfortunate that the
sites are far from being representative of the world's ecosystems. "The Biosphere Reserves project, like UNESCO's World Heritage Trust, is absurdly under-funded. As a minimum requirement
each of the earth's biogeographical provinces, almost 200 of
them, should feature at least one Biosphere Reserve: to date,
less than one-quarter of these provinces possess such a Reserve.
To establish a typical Reserve costs around $ 100,000, and to
maintain it another $ 50,000 per year: so a programme to set
up another 150 Reserves, and to run them until the end of the
century, would cost $ 165 million - the equivalent of a value-
added tax of 0.1 percent on internationally traded oil (as has
been proposed by Saudi Arabia) extended over a mere 20 months"
(126).

At the ninth session of the Senior Advisers to ECE government on Environmental Problems, representatives of Sweden and
The Netherlands took the initiative of promoting a network of
Representative Ecological Areas in the ECE Region (Eastern and
Western Europe and North America). The idea is in its first stage
of inception, with Sweden as the leading country, whereas The
Netherlands took the lead with the evaluation of international
measures for the protection of flora and fauna and their habitats in the ECE Region(127).

5.5 World Conservation Strategy

The WCS, launched in March 1980 by the International Union
for the Conservation of Nature and Natural Resources (IUCN) with
advice, cooperation and financial assistance from the World
Wildlife Fund (WWF) and UNEP, and in collaboration with FAO and
UNESCO, stresses that conservation - like development - is for
people: while development aims at achieving human goals largely
through the use of the biosphere, conservation aims at achieving

them by ensuring that such use can continue. The underlying idea is very much toward an international ecological order, as opposed to the UN debate on a New International Order in which there is no ecological foundation(128).

Conservation is a process to be applied cross-sectorally and is not an activity-sector in its own right. Living resources conservation has three specific objectives:

. to maintain essential ecological processes and life-support systems (climate, soil, water, minerals),
. to preserve genetic diversity,
. to ensure the sustainable utilization of species and ecosystems.

Hence, the goal of the WCS is to become integrated with the International Development Strategy for the Third United Nations Development Decade, to ensure that modifications to the planet indeed secure the survival and wellbeing of people and ecosystems alike. The first part of the strategy proper is the setting up of a framework for national and subnational strategies in nine subsequent steps, reaching from reviewing the objectives to administrative and legislative measures. In each of these steps and measures, four strategic principles should be taken into account:

. to integrate conservation with development
. to retain options for multiple and future uses
. to mix cure and prevention
. to focus on cause as well as symptoms.

Each of the principles is then worked out such that they include anticipatory and cross-sectoral policies, evaluation of the state of the ecosystems as well as assessment of environmental effects of development, existing and desired legislation and organization, the training and research needed for the capacity to manage at the national and subnational level, and finally participation and education to support conservation with special emphasis on rural development. The WCS concludes with a checklist of priority requirements for action strategies. For a more detailed analysis see(129).

5.6 World Charter of Nature

In June 1980 the government of Zaire attempted to include a proposal in the agenda for the thirty fifth session of the United Nations General Assembly for a World Charter of Nature, similar to and on equal footing with the Charter on Human Rights. The general principles read as follows:

. Nature shall be respected and its essential processes shall
   not be disrupted.
. The continued existence of all forms of life shall not be
   compromised: the population levels of all species must be
   at least sufficient for their survival and to this end
   necessary habitats shall be maintained.
. All areas shall be subject to the principles of conservation: special protection shall be given to unique·areas,
   representative samples of all ecosystems and the habitats
   of endangered species.
. Ecosystems and organisms which are utilized by man shall
   be managed to achieve and maintain optimum sustainable
   productivity, but not in such a way as to endanger the in-
   tegrity of the ecosystem and organisms with which they co-
   exist.
. Nature shall be secured against warfare and other hostile
   activities.

Following these general principles are the responsibilities
and requirements for implementation by governments, notably that
the principles set forth in the Charter shall be reflected in
the law of each state, and that conservation strategies shall
be essential elements of national planning.

The proposal of the government of Zaire is now under consideration for final adoption at the 36th session, 1981, of the
United Nations General Assembly(130).

It is to be hoped that it is not already too late to come
to the rescue of the world's ecosystems and the biosphere's life-
support system.

REFERENCES

1.   Crimshaw, N.S. & L. Taylor, 1980: Food; Scientific American,
      Sept. 1980; 74-84.
2.   Böll, W., 1980: Africa in Transition, the Tropical Issue;
      Probleme der Entwicklungsländer, Vierteljahresbericht 79,
      March 1980.
3.   Buringh, P., H.D.J. van Heemst & G.J. Staring, 1975: Computa-
      tion of the absolute maximum food production of the world;
      Agric. Univ. Wageningen.
4.   De Hoogh, J., M.A. Keyzer, H. Linnemann & H.D.J. van Heemst,
      1976: Food for a growing world population; Agric. Univ. Wage-
      ningen.
5.   F.A.O. 1978, The state of food and agriculture 1977.
6.   Curry-Lindahl, K., 1978: Technical Assistance with Respos-
      ibility: Environment and Development in Developing Countries;
      Report to the Swedish Government, 503 p.
7.   Curry-Lindahl, K., 1979: Is Aid to Developing Countries Des-
      troying their Environment?, Oryx 15: 133-137.
8.   Henriques, P.C., 1977; De Ingenieur 89 (36): 683-685 (letter
      to the editor).
9.   Marsh, G.P., 1864: Man and Nature; re-edited 1965 by David
      Löwenthal, Harvard Un. Press.
10.   Carson, R., 1962; Silent Spring; Dutch Version 1963, Brecht
      Amsterdam.
11.   Dorst, J., 1970: Avant que la Nature meure; Delachaux Niestlé,
      Neuchâtel; English version, Collins, London.
12.   Farvar, M.T. & J.P. Milton, 1972: The Careless Technology;
      Natural History Press, N.Y.
13.   Eckholm, E., 1976: Losing Ground; W. Norton Inc., N.Y.
14.   IUCN-WWF-UNEP, 1980: World Conservation Strategy; IUCN-Gland.
15.   "North-South: A programme for Survival", 1980, Report of the
      Independent Commission on International Development Issues
      (Brandt-Report), Pan Books Ltd.
16.   Dammann, E., 1979: The Future in our Hands; Pergamon Press.
17.   Chichilnisky, G., R. Falk & J. Serra, 1980: Authoritarianism
      and Development, a Global Perspective; IFDA-Dossier 19: 3-14.
18.   Ahmad, E., 1980: The Neo-Fascist State, Notes on the Patho-
      logy of Power in the Third World; IFDA-Dossier 19: 15-26.
19.   Coleman, A., 1976, Is planning really necessary?; The Geo-
      graphical Journal 142 (3).
20.   University of California Food Task Force Study, 1974: A hun-
      gry world.
21.   OECD, 1979, The state of the environment.
22.   The Global 2000, 1980: 34-35 (Report to the President of the
      United States).
23.   UN, 1978, Conference on Desertifi ation, round-up, plan of
      action and resolutions.

24. Bali, Y.P. & J.S. Kanwar, 1977; FAO Soils Bulletin 34.
25. Das, D.C., 1977, FAO Soils Bulletin 33.
26. Pereira, H.C., 1973: Landuse and water resources in temperate and tropical climate; Cambridge Univ. Press.
27. FAO, 1978: Forestry Paper 7.
28. Kendeigh, S.C., 1974: Ecology; Prentice Hall.
29. Hunt, C.B. 1976: Physiography of the United States.
30. From a television programme, NOS–Television 1976.
31. SCOPE–14, 1979: Saharan Dust, Wiley & Sons.
32. UN–Conference on Desertification, 1977.
33. Van der Schalie, H., 1974: The Aswan Dam revisited; Environment 16 (9).
34. Several newspaper articles, December 1980.
35. SMIC, 1971: Inadvertent Climate Modification, MIT–Press.
36. FAO, 1971: Environmental aspects of natural resources management: Forestry.
37. FAO, 1976: Fisheries Technical Papers 162.
38. FAO, 1977: Fisheries Technical Papers 172.
39. US–AID, 1979: Environmental Report on Sri Lanka.
40. FAO, 1975: Committee for Inland Fisheries in Africa, Technical Paper 3.
41. Ehrlich, P.R., A.H. Ehrlich & J.P. Holdren, 1977: Ecoscience; Freeman & Comp.: 161–166.
42. Semb, A., 1977; Water Air and Soil Pollution 6: 231–240.
43. Wright, R.F. & E.T Giessing 1976; Ambio 5 (5–6): 219–223.
44. Schofield, L., 1976; Ambio 5 (5–6): 228–230.
45. Knabe, W., 1976; Ambio 5 (5–6): 213–218.
46. Ottar, B., 1977, Ambio 6 (5) 262–269.
47. OECD–Observer 88, 1977: 6–8.
48. Ambio 5 (5–6) 1967, entire issue on SO_2 pollution, monitoring and impact.
49. Water, Air and Soil Pollution 6, 1976; entire issue on SO_2.
50. Ambio 9 (5) 1980, several articles on SO_2 and acid rains.
51. Proceedings of an international conference on Ecological Impact of Acid Precipitation, Sandefjord, Norway, March 1980; obtainable at the Norwegian Ministry of the Environment, Oslo.
52. ECE, 1980: First Session of the Interim Body for the Convention on Long-range Transboundary Air Pollution; 27–31 Oct. 1980, Geneva.
53. Vermeulen, A.J., 1978: Acid precipitation in The Netherlands; Provinciale Waterstaat Noord-Holland.
54. UNEP, 1980: Coordinating Committee on the Ozone Layer, Bilthoven, Nov. 1980.
55. Dethlevsen, V. & B. Watermann, 1980: Epidermal Papilloma of North Sea Dab (Limanda limanda); ICES-Special Meeting on Diseases of Commercially Important Fish and Shellfish, Paper 8.
56. Van der Wel, C., 1980: Vis uit de Noordzee, eet smakelijk; Nieuwe Revue: 49: 48–61.
57. Reynders, P.J.H., 1980; On the causes of the decrease in the Harbour Seal (Phoca vitulina) population in the Dutch Waddensea (doctor thesis at Wageningen Agricultural University).

58. Lasonder, D.M.J. 1981, Het Rijn proces tegen de Franse Kali-mijnen; Panda 17 (1): 9-13.
59. St. Reinwater, begunstigersberichten, Tweede Weteringplant-soen 9, 1017 ZD Amsterdam, (series of communications).
60. Schmit, H., 1980: Een tijdbom in de bodem; Trouw 6 Sept.1980.
61. Brandpunt-Television interview with Minister L. Ginjaar on 6 Dec. 1980.
62. Dalrymple, D., 1975; USDA Report 106, Washington D.C.
63. Brown, L.C., 1970: Seeds of Change; Preager, N.Y.
64. Glaeser, B. (ed), 1980: Factors affecting Land Use and Food Production; Sozial wissensch. Stud. zu intern. Probl. 55: 33.
65. FAO, 1975; Ceres, May/June.
66. Ehrlich, P.R., A.H. Ehrlich & J. P. Holdren, 1977: Ecoscience; Freeman & Comp: 339-341.
67. MIRCEN-News 1. 1980.
68. Schneider, S.H. & L.E. Mesirow, 1974: The Genesis Strategy.
69. Allaby, M, 1977: Miracle Rice Breeds Miracle Locusts; Ecologist 3 (5) May 1973: 180.
70. Wall Street Journal June 14, 1972 p. 1.
71. Mooney, P.R. 1980: Seeds of the Earth; Int. Coalition for Dev. Action, London: 43.
72. Perelman, M., 1977: Farming for Profit in a Hungry World; Landmarks series: 144.
73. Barnes, P., 1975 in R. Rodale ed.: New Food Chains.
74. Wilkes, G., 1977; Bull. of the Atomic Scientist, Feb. 1977:11.
75. Grubben, G.J.H., 1977; IBPGR: 7.
76. Kendrick Jr., J.B., 1977: Calif. Agric., Sept. 1977:2.
77. Eckholm, E., 1978, Disappearing Species, The Social Challenge 6.
78. Plant Genetic Resources of Canada, Newsletter 5, Oct. 1978:5.
79. IBPGR, 1977 Annual Report: 65-67.
80. Mooney, Seeds...: 21-22.
81. Mooney, Seeds...: 49.
82. Freistritzer, W.P. (ed.), 1976: Seed Industry Development, A Guide to Planning Decision Making and Operation of Seed Programmes and Projects, FAO: 28.
83. Mooney, Seeds...: 55-60.
84. Mooney, Seeds...: 61-63.
85. Mooney, Seeds...: 69.
86. IBPGR Newsletter 34, June 1978: 13.
87. Mooney, Seeds...: 71-72.
88. Mooney, Seeds...: 77.
89. US National Acad. of Sciences, 1978: Conservation of Germ Plasm Resources: An Imperative.
90. Mooney, Seeds...: 81-82.
91. Mooney, Seeds...: 84-85.
92. Whiteside, T., 1977: Tomatoes; The New Yorker, January 24:57.
93. Waterloo, C., 1978: Pesticides; what we don't know; Des Moines Sunday Register, April 23.
94. Science Council of Canada, 1977: Agriculture to the Year 2000, Draft report: 71.

95. UNEP, 1978: State of the Environment: Pesticides.
96. Half Million Poisoning Blamed on Third World Pesticides Use;
     The Western Producer, March 23, 1978: 37.
97. Saskatchewan Dept. of Agric., 1978: Science.
98. Mooney, Seeds...: 91.
99. Mooney, Seeds...: 103-107.
100. UNIDO, 1978: Report on the Technical Consultation on Produc-
     tion of Drugs from Medicinal Plants in Developing Countries,
     Lucknow, India, March 1978, ID/WG271/6.
101. IUCN - Red Data Book, 1975.
102. IUCN - Plant Red Data Book, 1978.
103. Myers, N., 1979: The Sinking Ark; Pergamon Press.
104. Raven, P.H., 1980: Statement before the Committee on Foreign
     Affairs, Subcommittee on International Organizations, U.S.
     House of Representatives, May 7th: 2.
105. Brown, L.R., Food or Fuel, World Watch Paper 35, March 1980:
     29.
106. U.S. Nat. Acad. of Sciences, 1978: Conservation of Germ Plasm
     Resources: An Imperative: 30.
107. U.S. National Research Council, 1980: Research Priorities in
     the Tropics.
108. Myers, N., 1980: The Present Status and Future Prospects of
     Tropical Moist Forests; Environmental Conservation 7 (2):
     104-108.
109. Myers, idem, 108-109.
110. Myers, idem, 110-112.
111. Grainger, A., 1980: The State of the World' Tropical Forests;
     Ecologist 10 (1-2) Jan.Feb.
112. World Bank, 1978: Forestry, Sector Policy Paper.
113. FAO, 1979, Trade Yearbook 1976.
114. IUCN-Bulletin, May 1980: 31.
115. Eckholm, E., 1979: Planting for the Future: Forestry for
     Human Needs; World Watch Paper 26:7.
116. Overhaart, J.C., 1980: The Family Cooker; Technical Univer-
     sity Eindhoven, Division of Appropriate Technology, Technical
     Guide.
117. The Family Cooker tested in Ghana, Ecoscript 11, Foundation
     Mondiaal Alternatief, PO box 168, 2040 AD Zandvoort, Nether-
     lands (in prep.).
118. Eckholm, idem: 48-56.
119. De Weille, G.A., 1977: Desertification and ecological meas-
     ures to combat it; Ecoscript 1 (Foundation Mondiaal Alter-
     natief, Zandvoort).
120. Korem, A., 1979: Influence of indiscriminate burning on
     desiccation and desertification - observations in Northern
     Ghana; Ecoscript 9.
121. De Weille, G.A., 1977: A suggestion for national plans to
     halt desertification: Ecoscript 2.
122. De Weille, G.A., 1978: Greening measures versus anthropogen-
     ic ecosystems deterioration in the Tropics; Ecoscript 8.

123. De Weille, G.A., 1978: Plans for Transnational Cooperation for Planned Regenerative Land Use Reform; Proc. Regional Conference of the International Geographical Union, Lagos.
124. UNESCO, 1973: MAB–Report 12.
125. UNESCO, 1978: MAB–Report 48 chapter 6 and annex with list of Biosphere Reserves.
126. Myers, N., 1980: The problem of disappearing species, what can be done?; Ambio 9 (5): 229–235.
127. ECE–ENV/R 125, 11 Nov. 1980 (Geneva).
128. De Weille, G.A., 1978: A Plea for a New International Ecological Order; Ecoscript 7.
129. Allen, R., 1980: How to Save the World; The Ecologist 10 (6–7) Aug./Sept. 1980.
130. U.N.–General Assembly, 1980: Proposal for a World Charter of Nature, by the Government of Zaïre; U.N.G.A., A/35/141, 11 June 1980.

DISCUSSION

A participant said that there has been an increasing interest in long-range weather prediction in Australia, on the order
of months to seasons and more. He asked Dr. Hallsworth if this
research interest has been translated into any sort of advice
to agriculturalists about rainfall. Dr. Hallsworth replied that
no advice of that sort has come forward.

A participant said that an alternative to exploiting new
arable land is to make better use of land already under cultivation. By any of today's standards the southeastern United States in the 1920s and 1930s was a developing country. Since that
time, agricultural production in the region has increased on
a substantially smaller area. One of the early extension programs
taught farmers how to cultivate corn on steep slopes without
causing major erosion. This was a significant program at the
time, but now it is useless because that kind of land is not
being used to grow crops. Dr. Hallsworth replied that soil scientists talk about permissible losses of 10 t/ha. This means that
6 inches of soil is gone in 250 years. That is not much longer
than most of the southern United States has been settled, and
much of that area has lost its topsoil. The same thing has happened in the south of England, only they have been farming there
for 1500 years and they are down to the rock. In the city of
Bath, the Roman baths were used until the year 500, then they
were lost until around 1900 when they were excavated from under
7 feet of silt.

Dr. Klingauf was asked about the reason for the response
of insects before a change in atmospheric pressure. He said that
this response has been observed one or two days before a change
in pressure, but that they do not know what causes it. Ionization
has been suggested as a possibility. Asked if control practices
can be matched to life cycles of pests and to climatic cycles
to set up schedules that will make them more effective, Dr. Klingauf replied that if pesticides are applied by schedules based
on climatic conditions and insect life cycles, side effects can
be controlled better. Integrated pest management can minimize
pesticide applications. To develop these schedules, it is necessary to monitor weather conditions in individual fields, particularly leaf wetness and temperature. Dr. Klingauf was asked
how this sort of information is transferred to farmers and whether the farmers use it. He replied that the German Agrometeorological Service in Frankfurt supplies the Plant Protection Ser-

W. Bach, J. Pankrath, and S. H. Schneider (eds.), Food-Climate Interactions, 361–365.

vice with weekly maps showing danger zones for Phytophthora, and
this information is disseminated to farmers.

There are problems in using pesticides for agriculture versus
public health hazards. Some pesticides used for agriculture have
caused mosquitoes to develop resistance, increasing the malaria
threat. It has been proposed that some pesticides be reserved for
public health use. If you increase food production but cause more
malaria, you are working against yourself in terms of human life.
Dr. Klingauf replied that it is forbidden to use antibiotics for
plant protection in Germany; they are reserved for medical use with
humans and animals. He said that it would be good to reserve some
pesticides, but it would cause an economic problem. For example,
beet flies and mosquitoes are closely related species and must be
controlled with the same pesticides.

In the United States, about ten universities and extension
service offices use computer networks linked to terminals in county extension agents' offices or even the farmer's own home to relay information rapidly to help the farmer assess all alternatives
and use pesticides only when they are really needed and are most
effective. Asked if mechanical control of pests is ever effective,
Dr. Klingauf replied that mechanical methods of pest control are
effective only in stored products such as grain, but they are not
used much because pesticides are cheaper.

In response to Dr. Pimentel's paper, a participant said that
eco-farming uses much less energy than conventional farming, suggesting a strong argument for eco-farming in both developing and
developed countries. Dr. Pimentel said that crop rotation substitutes land for energy by putting part of the land in crops like
legumes half the time. He said that we should manage our land resources intelligently and not go on mining the soil as we have been
doing.

A participant said that Dr. Pimentel had shown a relation between population growth and energy growth. From 1860 up to World
War I, the energy growth was approximately 4 to 4.5% per year. Between the two wars it was about 2% annually. From 1948 to 1973, it
was about 4.3%. For the last 7 or 8 years, since the beginning of
the oil crisis, it has slowed down to 2 or 3%. Up until a few years
ago, population growth was about 2% annually, which means a doubling time of about 35 years. This means there is practically no per-
capita energy growth. He suggested that maybe this is an indication of a decoupling of economic growth and energy growth. Dr. Pimentel replied that the amount of energy that goes into producing
food would be a better measure, as well as the amount of energy
that goes into protecting people from diseases. There is no doubt
that energy is needed to produce food above the subsistence level
and that energy is needed to purify water and protect man from di-

seases. These are important and are related to population growth, but a lot of energy use is not.

Dr. Pimentel said that a lot more information is needed on carbon cycling. He believes that many estimates made by ecologists have been too optimistic. He said that he is concerned about the impacts on agriculture by using crop residues in biomass energy production. This organic matter is crucial to good and productive agriculture.

A participant said that a conference on energy held in Germany in 1979 concluded that the energy crisis in developing countries is caused by the inefficient use of energy. He said that most of the energy used in developing countries is for cooking and relatively little is for agriculture. Dr. Pimentel said that too little energy is being used in agricultural systems in developing countries and that a lot more fertilizer should go into the systems. About 60% of the energy consumed in developing countries is used for cooking, 30% for agriculture and 10% for other uses. Most of the energy sources used for cooking are renewable and most of the energy for agriculture is from fossil sources. Agriculture is a net energy producer, even in U.S. and European grain production, where we get about 2.5 kcal out for each kcal of fossil energy input. With fruit or vegetables, it takes about 2 or 3 cal of input for each cal of output. It is important to consider what kind of energy is going in and out of these systems.

A participant pointed out that the threefold increase in energy use from 1945 to 1975 was in a period of declining real cost of energy. Given higher energy costs, he asked, what would that energy increase have been? If you held labor costs constant and increased energy prices fivefold, Dr. Pimentel replied, you probably would not have gotten that sort of increase in energy use, because labor would have been substituted for mechanization.

Responding to Dr. Hekstra's paper, a participant said that in our analyses of species, we have neglected the little ones, i. e. the insects and microorganisms. Environmental impact statements that are produced in the United States emphasize birds, mammal, fishes and so on, he said. But in terms of biomass, the insects amount to about 1000 kg/ha, the earthworms about 1000 kg/ha and fungi and bacteria about 2000 kg/ha. This compares to a human biomass that averages only 16 kg/ha in the United States and 80 kg/ha in Bangladesh. We could probably do without the birds and mammals, but we would be in real trouble without the insects and other little species. Yet we know very little about these small organisms and the really vital role that they play in agriculture and the biosphere itself. Environmental impact statements often concentrate on birds and mammals because they function as indicators. They are at the top of food-chain pyramids.

A participant said that we should look at the quality of species as well as the quantity. However, as we have not investigated the characteristics of every species in detail, perhaps it is safest to keep as many as possible.

Another participant said that we must recognize the need to manage a large portion of our ecosystem. We probably cannot do this and still maintain the full complement of species that would be found in a natural ecosystem. Dr. Hekstra added that good agricultural land should be used to raise crops and not squandered on highways. The land should be managed as a good agricultural ecosystem, and as many wild species as possible should be maintained as a genetic resource. The oil palm, for example, has gone from a wild species to a crop in only 50 years. A participant said that many developing and developed countries do not have good collections of germ plasm. They do not collect and store the material and have it available for retrieval. He said that there is a need for a concerted effort supported by the international community. Each nation should have its own store, but such collections are difficult and expensive to build and maintain. An international effort is necessary.

One participant took issue with claims that plant genetic resources have been disappearing as a result of the Green Revolution. He said that there has been a magnificent effort to collect practically all of the available germ plasm for the major crops that have been the focus of work at the international centers. These collections include near and wild relatives of the species with which the centers work and are more complete than they have ever been. He said that there are also efforts to add some vegetable, tree and root crops to the collections, and asserted that we are better assured than ever before that the germ plasm is in safe hands. Dr. Hekstra said that 400 old strains have been lost each year for the last 5 to 10 years in the European Community. Eight agencies are supporting European Community seed banks because the individual governments do not provide sufficient financing. All over the world there is a tendency for the best seed banks to be transferred from national agencies to corporate agencies, such as in the case of Royal Dutch Shell who owns 24 seed banks. One participant questioned the statement that genetic stock is in the hands of a few grasping companies. In the United States the American Seed Trade Association includes about four major seed corn companies, but at least 500 small companies get their genetic inbred stock from state universities and agricultural experiment stations, which are publicly supported. At least half of the hybrid seed planted in the United States comes from this second group.

A representative of the West German Ministry of Economic Cooperation said that his agency is supporting two genetic resource centers and the European Community also has two centers. He urged governments to take responsibility for maintaining genetic resources instead of leaving it to corporations.

Another participant said that some current figures on estimates of arable land are alarmist. In the Near East, for example, land that is actually used for farming amounts to 110% of the amount often estimated to be suitable for long-term agricultural use. Looking ahead to the year 2000, FAO envisages a 23% increase in arable land whereas the Global 2000 report projected only 4%.

BASIC DATA REQUIREMENTS - EXPERIENCE WITH THE WORLD WHEAT EXPERIMENT OF THE WORLD METEOROLOGICAL ORGANIZATION

Karsten Heger

German Weather Service, Offenbach
Federal Republic of Germany

1. INTRODUCTION

During the fifties and sixties the climatic conditions were so favourable in the principle wheat producing countries that there were hardly any negative effects on the yield. During both decades yields increased due to improved cultivation techniques. The yields reached such a dimension that in North America arable land had to be left uncultivated in order to balance supply and demand. This situation changed in the seventies. In 1972 the wheat cultivation in the USSR, China, India, Australia, South East Asia and the Sahel was impaired by unfavourable weather conditions. The world production of wheat and rice decreased by 33 million tons (1). This decrease in production could only be compensated by consuming almost all reserves. The crop failure in the wheat and rice production was alarming, as the nutrional situation would become disastrous (1) in the world if unfavourable or catastrophic weather conditions would occur in the wheat producing areas several years in succession.

People became aware again that the production of food depended not only on factors like variety, fertilising, cultivation, pests or intercompetition of weeds but essentially on the weather conditions during the entire vegetation period. According to the climatic conditions of a region single or several meteorological elements like precipitation, evaporation, air temperature, wind, radiation with differing intensity, affect positively or negatively the different phases of development. Some elements may act as restricting factors - water in arid areas or the number of days without frost during the period of growth in areas near the Pole. Due to the great variability of the weather within the different climatic regions, it seems obvious

367

W. Bach, J. Pankrath, and S. H. Schneider (eds.), Food-Climate Interactions, 367–382.

to ask which yield or changes in yield might be expected. This question
may refer to a certain year or a longer period of twenty or more years. In
the first case it is a matter of forecasting the yield in the corresponding
year. In the second case the question refers to the yield potential of an
area and to the effects of the climatic variability on the yield (i.e. the
variability of the yield). The annual fluctuations in the yield are superimposed on the general trend towards an increase of the yield, which has been
achieved in the past three decades due to improved production techniques.

The development of more sophisticated computers made it possible to
describe the dependency of the yield on the weather not only by purely
statistical models as in the previous decades. On the one hand these models
were valid for the climatic region only for which they had been developed
on the basis of the corresponding meteorological data. Thus the statistical
models are restricted to the site, the corresponding period, i.e. the limited
conditions of the possible weather fluctuations of the period under study and
restricted to the specific cultivation conditions. On the other hand it cannot be expected that the statistical model shows the interaction of physical
and biological laws in the complex system plant–atmosphere–soil. Consequently the yields calculated were not very satisfactory when compared
with the actual yields obtained. A better conformity is expected by using
simulation models in which account is taken of the biological and physical
laws of the plant growth. It was expected that such dynamic models could
be applied to other cultivation areas if the most relevant biological and
physical processes were described mathematically as accurately as possible.
Such models have been developed and published by numerous authors, among
them de Wit, Brouwer and Penning de Vries (2), Goudriaan (3) or van Keulen
and Louwerse (4) who verified their models on the basis of experiments which
had been performed under controlled conditions. However, these experiments
were restricted to certain regions so that it was not possible to verify if they
were transferable to other climatic regions. This situation just called for an
international experiment, providing values of biological, meteorological
and soil–physical parameters of different climatic regions according to
methods common to all participants.

2. DEVELOPMENT OF THE WORLD WHEAT EXPERIMENT OF THE WORLD METEOROLOGICAL ORGANIZATION (WMO)

Planning and performance of a global wheat experiment were initiated
by the Food and Agricultural Organization (FAO). Already at the beginning
of the seventies FAO conceived an agrometeorological research programme
in order to obtain meteorological observations from different vegetative
stages of agricultural crops at the same period. The reactions of the crops
to the meteorological conditions in different climatic regions were to be

studied. The activities envisaged aimed at establishing, on the basis of the data obtained, statistics of the close relationship between plant production and climate. Apart from this it was planned to use the data for the preparation of yield models based on biological and physiological laws, i.e. for the preparation of simulation models or dynamic models.

This FAO experiment was taken up by the Commission for Agricultural Meteorology (CAgM) of the WMO. At its fifth session in Geneva, 18 to 29 October 1971, a working group was established.

At their session the representatives of the working group stated that the basis of the experiment was the performance of meteorological and biological observations according to methods common to all experimental sites. The data collected were to be used for the establishment of bio-physical yield models replacing the statistical relationship between yield and weather connected with site and time. Wheat was chosen as a test plant as it grows under climatic conditions which differ considerably from place to place. Already at this stage of the preparations of the experiment it was clear that the biological reaction of the plants - in this special case of wheat - on the physical environmental conditions in the atmosphere near ground level and in the soil represented an extremely complex system. For this reason the participants in the working group agreed to proceed gradually in evaluating the experiment. First of all attempts were to be made to establish a simple simulation model. This seemed feasible as the most important elements for the growth of the wheat and its yield, i.e. temperature and soil humidity, were known. Such a simple simulation model could already be used as a basis for agroclimatological activities in which e.g. the eventual yields under extreme climatic conditions, can be estimated. This is a problem of particular interest to FAO in order to estimate the possible food reserves.

As another future objective of the experiment the preparation of a dynamic model was considered showing as accurately as possible the growth of wheat by considering physical effects and corresponding biological reactions of the plants during their entire life cycle. Such a model offered several possibilities of application:

- It is possible to estimate and forecast at the different stages of development in the life cycle of the plant the expected yield on the basis of the weather occurred. However, an exact prognosis of the yield cannot be made under present conditions as it is not possible to forecast the weather several weeks and months in advance.

- The dynamic model serves as a basis for agrometeorological computations of the potential production on a global scale and provides better founded results than the simple model.

- Expected climatic changes can be simulated in the model. These simu-
lations show the effects on the yield if the climatic conditions change
but the biological conditions for the wheat variety used remain the same.
This is an aspect which has become rather topical due to the World Climate Programme.

Based on the above objections the working group set the following
terms of reference:

a) Develop techniques for taking observations on wheat and its environment
for evaluating the effect of weather/climate on its potential production
and quality on an international basis.

b) Define the experimental observations and procedures necessary for the
conduct of these trials.

c) Arrange for the conduct of trials in a number of countries with differing
climatic regimes.

d) Make suitable arrangements for the recording of the agrometeorological
and biological data, and for their collection centrally.

3. EXPERIMENTAL PROGRAMME

The experimental programme was conceived in such a way that the
largest possible variety as regards climatic conditions was given. This was
achieved by the participation of the eight countries|listed in Fig.1.The ex-
periment was planned so that only the climatic conditions varied but not
the other factors having an influence on the growth of the plant (apart from
certain variations with respect to fertilization and date of sowing).

The Mexican variety Siete Cerros developed by the International
Maize and Wheat Improvement Centre (CIMMYT) was sown at all experimental sites. This variety is marked by a photoperiodic indifference. In
addition a home variety was to be grown. Its yields would show whether
the weather of the corresponding vegetation period had any positive or
negative effects compared with preceding years.

The experimental site had to meet the following requirements:

a) the soil is representative for the area

b) the soil is fertile and free of stones

c) the soil is without impermeable layers (hard pan, rock layers) to a depth
of 150 cm

d) the orography is such that there will be no mesoscale bias, i.e. the climatic pecularities of the soil are not too marked.

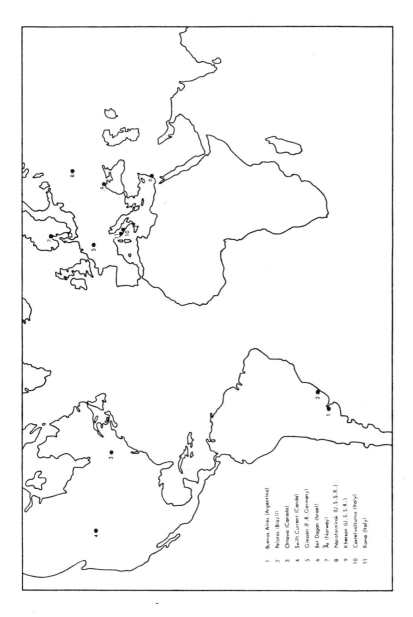

Figure 1: Position of the stations of the World Wheat Experiment

1   Buenos Aires (Argentina)
2   Pelotas (Brazil)
3   Ottawa (Canada)
4   Swift Current (Canada)
5   Giessen (F.R. Germany)
6   Bet Dagan (Israel)
7   Ås (Norway)
8   Narotominsk (U.S.S.R.)
9   Kherson (U.S.S.R.)
10 Castelvolturno (Italy)
11 Roma (Italy)

Figure 2 shows the arrangement of the plots where eight varying tests were made. Each plot was 8 x 4 m and the plots were separated from each other by rows of 0,5 m width. The border area was 10 m wide where the home variety had been grown. When 4 replications were made ca. 0,3 ha were necessary including the border area.

In this experiment, to be repeated every 4 years, the factors were to be kept constant. For this reason the same plot had to be used.

As "optimum fertilizer" the following applications are considered:

Nitrogen   (as water soluble N)   6 - 12 ppm

Phosphorus   (as water soluble P)   6 - 12 ppm

Potassium   (as water soluble K)   20 - 60 ppm

ph of soil should be within a range of 6.4 to 7.4.

The normal quantity of fertilizer was applied during the final stages of the site preparation. The additional quantity of fertilizer of 15 % was given during the jointing stage for autumn plantings and at seeding time for spring plantings. Seed was planted at a rate of 110 kg/ha, spaced in rows 15 cm apart and planted at a depth of 2 - 5 cm.

The meteorological measuring programme was conceived in such a way that only those parameters were recorded which are internationally required and exchanged for general meteorological purposes. This is important for subsequent applications of the growth and yield model aimed at international planning, as only these data are operationally available. Air temperature and relative humidity were recorded hourly at a height of 2 m. Hours of bright sunshine and daily global radiation, average wind speed of the day and of the night, daily precipitation maximum and minimum temperature and soil temperature values at 5 cm, 10 cm and 20 cm depth were measured. The experimental site was to be located near (maximum distance 100 m) a climatological station with long-term observations.

The weekly measurement of the soil moisture in layers of 10 cm to a depth of 100 cm was to document the water status of the crops as the soil moisture is a factor which considerably restricts the yield, especially in arid areas. Apart from some exceptions, the observations were made according to the gravimetric method.

The biological observations from which conclusions could be drawn on the growth stages of the cultures, were of great importance. The phenological observations were:

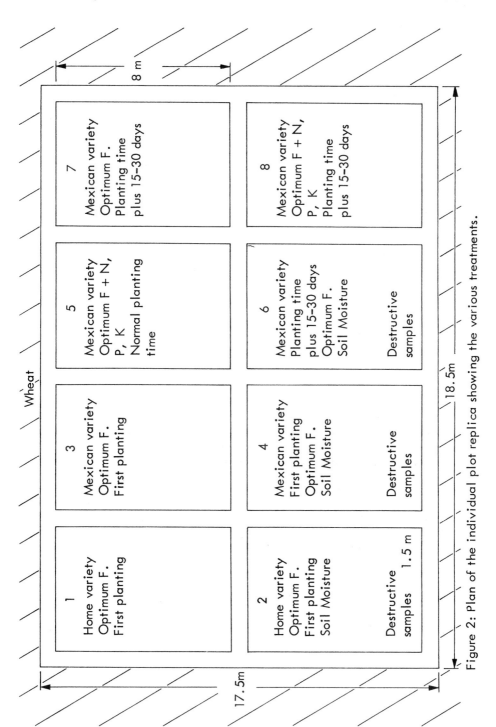

Figure 2: Plan of the individual plot replica showing the various treatments.

date of sowing,
> of emergence,
> of jointing,
> of heading (when 50 % complete ear emergence is observed from 4
> samples of 20 plants each),
> of flowering,
> of milk stage,
> of soft dough stage and
> of total ripeness.

Additional observations during the growing period were to show ir-
regularties which might account for variations in yield. They could be
caused by winter kills due to severe frost without snow cover or due to
lack of oxygen under iced snow cover, heavy plant diseases or noxious
animals.

At the end of the vegetation period the yield was determined at the
different plots:

a) total weight of grain (kg/ha)

b) weight of a 1000 kernels (gm)

c) kernel-water precentage and/or dry matter of the kernels

d) dry matter of straw.

As soon as the home variety or the Mexican variety sown at the two
different periods had reached the jointing stage, the plants were cut every
two weeks 5 cm above ground from 4 different rows at a length of one meter.
Then they were dried in order to determine the dry matter. The average
value of the four samples was transferred to an area of 1 m^2.

In addition samples of the home variety and the Mexican variety for
both normal and delayed planting dates of 2,5 kg sizes have been sent to
the Bundesanstalt für Getreideforschung in Detmold (Federal Research
Authority for Grain Utilization), Federal Republic of Germany, to determine, according to a homogeneous method, protein content, baking quality,
ash constituent and starch content.

4.   APPLICATION OF A GROWTH MODEL TO THE DATA OF THE
WORLD WHEAT EXPERIMENT

4.1 Model description

After the data compiled during the World Wheat Experiment had
passed a quality control, my colleague Dannecker applied a model published by Hodges and Kanemasu (5) on these data. Both authors had
measured at different sites near Riley, Ellsworth and Finney Counties,

Kansas (USA) with infrared gas analyzers (URAS) the exchange of CO_2 in chambers with 200 or 400 wheat plants each. At the experimental sites the leaf area index LAI was measured with an optical planimeter and the dry matter of the plants 2 to 4 times per month. Apart from the photosynthetic active radiation PAR the reflected PAR (RPAR), the PAR (TPAR) transmitted within the crop and the PAR (RSPAR) reflected by the soil were registered with different sensors. Air temperature and relative humidity were recorded with thermographs. An evapotranspiration model by Kanemasu et al. (6) for sorghum and soybeans was used for the determination of the soil water status.

The measurements of the carbon dioxide exchange and the meteorological and phenological parameters served to

a) investigate whether photosynthesis and respiration depend on the phenological stage and to

b) develop equations describing photosynthesis, respiration, dry matter production as a function of the leaf area index LAI and the meteorological variables.

The gross photosynthesis deviates between emergence and jointing from the one between jointing and maturity and is a function of the intercepted photosynthetically active radiation IPAR and a water stress factor dermined on the basis of the available soil water. The night respiration is a function of the night length, the daily net carbon exchange and the accumulated dry matter. The daytime respiration depends on the length of the day, gross photosynthesis, accumulated dry matter and air temperature. The net carbon exchange of a whole day as a sum of the gross photosynthesis and respiration is converted into the dry matter production. The constants appearing in the different equations have been determined by multiple regressions. Thus these equations have been determined by pecularities of the site, time and variety during the experiment in Kansas so that the transferability is problematic.

4.2 Application of the model to the World Wheat Experiment data

In spite of scruples with respect to the transferability of the model on the dry matter production of winter wheat prepared under the climatic conditions of Kansas to other climates it was used for the data of the World Wheat Experiment. Two results are possible:

a) The conformity between calculated and observed yields is satisfactory. Then the World Wheat Experiment would have served to prove the transferability of the model to any climate, as the climatic differences of the stations participating in the experiment are sufficiently large. There would be no objection to applying the model for determining the potential yield.

b) The conformity between observation and calculation of the yield is not
satisfactory. There is then at least the positive result that experience
has been gained for the preparation of another model. Apart from this
it is easier to see whether a sufficient amount of data has been provided
by the World Wheat Experiment as far as extent and quality are concerned
or whether it is necessary to continue the experiment with changed
conditions.

Table 1 shows the yields achieved at the stations Ås, Giessen, Rome,
Kherson and Buenos Aires between 1973 and 1976. In the first and second
line the calculated and measured dry matter of Siete Cerros are compared,
line three shows the percentage of the change. The stations Ås, Giessen
and Rome are a cross-section of Europe. Apart from one exception each at
Giessen (1974), Rome (1976) and Swift Current (1973) the conformity between calculated and measured dry matter is relatively good. With the very
low yield at Kherson the deviations up to 356 % are extremely high. They

Table 1: Measured and calculated dry matter $[\mathrm{mg}\ \mathrm{cm}^{-2}]$ of the Mexican
variety Siete Cerros. First line measured value, second line
calculated value, third line percentage of the change.

Station	1973	1974	1975	1976
Ås	94	80	82	44
	107	105	64	28
	+ 16%	+ 31%	- 22%	- 39%
Giessen[+)]	78	84	81	87
	111	155	87	80
	+ 42%	+ 84%	+ 7%	- 8%
Rome	--	132	117	83
	--	131	99	142
	--	- 1%	- 16%	+ 71%
Kherson	--	36	14	40
	--	91	64	60
	--	+153%	+357%	+ 50%
Buenos Aires	59	90	75	77
	74	84	80	124
	+ 25%	- 7%	+ 6%	+ 61%
Swift Current	36	39	45	86
	16	47	37	84
	- 56%	+ 21%	- 18%	- 2%

[+)] winter wheat (plot 2)

might be attributed to the following facts: The wheat-growth-model developed by Hodges and Kanemasu only on the basis of experiments in the climatic area of Kansas, does not take sufficiently into account the especially dry conditions in Kherson during the World Wheat Experiment. Apart from this lower yields might be due to cultivation methods at Kherson differing from those in the USA.

The conclusion has to be drawn that the application of the simple wheat-growth-model developed by Hodges and Kanemasu did not yield the desired result to be used as a basis for agroclimatic studies showing the possible wheat production of different regions.

5. POSSIBILITIES AND LIMITS OF THE WORLD WHEAT EXPERIMENT

The four years' World Wheat Experiment being terminated, some experience has been gained and the possibilities and limits of such an experiment may be discussed.

Some reductions have to be made as regards the first objective, i.e. the collection of meteorological and biological data in different climatic regions under homogeneous test conditions. The request for a uniformity of the cultivation conditions has not always been achieved. The great differences in the production of the stations cannot only be explained by the deviating climatic conditions. Table 2 shows a comparison of the dry matter (kernels and straw) of the variety Siete Cerros and of the home variety at all experimental sites. Especially obvious are the great differences in the production and the small yields of the two Russian stations.

Table 2: Total dry matter at maturity of the home variety / the Mexican variety Siete Cerros $\left[\text{mg} \cdot \text{cm}^{-2} \right]$

Station	1973	1974	1975	1976
Ås	101/ 94	89/ 80	99/ 82	61/ 44
Rome	-/ -	132/132	120/117	83/ 83
Castelvolturno	-/ -	118/124	124/139	-/ 95
Kherson	-/ -	66/ 36	13/ 14	32/ 40
Naroforminsk	36/ 19	99/ 48	57/ 22	116/ 83
Bet Dagan	137/108	-/ -	(70/ 65)	-/ -
Swift Current	38/ 36	38/ 39	43/ 45	92/ 86
Ottawa	-/ -	45/ 48	32/ 21	-/ -
Buenos Aires	53/ 59	68/ 90	42/ 75	60/ 77
Pelotas	58/ 54	96/ 88	72/ 62	60/ 61

Table 2 also answers the question how far the selection of the Mexican variety Siete Cerros has proved successful. In general there is no need to be afraid of a comparison with the home varieties. Solely in Buenos Aires the yield was considerably higher, whereas it was considerably lower in Naroforminsk.

Table 3 shows the effects of the differences in the fertilization as well as the differences in the production in mg/cm^2 for the different years and the total of the annual deviations at the individual stations. There is a tendency towards higher yields when the common quantities of fertilizer-admissible in given limits - N, P and K are increased by 15 %. Larger increases in the production were achieved in Naroforminsk and Pelotas. This may be interpreted thus, that there is too great a margin as far as the definition of the so-called optimum fertilizer is concerned. But there might also have been differences in the technical conditions and measures of cultivation so that the requests for homogeneous test conditions of the World Wheat Experiment have been met only partially. The question remains unanswered whether a uniformity is at all possible when account is also taken of the natural differences in the soil-physical conditions between the different experimental sites.

Which experience might be gained with the available data from the tests with the growth model? A significant obstacle in the application of the model is that there are no indications with respect to the leaf area index. For the development of a complicated growth model no areas are available to which assimilation and respiration can be related. Even if at two weekly intervals during the vegetation period of the World Wheat

Table 3: Differences of dry matter $\left[mg\cdot cm^{-2}\right]$ of the Mexican variety Siete Cerros between higher fertilizer and normal fertilizer.

Station	1973	1974	1975	1976	1973-1976
Ås	+ 4	- 1	+ 7	0	+10
Giessen	+12	+ 7	- 7	+ 5	+17
Rome	-	+ 9	+ 2	+ 4	+15
Castelvolturno	-	- 1	-14	+16	+ 1
Kherson	-	-	- 4	- 3	- 7
Naroforminsk	- 5	+19	+ 3	+22	+39
Swift Current	- 3	+ 4	+ 2	+ 2	+ 5
Ottawa	-	- 7	+ 5	-	- 2
Buenos Aires	-21	+18	+ 8	+ 3	+ 8
Pelotas	+20	+12	+16	- 4	+44

Experiment the dry matter was determined, a conversion of dry matter into the leaf area index is very doubtful as no account is taken of the differences in the thickness of the leaves. In order to be able to describe sufficiently accurately the water consumption, it will be necessary to develop a model of the heat balance of the transpiring leaves. This can only be realized if measurements of temperature, ventilation and moisture condition within the crop and in its direct vicinity – carried out with sophisticated instruments – are available at intervals of a least one hour. For this purpose also hourly measurements of the radiation fluxes are necessary. But too sophisticated instruments are in opposition to the request that as far as possible only those parameters should be used in the growth model which are part of the normal observations of a meteorological station. Here we see the conflict between the measuring techniques necessary for the preparation of a complex simulation model and its possible applications. This applies to the fields of application mentioned at the beginning, that is a prognosis of the yield, the determination of the production potentials and effects of climatic changes.

The experience gained during the World Wheat Experiment may be summarized as follows:

1. The meteorological and biological data are suitable to verify whether not too sophisticated wheat growth models provide realistic results under different climatic conditions.

2. More extensive experiments should be made in view of testing simulation models describing during the entire growth period the interaction be-tween physical conditions of the environment and biological reactions of the plant. This should be made without having to resort to sophisti-cated instruments which are not available in the meteorological routine operation.

This fact has been realized by the members of the working group and they are willing to continue the experiment with an extended measuring programme at a reduced number of stations. Italy, Israel, the Federal Republic of Germany and Norway have expressed their intent to participate.

Independently of the envisaged continuation of the experiment by the CAgM, the Central Agrometeorological Research Station of the German Weather Service in Braunschweig has restarted in 1980 experiments with the variety Siete Cerros. The experiments have been extended by the in-stallation of a heat balance station within the crop. This station measures vertical heat and water vapour fluxes and the different radiation components. When preparing the measuring programme use could be made of experiences gained in connection with a research programme on the water status of maize (Hoyningen-Huene, Braden (7); Schrödter, Hoyningen-Huene,

Braden (8)). The extensive observations with solution in time and space of parameters like air temperature, relative humidity, air motion and radiation, are supplemented by the weekly determination of the leaf area index and the plant mass. From 1981 an experimental site will be available where a lysimeter (described by Hoyningen-Huene and Bramm (9)) has been installed for a better determination of the water status in the soil. In this way it will be possible to study the heat balance within the crop and thus closely related to the water status of the plants.

6. CONCLUSION

The preparation and performance of the World Wheat Experiment aimed at obtaining meteorological, biological and soil-physical data of different climatic regions in order to develop a generally valid model of the wheat growth. Generally valid means that reliable results of the yield are obtained from the model under warm an cold, dry and humid conditions of individual years of different climates. This is feasible when the growth of the plants is mathematically described by means of physical and biological laws which can be applied to any place. The difference in the yield between two different sites is the result of the differences in the environmental conditions. This project is a very ambitious goal as a complex system is to be represented as accurately as possible, in which plants are reacting on multiple outside influences and in which the physical environmental conditions are modified by the plants.

The first observational phase of the World Wheat Experiment being concluded, the present status may be described as follows. The quality-controlled data have been made available to the participating states for developing their own models. The Deutscher Wetterdienst was the first to verify the model of Hodges and Kanemasu at all experimental sites which were located in the different climatic regions. Even if it was not possible to prove the transferability of the model to other climatic regions, it will have to be seen whether a similar experiment with a model, published by de Wit et al. (2), will provide a better result. The usefulness of the World Wheat Experiment has already been proved in verifications of existing models. All models known to refer to the growth of plants - and this does not only apply to wheat - had the disadvantage of having been developed on the basis of data of experiments made in restricted areas and thus being characteristic of the climate of the experimental site. The World Wheat Experiment, however, provided data of four years and eleven experimental sites located in differing climates. There has never been any precedent. It is not yet known whether generally valid wheat growth models can be developed. But it is obvious that the World Wheat Experiment should be continued, as it is necessary to expand the measuring programme in order to study intensively certain aspects like heat balance in the crop and on the individual leaf and the closely related water balance of the plants. An important aspect is the measurement of the dry masses and the leaf area index in short intervals in order to determine photosyntheses and respiration during the different stages of development.

The WMO Commission for Agricultural Meteorology (CAgM) intends to perform experiments with other important cultures based on the experience gained during the World Wheat Experiment. In this context the proposal was discussed to find an international institute to perform this task. However, this proposal was not pursued as the difficulties of finding a corresponding sponsor seemed insurmountable.

Literature

(1) Baier, W.:
     Crop-Weather Models and their use in yield assessments.
     Tech. Note World Meteorol. Organ   Nr. 151,
     WMO-No 458 (1977)

(2) De Wit, C.T.; Bronwer, R.; Penning de Vries, F.W.T.:
     A dynamic model of plant and crop growth. In: P.F. Wareing and
     J.R. Cooper (Eds.) Potential Crop Production,
     A Case Study, Heinemann Educational Books, London (1979),
     117 - 142

(3) Goudriaan, J.:
     Crop micrometeorology: a simulation study, Wageningen Centre for
     Agricultural Publishing and Documentation (1977). Simulation Mono-
     graphs.

(4) Keulen, H. van; Louwerse, W.:
     Simulation models for plant production. Agrometeorology of the wheat
     crop-proceedings of the WMO
     Symposium, Braunschweig
     WMO-No 396 (1973), 196 - 209

(5) Hodges, T. and Kanemasu, E.T.:
     Modeling Daily Dry Matter Production of Winter Wheat
     Agron. J. 69,6 (1977)

(6) Kanemasu, E.T.; Stone, L.R.; Powers, W.L.:
     Evapotranspiration model tested for soybean and sorghum
     Agron. J., 68, 569 - 572 (1976)

(7) Hoyningen-Huene, J. v.; Braden, H.:
     Bestimmung der aktuellen Evapotranspiration landwirtschaftlicher
     Kulturen mit Hilfe mikrometeorologischer Ansätze. Mitt. Dt.
     Bodenkundl. Gesell. 26 (1978), 5 - 19

(8) Schroedter, H.; Hoyningen-Huene, J. v.; Braden, H.:
Ergebnisse experimenteller und modelltheoretischer Arbeiten zum
Problem des Wasserhaushaltes im System Boden – Pflanze – Atmosphäre
Beitrag z. Agrarmeteorologie Dt. Wetterd., Agrarmeteorol.
Forsch.-Stelle Braunschweig Nr. 63 (1978)

(9) Hoyningen-Huene, J. v.; Bramm, A.:
Die wägbare Unterdrucklysimeteranlage in Braunschweig-Völkenrode –
Aufbau und erste Erfahrungen
Landbauforschung Völkenrode, 28 (2), 95 – 102
Braunschweig (1978)

THE TECHNOLOGY OF CROP/WEATHER MODELING

Clarence M. Sakamoto

Chief, Models Branch, NOAA/Environmental Data and
Information Service; Center for Environmental Assessment Services; Federal Bldg., Columbia, Missouri USA

ABSTRACT   This paper discusses the technology of crop modeling
including its applications, problems and needs that can assist
in providing a kind of early warning tool to help stabilize
the world food security problems. Illustrations are provided
to suggest that depending on the use of the model as well as
the availability of data, a range of technological development
in using weather data can be useful.

    Several criteria for model selection should be considered.
In addition to the objective of the application, accuracy, data
availability, cost, timeliness and consistency with scientific
knowledge should be considered.

    Future prospects in problem and potentially fruitful areas
are briefly discussed, including the need to establish a basic
soil climate and crop data bank; assessment of catastrophic
events, aggregation of meteorological data; management of technology and the potential application of remote sensing.

INTRODUCTION

    The history of technology can be divided into three stages
of evolution. The first stage is the development of crude
implements. The second is the machine stage of the industrial
revolution where the concern was to produce labor-saving devices
for accuracy and precision. Today, we have been thrust into
the third stage, the cybernetics stage, where profusion of data
and the processing of these data into information is becoming
common for users. Similarly, the modeling of crop yield can be
divided into these three stages depending upon data availability

W. Bach, J. Pankrath, and S. H. Schneider (eds.), Food-Climate Interactions, 383–398.
Copyright © 1981 by D. Reidel Publishing Company.

and the sophistication of hardware available. The crude implement stage uses basic observations (e.g. temperature and precipitation), as indices; the machine stage involving simple
models (e.g. regression) which short-circuits the causal
approach; the third stage, the process-oriented modeling
approach involves a system approach to the problem. However,
the realization is that as a technology, crop modeling can be
applied successfully only if the advantages and limitations are
recognized and considered for assessment purposes. This, in
turn, has led to the need to deal with systems and the quick
realization that the system is indeed very complex.

This paper discusses the state of the art in crop modeling,
its implications, problems and needs that can assist in providing a kind of early warning to help stabilize the world food
security problem. For the purpose of this paper, crop/climate
or crop/weather models are mathematical relationships that
relate the measured response of crops to the environment.

USES OF CROP/WEATHER MODELS

The development of crop/weather models serves several purposes. In any development, it is important to address the
objective and scope of model building. Associated with this is
the identification of the user. These factors, in turn, affect
model strategy; that is, what kind of assessment model should be
used? What are the data available for analysis, and, just as
important, what types of data will be accessible for real-time
assessment?

Basically, there are two types of crop/weather models: (a)
the process-oriented, and (b) the statistical-empirical. In terms
of the stages of technology alluded to earlier, the statistical-
empirical model is anologous to the tool stage, a stage much
more advanced than the crude implement period, while the process-
oriented (physiological, simulation, deterministic models) can
be linked to the machine-cybernetic stage. In the latter case,
one looks at a series of sub-model systems which are based on
data from field experiments or from the literature.

Figure 1 illustrates the types of models relative to
spatial and temporal scales. The second and third generation
models are in the realm of process-oriented.

Baier (1977) outlines at least two roles for crop/weather
models: (a) the assessment of crop production from current
weather data, (b) the evaluation of the impact of climatic
change on the variability of crop production (NDU, 1978). In
addition to these, (a) these models serve as a platform to investigate composite effects of the environment on plants.

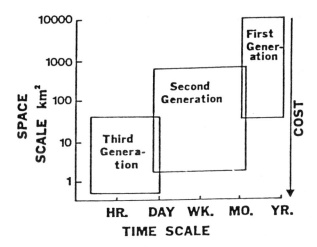

FIG. 1. Spatial and temporal data required for various
         classifications on yield models.

This is to say that environmental factors can be viewed through
the dynamic soil-plant-atmosphere continuum, recognizing that
several factors are involved, possibly simultaneously; (b) these
models allow the systematic quantification of risk analysis.
An example of such an analysis is shown by Ritchie (1979). Using
a process-oriented but semi-empirical wheat model, Ritchie was
able to provide estimates of probabilities (and hence, risk) of
available soil moisture in wheat fields as a function of planting time, using historical weather data (Figure 2); (c) these
models serve to aid in the development of new knowledge. As
modelers seek to quantify the biological response of a system to
the environment, new methodologies, and new information evolve;
(d) models serve as a management tool. Effective and useable
models permit users to play "what-if" games and address options
that might be available to them. Finally, the process-oriented
models (e) serve as an organizational tool. Available information on a system can be organized into a coherent, useable framework. Interrelationships between systems are more clearly
revealed and necessary contributions from each discipline
are more easily identified.

     Statistical or correlative models are also very useful.
This black-box approach to crop/weather modeling has been
critized, perhaps justifiably; however, the author is of the
opinion that such models, when judicially used can provide much
information in the absence of proven, operational modeling of
the more sophisticated process-oriented type.

EFFECT OF MAX. SOIL WATER STORAGE CAPACITY

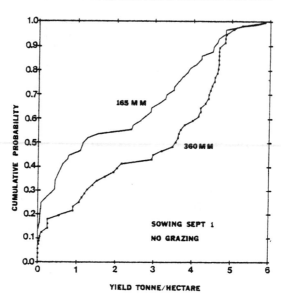

FIG. 2.   Example of the use of a process-oriented model to study
impact of soil moisture on yield (after Ritchie, 1979).

YIELD MODELING IN LACIE

In 1974, the United States embarked on an experiment to
test the capability of current technology to support real-time
monitoring of global crop production.   This experiment, called
LACIE (Large Area Crop Inventory Experiment) ended in 1978.
This experiment demonstrated the capability that much information about the crop over large areas can be inferred before
harvest from meteorological data (Hill, et al. 1980).   Figure 3
is an example of the regression approach used with a historical
time series data set.   In the example, an increase of yield is
indicated for the late 1940's through the 1960's.   The increase
is assumed to be associated with improvements in the management
of technology such as better varieties, increased fertilizer
application, increased irrigation practices, herbicide and
insecticide controls, etc.   The variations about the technology
effect is assumed to be associated with the effects of weather.
One should expect that these models tested with independent data
may provide large errors during specific years but when the
results are aggregated over large areas, the errors diminish.
To be useful and effective, the variables need to be reasonable
with respect to the signs of the coefficient relative to the

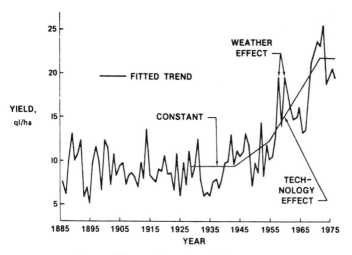

Reported Kansas winter wheat yields (source: USDA).

FIG. 3. A regression approach to crop yield modelling illustrating assumptions of weather and technology effects.

growth stage. When monthly data are used, the growth stage,
as well as the planting and the maturity dates can change, in
some cases as much as a month. Consequently, the results of
these models must be viewed with caution. Figure 4 shows the
result of the U.S. wheat models when compared to the reference
USDA yields. Unusually large departure years are sometimes
associated with what has been termed subitaneous (episodic)
events; that is, those non-modeled events that either occur
naturally over short periods (such as hail, freeze or flood)
or are non-natural (social or political). The fact is that
even technological changes are not often modeled or may be
considered inadequate because of their complex interactions
with weather, economics and social responses of man. This is
an area that is currently being addressed as a task in the new
project called AgRISTARS (Agricultural Resources Inventory
Surveys through Aerospace Remote Sensing) in the United States.

Following the LACIE experience with the first generation
model, it was hypothesized that because of crop calendar variability, improvements could be achieved if periods shorter than
monthly intervals were used. This shorter interval has been
referred to as the second-generation model. In LACIE, two
second-generation models were developed, one by Feyerherm (1979)
and the other by LeDuc (1979). In these models, daily meteorological data were used for the duration between selected phenological periods. The phenological periods were either modeled

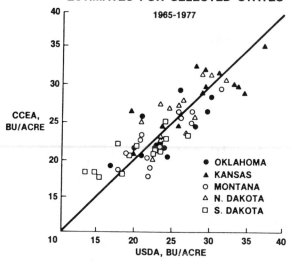

FIG. 4.   USDA wheat yield estimates vs CCEA model,
          estimates for selected states.

(Robertson, 1968) or observed by selecting the mean observation
date.   Feyerherm's model included technology as a function of
relative yielding ability, nitrogen applied and the proportion
of acreage under irrigation or fallowed the previous year.   A
soil moisture program after Baier and Robertson (1966) was used
to simulate soil moisture through several layers of soil.   In
spite of the detailed analysis, yield results did not improve
significantly over those models that utilized longer data
periods such as a month.   LeDuc, in her approach, minimized the
phenological date as a source of error by using observed data,
and aggregated daily meteorological observations into weekly
periods to include total precipitation, mean maximum and mean
minimum temperatures.   From these data, she was able to develop
a regression model that seemed, with limited independent data,
to capture the large variability in the yield (Figure 5).

ERRORS IN BIOLOGICAL SYSTEMS

     In science, we are aware of two types of numerical variables:
deterministic and stochastic (Thom, 1970).   In a biological
system for which crop growth and yield models are being developed, a deterministic variable is exactly deterministic in concept
only.   This is to imply that a deterministic state in process-
oriented models may not be reached because of the random

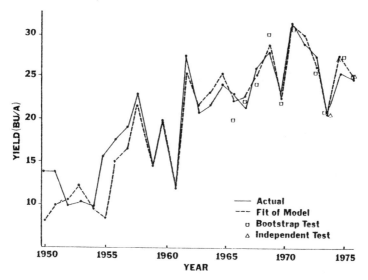

FIG. 5.   Yields for North Dakota hard red spring wheat, actual
          and estimated from alternate model.

component or "errors" with field or laboratory measurements and
because of differences in responses to the microenvironment
(Sakamoto and LeDuc, 1980).   In the process-oriented models,
the implicit assumption is made that the deterministic component
and the random component are not separable.

     The error concept can be used in large areas of crop fore-
casting by considering not only the mean, but also a measure of
variation.

     Table 1 from Hill, et al. (1980) shows the difference in
the number of days between the observed and estimated growth
stages as determined by Robertson's (1968) method.   It is clear
that the modeled error varies considerably, ranging from 10
days earlier to 10 days later at the jointing stage and from 8
days earlier to 10 days later at the critical heading stage.
Furthermore, for a particular crop district in a spring wheat
growing area of the state of North Dakota in·the U.S., (Figure 6)
the standard deviation of the heading date varied from 3 to 10
days or about 1 to 2 weeks (LeDuc, et al., 1979).   With this
natural field variability plus the possible modeled error, the
results strongly suggest that, at least for large yield estimates,
the meteorological data period of a week would suffice to capture
the major responses of the crops to the variability of weather.

     Periods longer than a month could also provide useful

TABLE 1 Difference (days) between spring wheat crop calendar
estimates and field observations of growth stage.

Test site location	Growth stage		
	Jointing	Heading	Soft dough
Hand, SD	−10	−2	*
Williams, ND	*	2	12
Hill, MT	10	6	15
Toole, MT	2	−1	*
West Polk, MT	−7	−2	*
Ft. Sask, Alta.	−1	−7	*
Lethbridge, Alta.	12	10	*
Melfort, Sask.	9	7	*
Swift Current, Sask.	9	−4	*
Torquay, Sask.	7	−2	*
Stony Mt., Man.	6	1	*
Starbuck, Man.	4	−3	*
Altona, Man.	3	−8	*

* Not observed.

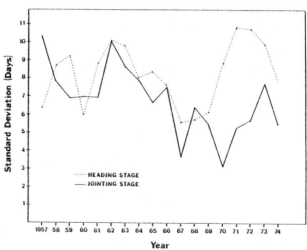

FIG. 6   Standard deviation of the heading
and jointing stage in North Dakota.

information. McDonald and Hall (1980) for example, showed a
close relationship between the May–June monthly precipitation
minus potential evapotranspiration and yield of the spring
wheat areas in the Soviet Union in 1977 (Figure 7).   In
Figure 8, the percent deviation from the trend (technology) was
estimated, which suggest the response to moisture stress during
May–June. A drought pattern emerged that clearly suggested a
serious problem.

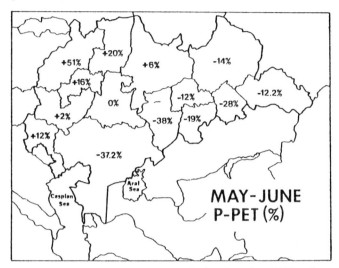

FIG. 7.   Departure of moisture deficit in the USSR wheat belt.

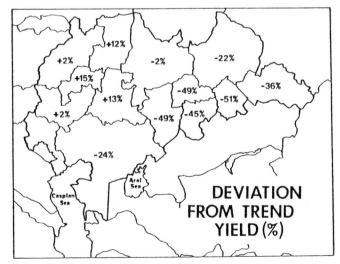

FIG. 8.   Impact of climate variability on the USSR wheat yield.

Meteorological data should be viewed as crude indices, however. When these data are used in soil moisture or potential evaporation, they are considered agroclimatic indices. These indices have been shown to provide better estimates of crop production than precipitation (and/or temperature).

PRODUCTION ESTIMATES IN DATA LIMITED AREAS

Illustrations have been provided where typically, and relative to lesser developed countries, the reporting meteorological network is much more dense and reliable. However, even in developed countries, the reporting network for operating some types of models in a systematic manner is less than desirable. Meteorological satellite data are, of course, one way that might be used to increase the spatial and temporal resolution of data. However, use of satellite technology is still at a stage where additional research needs to be accomplished before they can be routinely utilized. They are, however, useful in providing information as to percent cloud cover and with experience, indication of the types of clouds. More recently, satellites have been used to estimate solar radiation directly from satellite (Tarpley, 1978).

With limited data such as often exists in developing countries, crop production assessment can be made. Motha and Sakamoto (1979), for example, used the Crop Moisture Ratio, which is defined as the ratio of precipitation to potential evaporation and ranked these ratios with time. In Senegal (Fig. 9) these ratios clearly indicate the low ranks associated with reduced crop production. Assessment of the 1979 crop year with the method proposed by Frere' and Popov (1979) to show approximate planting periods together with observed precipitation and climatological information, indicate the practical and cost effective use of agroclimatic indices in crop yield assessment. (Dale, 1980)

CRITERIA FOR MODEL SELECTION

In previous sections, uses of models and results of real-time application of crop/weather models and agroclimatic indices were discussed. There are, however, several criteria which should be considered in any effort to use weather-based crop models. When a model strategy is developed, the purpose of the model is specified. In the context of this workshop, these models and indices can be used to provide a tool for early warning of potential food shortages. In this sense, the users of these models include local, national, as well as international decision-makers, banks, industry as well as farmers who might benefit from strategies that could be developed from playing "what-if" games from selected kinds of models.

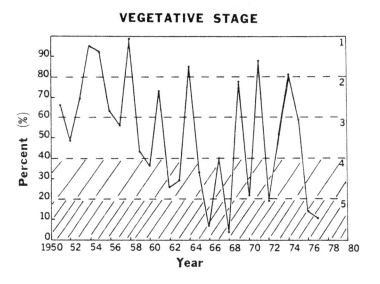

FIG. 9. Senegal millet model with yield estimates for 1974–1978, based on the crop moisture index.

It is desirable that the model provide an accurate estimate
of the "true" observation. In a biological system, the "true"
observation like the circle in the center of a target is not a
single point, but rather a range of values.

Data availability is obviously another serious consideration
in developing useable models. Recently, the author had the
opportunity to discuss potential application of a process-
oriented crop modeling effort in the Sahel with a scientist.
Data requirements include daily temperature and precipitation.
Thinking that his candidate model might have universal application, he attempted to test his model in the Sahelian countries,
only to be much disappointed because of data limitation. If
these areas are important in current world food assessment, as
I believe they are, we need to initiate systematic data observations. The cost of acquiring data is hypothesized to rise in
an exponential manner as one increases the temporal and spatial
scales (Figure 1). The operational cost involves not only the
acquisition, but it also involves quality controlling of the data
from the WMO Global Telecommunication Network.

The timeliness of a model with regard to its utility is
another factor that needs to be considered. The greater the
lead time, the greater is the available time for planning
options. In the LACIE models, reserve soil moisture before
planting as well as winter temperatures in the winter wheat
areas of the Soviet Union provided much information on the
likelihood of the condition of the crop. These early-season
models can be used in connection with probability analysis to
provide a distribution of yield based on historical data.
(Figure 10)

In addition to the above, the models should be consistent
with scientific knowledge. Statistical models attempt to short-
circuit the complex processes that exist in a plant-soil-atmos-
pheric continuum. Process-oriented models provide a systems
approach to the problem. Both types are useful, but limitations
of each should be known.

FUTURE PROSPECTS

The technology of crop/weather modeling is an evolutionary
process. The LACIE experiment, in response to the food crises
in the 1970's has kindled interest from many countries to deal
with the vagaries of weather. This monitoring tool is a small,
but effective way of dealing with the assessment of food shortages. Much can still be done to increase the utility and
effectiveness of crop weather models. The following are areas
that can be addressed.

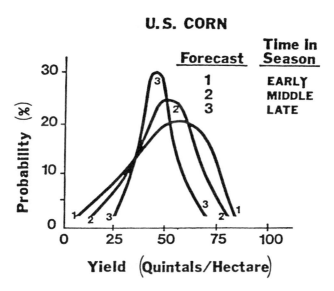

FIG. 10.   Probability distribution of early, middle and late
estimate of corn in the U.S.

Data

     There is a need to establish basic soil, climatic and crop
data which can be made accessible to modelers. In developed
countries, much repetitive effort is expended in obtaining
identical data bases by different individuals and organizations.

Model Testing

     A systematic procedure for model evaluation is needed
including selection of criteria under which these models can be
evaluated. Support should be given for developing, calibrating
and evaluating crop models. In a new program for the United
States, called AgRISTARS, attempts have been initiated to review
candidate models systematically under selected criteria.

Aggregation of Data

     The spatial agronomic and meteorological network from which
modeled estimates are derived need to be aggregated in a useable
form. This procedure should be accomplished in a manner that
will permit extraction of maximum information content from a
limited data network.

Assessment of Episodes

Methodologies that permit a quantitative assessment of weather and non-weather related episodes are needed. These may be acquired through a systematic analysis of historical data as well as field experiments that will allow crop modeled adjustments.

Management Impact

The impact of technology on crop models cannot be adequately assessed until data that affect technology of crop productions are recorded systematically. These management impacts are long-term as well as short-term. Examples of long-term practices include adoption of selected varieties and information transfer to farmers. Short-term includes fertilizer/energy supply, government programs that expand or curtail acreage allotments.

Remote Sensing

This new technology may be utilized as part of a system to assess crop condition, to indicate stress areas and provide indicators of yield and acreage estimates (McDonald and Hall, 1980).

SUMMARY

The evolving technology of crop/weather modeling can be applied in many ways to meet the needs for better planning to help stabilize world food security problems. The model or method selected will depend, to a large extent, on the data which are available to assess the situation. The need for data, including environmental and agricultural, are clearly revealed in developed, as well as developing countries. Systematic collection, model testing and techniques including use of remote sensing are suggested to better assess the problem.

REFERENCES

1. BAIER, W.: 1977, *CROP-WEATHER MODELS AND THEIR USE IN YIELD ASSESSMENTS,* WMO Technical Note 151, World Meteorological Organization, Geneva, Switzerland, 48 pp.

2. BAIER, W., and ROBERTSON, G.W.: 1966, *A VERSATILE SOIL MOISTURE BUDGET,* Canadian Journal of Plant Science, 46 pp. 299-315.

3. DALE, J.G.,: 1980, *EVALUATION OF THE 1979 GROWING SEASON IN THE SUB-SAHARAN COUNTRIES AS DETERMINED FROM CEAS ASSESSMENT TOOLS,* U.S. Department of Commerce, Federal Building, Columbia, Missouri, 65201.

4. FEYERHERM, A.: 1979, *RESPONSE OF WHEAT GRAIN YIELDS TO VARIATIONS IN WEATHER AND CULTURAL PRACTICE,* in Proceedings of the Crop Modeling Workshop, Columbia, Missouri, October 3-5, 1977. U.S. Department of Commerce, National Oceanic and Atmospheric Administration, Environmental Data Service, Federal Building, Columbia, Missouri, 65201.

5. FRERE', M. and G.F. POPOV,: 1979, *AGROMETEOROLOGICAL CROP MONITORING AND FORECASTING,* Food and Agriculture Organization of the United Nations, Rome, 64 pp.

6. HILL, J.D., STROMMEN, N.D., SAKAMOTO, C.M. and LeDUC, S.K.: 1980, *LACIE-AN APPLICATION OF METEOROLOGY FOR THE UNITED STATES AND FOREIGN WHEAT ASSESSMENT,* Journal of Applied Meteorology, 19, pp.22-34.

7. LeDUC, S.K., SAKAMOTO, C.M., STROMMEN, N.D., and STEYAERT, L.: 1979, *SOME PROBLEMS ASSOCIATED WITH USING CLIMATE-CROP YIELD MODELS IN AN OPERATIONAL SYSTEM:* An overview, presented at the Eighth Biometeorological Congress, Shefayim, Isreal, September 9-15, 1979.

8. LeDUC, S.K.: 1979, *CCEA SECOND GENERATION MODEL,* in proceedings of the Crop Modeling Workshop, Columbia, Missouri, October 3-5, 1977. U.S. Department of Commerce, National Oceanic and Atmospheric Administration, Environmental Data and Information Service, Federal Building, Columbia, Missouri, 65201.

9. McDONALD, R.B. and HALL, F.G.: 1980, *GLOBAL CROP FORECASTING,* Science 208, pp. 670-679.

10. MOTHA, R.P. and SAKAMOTO, C.M.: 1979, *THE PERFORMANCE OF CROP YIELD MODELS IN DROUGHT-PRONE COUNTRIES IN CENTRAL AFRICA*, presented at the 71st annual meeting of the American Society of Agronomy, Ft. Collins, Colorado, August 5-10, 1979.

11. NATIONAL DEFENSE UNIVERSITY: 1978, *CLIMATE CHANGE TO THE YEAR 2000*, Fort Lesley J. McNair, Washington, D.C., 20319.

12. RITCHIE, J.T.: 1979, *WATER MANAGEMENT AND WATER EFFICIENCIES FOR AMERICAN AGRICULTURE*, in Proceedings of the Symposium on Weather and Climate, October 1-2, 1979, Kansas City, Missouri.

13. ROBERTSON, M.G.W.: 1968, *A BIOMETEOROLOGICAL TIME SCALE FOR CEREAL CROP INVOLVING DAY AND NIGHT TEMPERATURES AND PHOTO PERIOD*, International Journal of Biometeorology, 12, pp. 191-223.

14. SAKAMOTO, C.M., and LeDUC, S.K.: 1980, *DISCUSSION TO "PLANT RESPONSES TO ENVIRONMENTAL CONDITIONS AND MODELING PLANT DEVELOPMENT,"* by D.N. Baker, J.A. Landivar, F.D. Whisler and V.R. Reddy in Proceedings of the Symposium on Weather and Agriculture Symposium, October 1-2, 1979, Kansas City, Missouri.

15. TARPLEY, J.D., SCHNEIDER, S.R., BRAGG, J.E. and WATERS, M.P., III: 1978, *SATELLITE DATA SET FOR SOLAR RADIATION STUDIES*, NOAA Technical Memorandum NESS 96, Washington, D.C., 36 pp.

16. THOM, H.C.S.: 1970, *THE ANALYTICAL FOUNDATIONS OF CLIMATOLOGY*, Arch. Met. Geoph. Biokl, Serial B 18, pp. 205-220.

STATE OF THE ART OF PREDICTING SHORT PERIOD CLIMATIC VARIATIONS

Jerome Namias

University of California, San Diego
Scripps Institution of Oceanography
La Jolla, California 92093

ABSTRACT. This paper offers an attempt to summarize, compare and critically examine methods of long range weather forecasting and to suggest avenues where greater progress might be found.

It starts out with some general remarks about the extremely complex nature of the prediction problem on time scales of months, seasons and years, which explains in large part the low degree of skill presently achieved. Also discussed are the poor quality and quantity of available global data, a deficiency which is frequently and erroneously advanced as the principal cause of forecast failures. But it is concluded that the root of the problem and the main obstacle in the quest for better long range forecasts is a lack of understanding of the physics governing these large scale phenomena.

There follows a treatment of methods used which involve statistical, synoptic and physical (conceptual) procedures.

Among the methods are:

1. The use of analogues (similar past situations which can be used as guides for the future).
2. Long period trends in atmospheric data.
3. Teleconnections, or interactions between remote anomalous weather-producing circulations.
4. Regression equations and adaptations of empirical orthogonal functions and other devices designed to arrive at significant lag relationships.
5. Climatological contingencies indicating probabilities of monthly or seasonal surface temperature and precipitation

399

W. Bach, J. Pankrath, and S. H. Schneider (eds.), Food-Climate Interactions, 399–422.

following given observed conditions.

6. Use of the anomalous character of surface boundary conditions like snow cover, sea surface temperature and soil
moisture to determine probable forcing and stabilization of
anomalous atmospheric patterns (including storm tracks, etc.).

In practice the above methods are usually used jointly,
with a weighting which changes according to situation. This
weighting, largely subjective, is often the reason for both
forecast success and forecast failure.

Some evaluations of forecast skill are presented which can
be viewed as encouraging or discouraging, depending on the re-
ceptivity of the reader.

Finally, an optimistic view of the future is presented
wherein greater understanding of the physics of large space and
time scales may permit numerical (dynamical) long range simula-
tions and predictions for months or seasons and conceivably open
the door to predictions of anomalous climate regimes for years
and decades. This happy event defies setting a time frame, so
that prediction of the state of the art years hence becomes as
difficult as a long range weather forecast a season in advance is
at present. However, some recent encouraging dynamical predictions out to a month or so give some cause for optimism.

INTRODUCTION

Scientists who work in long-range weather forecasting encounter great difficulties, not only in the intricacies of their
chosen field but also in getting across to other scientists and
the lay public the essential nature of their problem and the
reasons for their painfully slow progress in the modern-day
milieu of satellites, computers, and atomic reactors. When solar
eclipses can be predicted to fractions of a second and the position of a satellite pinpointed millions of miles out in space, it
is not readily understandable why reliable weather predictions
cannot be made for a week, month, season, or years in advance.
Indeed, eminent scientists from disciplines other than meteorology, underestimating the complexity of the long-range problem,
have tried to solve it only to come away with a feeling of
humility in the face of what the late von Neumann used to call
"the second most difficult problem in the world" (human behavior
presumably being the first).

And yet, the potential economic value of reliable long-
range forecasts probably exceeds that for short-range (daily)
forecasts. Many groups need as much as a month or a season or
more lead time to adjust their plans. These include such diverse

types as manufacturers (e.g., summer suits, raincoats, farm implements), fuel and power companies, agriculturists, construction
companies, and commodity market men, to say nothing of vacationers.
Aside from this, long-range forecasting, by setting the climatic
background peculiar to a given month or season, is of distinct
value to the short-range forecaster. For example, it can alert
him to the likelihood of certain types of severe storms, including hurricanes, intense extratropical cyclones, and even broad
areas most frequently vulnerable to tornadoes.

Perhaps the most urgent needs for longer-range forecasts
have emerged during the present decade because wheat and other
grain supplies depend on climatic conditions, and demand for
energy resources depends on short-period variations in both
winter and summer temperatures. Many of the needs for long-
range forecasts cannot be met at present, however, because of the
low skill of predictions or the inability to predict anomalous
weather at sufficiently long time ranges. Why is the problem so
intractable?

Before attempting to answer this question, we shall first
describe—as background—the relationship frequently found on
daily weather maps between cyclones and anticyclones (and their
associated fronts) and the larger-scale pressure and wind patterns. Figure 1 shows a schematic diagram on which the cyclones
are represented as waves along the polar front, moving from
southwest to northeast and developing and occluding as they progress. These waves are embedded in a broad low-pressure trough
between two extensive high-pressure areas - one to the northwest
composed of cold air masses and the other to the southeast composed of warm air masses. Associated with the group of shortwave
disturbances is a long (or "planetary") wave in the westerly winds
aloft, which has a length of the order of four times that of the
cyclone waves. There may be about four to six of these waves
present around the hemisphere on any one-day map for the 500 mb
level. The long waves and the cyclone waves provide the atmosphere's mechanism for exchange of heat, momentum, and water
vapor between the tropics and polar regions. Long waves provide
a means for deploying cold and warm air masses into the temperate
latitudes where cyclones and anticyclones are generated. It must
be constantly borne in mind that both systems, the long waves and
the short waves (the cyclones) are interactive so that they cannot be treated independently.

Even if one averages the upper-level or sea-level pressure
and wind distributions over periods of weeks, months, or seasons,
the long waves do not disappear. They assume positions and
amplitudes that vary from one period to another, and thus the
prevailing air masses, storm tracks, and associated weather
conditions may be inferred from the mean configuration of the
long waves.

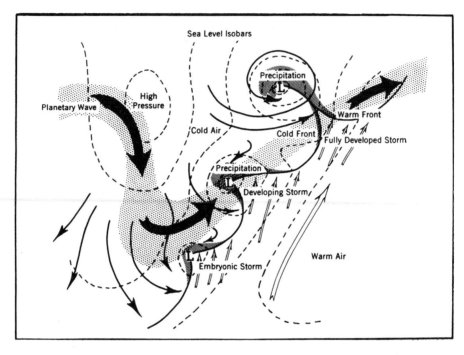

Figure 1. Schematic illustration of the relationship
between planetary waves in the middle and upper troposphere and
air masses, fronts, cyclones, and anticyclones customarily seen
on daily weather maps. Note that the wavelength of the planetary wave is several times the size of cyclone waves along the
surface polar front.

THE GENERAL PROBLEM

    In the first place, long-range forecasting requires routine
observations of natural phenomena over vast areas—and by vast we
mean at least hemisphere-wide coverage in three dimensions. More
probably, the entire world's atmosphere, its oceans, and its continents must be surveyed because of large-scale interactions
within a fluid that has no lateral boundaries but surrounds the
entire earth. In contrast to the physicist, the meteorologist
has no adequate laboratory in which to perform controlled experiments on this scale, although some recent work with electronic
computers holds out hope for useful simulation.

Inadequate Observational Networks

    When the immensity of the scale of the atmosphere is

comprehended, it becomes clear that the present network of meteorological and oceanographic observations is woefully inadequate. Even in temperate latitudes of the northern hemisphere, relatively well covered by surface and upper-air reports, there are "blind" areas of a size greater than that of the United States. The tropics are only sparsely covered by reports, and the data coverage in the southern hemisphere is poorer still. The World Weather Watch Program of the World Meteorological Organization is designed to ameliorate this situation.

In the southern hemisphere, a moat thousands of kilometers in diameter separates the data-rich Antarctic continent from the temperate latitudes, making it virtually impossible to get a coordinated picture of what is occurring now, let alone what may occur in the future. The "secrets of long-range forecasting locked in Antarctica"—a cliché often found in press articles—are indeed securely locked. Of course, wind and temperature data from observations of clouds and radiation by geostationary and polar-orbiting satellites are assisting to an ever-increasing degree, but improved methods are urgently needed.

Inadequate Understanding

Even if every cubic kilometer of the atmosphere up to a height of 20 km were continuously surveyed, however (and there are 10 trillion such volumes), reliable long-range forecasts would still not be realizable. Regardless of their frequency and density, observations are not forecasts; they merely provide "input data" for extended forecasting. Meteorologists have yet to develop a sufficient understanding of the physics of the atmosphere and the ocean to use these input data effectively in long-range forecasting, although this understanding is unlikely to come about in the absence of such data. The second objective of the Global Atmospheric Research Program (GARP) is designed to obtain this understanding. This second objective is to obtain "an understanding of the factors that determine the statistical properties of the general circulation of the atmosphere which would lead to better understanding of the physical basis of climate." Indeed, this effort, which involves sophisticated numerical modeling, holds out the best hope for physically based long-range forecasts. Thus far it is only a hope and far from realization.

THE PRESENT SITUATION

The Data Base

Today the data and facilities for making long-range forecasts, inadequate as they may be, are far better than ever. In

addition to about 25,000 surface weather reports (22,000 over
land and 3000 over sea) available each day at a center like
Washington, there are 900 balloon observations of wind direction
and speed and 1500 radiosonde observations of upper-air pressure,
temperature, and humidity and, frequently, wind. In the same
24-hour period, data in the form of about 1300 aircraft reports,
hundreds of indirect soundings of upper-air temperatures made by
the Nimbus-SIRS satellite system, and hundreds of cloud images
from geostationary and polar-orbiting satellites are collected.

While these figures are impressive, they are inadequate,
especially because they represent a most uneven geographical array
of observations and neglect proper surveillance of the ocean. The
vast blind areas are, unfortunately, located in such important
wind and weather-system generating areas as the northern Pacific
Ocean, the tropics, and parts of the southern hemisphere. These
systems, once generated, soon influence weather in distant areas
around the world, their complex effects often traveling faster
than the storms themselves. Hence, if an area is especially
storm-prone during a particular winter, the storms will persis-
tently influence other areas thousands of kilometers distant,
sometimes leading to floods or droughts. Obviously, if the wind
and weather characteristics in the primary generating area are
imperfectly observed, one cannot hope to predict the distant responses. The responses and generating areas are almost always of
a kind in which weather extremes over one area are compensated
by opposite extremes in another. For example, when the western
United States is very cold, the east is unusually warm. A similar
compensation often occurs in connection with precipitation abnormalities. Thus drought in one area is almost invariably
associated with heavy rains and possibly floods in other adjacent
and sometimes remote areas. This compensation obscures the meaning of global average variations and, of course, makes these
variations very small.

As pointed out earlier, data alone, regardless of how extensive in space and how frequent in time, are not sufficient to
insure reliable long-range forecasts. It does appear, however,
that more data of special kinds and accuracy are required if a
successful solution is to be obtained. The kinds of data required and a rough estimate of the density will be discussed
later.

State of the Art

Forecasts can be made for future days by using elaborate
numerical-dynamical models with high-speed computers. Using
these methods, one predicts various meteorological elements at
many levels for successive time-steps. The approach always begins with the initial conditions observed at many levels at a

certain time over a large area like the northern hemisphere and forecasts for time-steps of about 5 min. Each iteration starts from the last prediction, and the forecast is carried forward for many days.

Numerical predictions of this kind form the basis for the extended (5-day) forecasts made by the national weather services, an additional component being supplied by the experience of the forecaster. More specifically, the core of the extended forecasts (3-5 day) consists of numerically produced wind forecasts for different layers in the atmosphere with the help of models embodying fluid dynamics and thermodynamics that are solved numerically for a grid covering most of the northern hemisphere. These predictions indicate storm paths and new developments, and they also provide means for translating the wind systems into temperature and precipitation forecasts. These translations into weather are greatly assisted by the use of statistical equations derived from a few decades of past data. Finally, the human forecaster's experience frequently allows him to modify the objective predictions and come out with a still better product.

How accurate are these 3-5 day detailed forecasts? Figure 2, supplied to me by Donald Gilman of the National Weather Service's

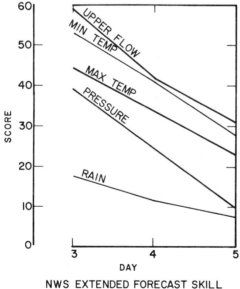

NWS EXTENDED FORECAST SKILL
(AFTER GILMAN)

Figure 2. National Weather Service skill scores for forecasts of various elements from 3 to 5 days in advance (courtesy of Donald Gilman). A score of 0 represents climatological probability; 100 would be perfect.

Long-Range Prediction Group, gives an idea of the accuracy for
various elements. While the score is not identically computed
for each element, they are comparable in that a perfect forecast
would score 100. A forecast no better than randomly selected
maps would score zero. The first thing to note is that forecasts
for all the elements have positive skill from the third to the
fifth day. Second, is that the skill of all predictions decays
rapidly with time, and, therefore, there is a limit to the
economic usefulness obtainable from such predictions. Third, the
temperature and upper-wind-flow predictions are apparently better
than those for rainfall. Part of the reason for this is that
temperature anomalies occur over broad areas while rainfall patterns are much more patchy (small scale).

From Figure 2 it is clear that detailed predictions made by
numerical iterative methods are unlikely to be of value for forecasts beyond a couple of weeks. In fact, rigorous theoretical
calculations show that, even with perfect data, the limit of detailed predictions is at most two weeks or so. Some recent work
at the NOAA Princeton Geophysical Fluid Dynamics Laboratories indicates that in the foreseeable years to come predictions of this
sort might have some positive although small skill out to 10 days.

While these results appear to sound the death knell of long-
range forecasts for a month, a season, and beyond, this is not
necessarily true. In the first place, forecasts of the general
character of weather for the coming month and season have already
been developed that have some positive skill over climatological
probability and, used intelligently, can be of economic value. It
is these general forecasts, not specific in space or time and out
from a month to a season and more, with which we shall be concerned for the rest of this chapter.

METHODS OF LONG-RANGE FORECASTING

If the statistics gathered from daily numerical predictions
(such as the mean and the variance) can be shown to have skill
above zero after their day-by-day demise indicated in Figure 2,
then the averages of computerized forecasts for a month or season
in advance could turn out to contain economically valuable information. Unfortunately, there is no evidence that this is so, but
the hope exists that better and more observations combined with
more knowledge of atmospheric modeling will result in this advance.
A tremendous effort in this direction is currently proceeding at
numerous meteorological centers in the world including the Soviet
Union, Great Britain, Japan, and Germany, as well as the United
States. (See (3), (4), (6), (8), (9), (14) and (15) for details).
On a pragmatic basis, many countries in the world support operational and research units that issue monthly and seasonal forecasts.

These forecasts, even though highly imperfect, are requested and used by many industries and for many purposes. The primary elements predicted are temperature and precipitation. In the United States, the present skill at forecasting departures from normal of average temperatures at 100 cities over the United States for 5-day, 30-day, and seasonal forecasts may be roughly given as 75, 61, and 58 percent, respectively, if chance is defined as 50 percent, (See 7, 9 and 14). Similarly, for precipitation, 5-day, 30-day, and seasonal forecasts average about five percentage points lower. While these skills are far from perfect, they do indicate that the methods contain some understanding of long-term atmospheric behavior and appear to be of definite economic value, judging from hundreds of users and from their reaction when the forecasts are not received on time (see (10) and (12)).

Similar skills are indicated in the forecasts made in Great Britain, the Soviet Union, Japan, Germany, Sweden, and probably other countries. The methods used, while varied, have as their common core a statistical-physical-synoptic (synoptic here meaning an overall view with the help of maps) approach. It would appear that since the forecasts have some skill, numerical modelers might profit by greater familiarization with operational long-range forecasting methods.

In the following paragraphs, we shall discuss several methodologies listed below but not in order of their relative importance: analogies and translations of wind patterns into weather; trends in features of the general circulation of the atmosphere (kinematics and extrapolation); teleconnections between parts of the general circulation; regression analysis (stepwise multiple regression and empirical orthogonal functions); contingencies based on historical climatological data for scores of years; and air-sea interactions and other boundary influences.

Analogue Methods

The so-called analogue method employs the philosophy that the weather has behaved in the past in such a way that the present initial conditions, if found to be similar to a past situation, will repeat just as in the earlier situation. This procedure requires similarity of many meteorological elements in three dimensions and preferably over much of the northern hemisphere. It obviously also requires a long file of historical meteorological records. Since the weather over large areas is never precisely the same, it is obviously not easy to find "good" analogues. Nevertheless, this procedure is used to some extent in many countries of the world. Perhaps the most systematic and thorough use is carried on in Great Britain (4), where an analogy is found with the help of computer selection. One selection is

for the monthly mean-temperature anomalies over much of the
northern hemisphere. For example, an analogue for October is required for the preparation of a November forecast—all the
previous Octobers are examined, not only surface temperatures are
used but also layer average temperatures of the troposphere.

The second consideration seeks an analogue for the monthly
mean-pressure anomaly, again over the hemisphere or as much of it
as available.

The third element involves matching by means of weather
types, classified largely on the basis of wind systems and storms
affecting the British Isles. A new twist involves the selection
of sea-surface temperature over the North Atlantic.

In some countries, the phase of the sunspot cycle is considered.

After the selection of a number of possible analogue years,
some poorer matches are rejected and a short list of acceptable
ones are used. The sequences of charts during these years is
then examined to help make the final selection.

Finally, a conference of participants employs parts of other
techniques described below to make the final forecast as to
whether the coming month will be much above, above, near, below,
or much below normal for the time of year. Rainfall is similarly
forecast in terms of above, near, or below average. At times,
subdivisions for the British Isles or for the monthly period are
indicated, and special indications of elements such as fog, frost,
gales, thunder, and snow may be stated. The forecasts are couched
in worded form along with a weather survey for the preceding month.

The British claim that about 74 percent of their forecasts
of mean temperature show excellent, good, or moderate agreement,
and rainfall about 62 percent.

Recently in the U.S.A. the selection of analogues has been
made more objective and faster by the use of eigenvectors which
remove redundancy from vast data banks and which can combine
atmospheric variables with oceanic variables (see (2) and (5)).
Some encouraging skill has lately shown up, but more understanding
is required to evaluate the source of skill from multidimensional
analog fields.

Trends

Even over periods of two or three months, there are frequently long-period slowly evolving trends in the features of the
general circulation of the atmosphere. These trends can be

brought to light in numerous ways. One method is to track singular map features or anomalies from month to month or between overlapping monthly means prepared twice a month. The slow evolution
of patterns on such maps is illustrated by a case showing the
gradual movement of an upper-level anticyclone over the period
from mid-November 1949 until March 1950 (Figure 3). The increasing chaos of circulations averaged for shorter periods of time
(5 days) is also shown in Figure 3. If the positions of component
high-pressure centers on daily maps were plotted, the chaos would
increase still further. Statistics of the motions of singular
features of maps show that when an averaging process is performed,
perturbations in the prevailing westerlies do not vanish but,
surprisingly, remain in sharp focus. Indeed the standard deviations of 30-day mean values taken for the same month of many years
may approach and at times equal the standard deviations of the
day-to-day values within the month. This characteristic is partly associated with the serial correlation present in day-to-day
meteorological elements but is greatly enhanced by the recurrence
of similar circulations in the same geographical areas.

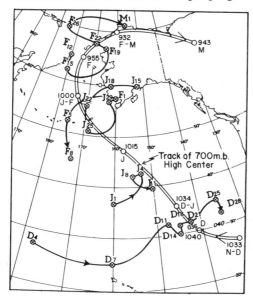

Figure 3. Track of a great anticyclonic cell as determined from 30-day mean charts from mid-November to mid-December
(N-D) 1949 to March (M) 1950. Intensity of center is indicated in
tens of feet for 700 mb, in excess of 1000 for sea level. Five-
day (crossed circles) positions of eastern Pacific anticyclone at
700 mb; J_1 refers to the 5-day period ending January 1, J_{1-15} indicates the period January 1-15, etc.

   The kinematic methods may suggest new developments due to
changes in the wavelength of the long waves as the seasonal forcing of the westerlies changes. For example, if a trough and the
upstream ridge are found to be moving in opposite directions (as
in Figure 4), a new pertubation in the westerlies may be suggested
and other indications are studied for clues as to just where the
new development will occur. This difficult decision is assisted
by (1) monthly or seasonally stratified climatological frequencies
of troughs and ridges and (2) areas of strong thermal contrasts
suggested by snow boundaries, oceanic temperature contrasts, or
coastlines. These new developments constitute one of the most
vexing of long-range forecasting problems.

   Some physical considerations involving dynamics are that the
trend methods described above, as well as other methods described
below, are always considered against the background of dynamical
concepts. Thus, the forecaster has at his disposal (1) stationary-
wavelength studies stratified according to area and season, (2)
studies of interactions between height anomalies in different
areas and seasons (shortly to be described), (3) studies of vorticity redistribution as implied by the upper-level contour
patterns, (4) correlations between the zonal wind systems (e.g.,
position and the strength of the Westerlies interacting with the
Northeast Trades), and (5) modifying effects exerted by differing
lower-boundary conditions.

   It is the ultimate aim of long-range forecasting to incorporate the above and other factors in an entirely objective
fashion with the help of numerical models. Lacking these, sta-
tistical-synoptic procedures have to suffice.

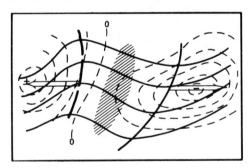

   Figure 4. Schematic example of mean flow patterns in
midtroposphere (solid lines) and 10-day centered tendencies (broken
lines) with arrows representing computed kinematic displacements
for one month. Shaded area suggests possible new trough because
of expanding wavelength, with tendency to become superstationary.

Teleconnections

    One of the most important tools used in long-range forecast-
ing involves "teleconnections" or the influence of one perturba-
tion on other remote atmospheric circulations. An example from
a recently constructed set of charts showing teleconnections in
midtroposphere is shown in Figure 5. Here the 700 mb height
anomaly at 45°N 145°W (at the + sign in the east central Pacific)
for all winters from 1948 through 1975 has been correlated with
484 points in the hemisphere north of about 15°N having a separa-
tion of 5° of latitude and 10° of longitude. Figure 5 indicates

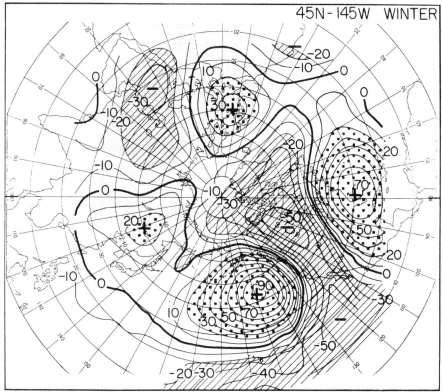

TELECONNECTIONS OF 700 DM

    Figure 5. Cross correlations from 700 mb height
anomaly center at 45°N 145°W with remainder of hemisphere.
Centers are labeled and lines are drawn for each 0.10. Values
exceeding ±0.20 are stippled (positive) or shaded (negative).

that positive anomalies at 45°N 145°W in winter are strongly
correlated in a positive sense with those over and off the southeastern United States coast but negative with those over western
North America. These teleconnections are largely manifestations
of the influence of one wave trough or ridge in the eastern Pacific
on the downstream flow via vorticity redistribution. It also includes the net effect of other forcing actions produced by
mountains and coastlines. A series of over 1900 charts like
Figure 5 have now been constructed for each of the 484 points
used in the hemisphere stratified by season. In short, these
teleconnections enable the forecaster to make a good estimate of
the anomalous pressure and wind flow over many areas, if he
chooses the most likely seats where strong pertubations are to be
found.

        Perhaps further comment on Figure 5 would clarify the usage.
If the eastern Pacific upper-level anticyclone is unusually strong,
one can quite confidently predict a strong trough extending from
west of Hudson Bay into the U. S. Rockies and also a strong Ber-
muda High. In this case, cold arctic air floods the West, but
unusually warm air dominates the East. Between the cold and warm
air masses appreciable storminess and precipitation occur. More
will be said about weather aspects as functions of circulation
later on. Note in Figure 5 that the anomalous wave train from
the Pacific carries on into the Atlantic, Europe, and perhaps
Eurasia, although in a weakened form. Naturally teleconnections
are more reliable when used in areas not too remote from the site
of the chosen pertubation.

Regression Analysis

        For both monthly and seasonal predictions regression analysis
has been found helpful. There are several methods that can be
employed, including stepwise multiple regression (called screen-
ing) or perhaps empirical orthogonal functions. These methods
may be employed to forecast the pressure fields at one or more
levels or to forecast one element—let us say temperature—from
another—pressure. Perhaps the simplest method involves the
computation of linear regressions of midtroposphere height for
each point in a grid whose resolution is about 5° of latitude and
longitude. These regressions may be worked up for the coming
month or coming season, based on the analysis of roughly 30 years
of data. All this work is greatly facilitated by modern computing
and plotting routines. While the use of eigenvectors is being
explored in many places in the desire to extract maximum information from limited data, these methods are not yet used operationally; except in the case of analogues discussed earlier.

Climatological Contingencies

For monthly and seasonal predictions, climatological contingencies, which give the probability of future temperature
anomalies based on antecedent temperature anomalies, have been
found useful. In the United States, this is done with the help
of contingency charts such as illustrated in Figure 6, where data
are shown by states whose average temperatures were taken from a
60-year record. From these data, 3 x 3 contingency tables are
worked up for each state and for each pair of adjacent seasons.
Thus each state's seasonal mean temperature was categorized into
three equally likely classes denoting cold, normal, and warm. Of
course, the percentage expected by chance in each class would be
33 1/3 percent. In Figure 6, the numbers indicating the probabilities greater than or less than chance are given for each
state in fall following a cold, warm, or normal summer. For example, if summer is cold in California, the upper chart indicates
that there is a 55 percent chance that the fall will be cold (33
percent chance + 22 percent from chart) but only a 15 percent
chance that the fall will be warmer than normal following a cold
summer. It is obvious from Figure 6 and other similar charts
that persistence is geographically and seasonally variable. Contingencies are always used in conjunction with the pressure and
wind-circulation indications described earlier. Occasionally, an
attempt is made to strengthen contingency indications by considering whether deep snow cover is present during the cold season or
if heavy rains (wet soil) or lack of rain (dry soil) are present
in the spring and summer.

Large-Scale Air-Sea Interactions

During the past several years, a great deal of information
has come to light through the study of large-scale air-sea interactions, (see (1), (11), and (13)). It is now well known that the
atmosphere and the ocean operate as a coupled system. However,
these two media operate on entirely different time scales—the
ocean on a scale that is about an order of magnitude slower in
time scale shows up vividly in the statistics of the persistence
of thermal anomalies in the upper layers of the ocean relative to
the atmosphere. Because of these different time scales, the ocean
frequently serves as a heat reservoir for the atmosphere, even
though its thermal structure is greatly influenced by the antecedent and contemporary atmospheric systems. The contemporary coupling is sufficiently clear so that given the sea-surface temperature
for the month or a season, the pressure and wind distribution
overlying it are reasonably well specified. This philosophy holds
out the hope for better long-range forecasts by considering the
ocean as well as the atmosphere.

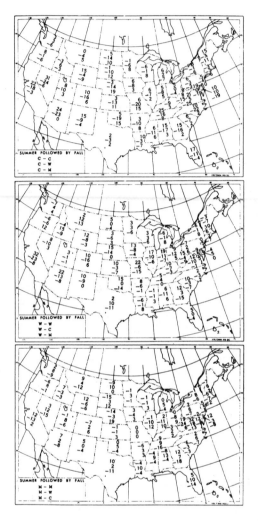

Figure 6. Contingencies of seasonal temperature anomalies by state. Numbers give excess or deficit, in percent, from
expected (33 1/3 percent) in the three classes, cold (C), warm
(W), and moderate (M), following the class observed in the antecedent season. For example, using the key in the upper chart, it
may be seen that cold summers in California tend to be followed
by cold falls 55 percent (33 percent chance + 22 percent from
chart) of the years, by warm springs 15 percent, and by moderate
springs 30 percent.

These air-sea concepts are being employed in a number of
countries. For example, they enter into some of the monthly forecasts made in Great Britain by consideration of the sea-surface-
temperature anomaly patterns over the North Atlantic.   In some
experiments, carried on at the Scripps Institution of Oceanography,
the starting point for seasonal experimental predictions is the
analysis of sea-surface-temperature patterns over the North
Pacific (11). These patterns are projected by advective means
(using normal large-scale ocean currents), and stepwise multiple
regression equations allow an estimate of the forthcoming season's
North Pacific airflow patterns.   Obtaining the latter patterns
provides a base for the utilization of the atmospheric teleconnections described earlier.   The predicted atmospheric flow patterns
can then be translated into weather.   This is done by using well-
known methods, which in effect translate midtropospheric circulation into associated temperature and precipitation patterns.   Indeed, this particular step can now be done with the help of
stepwise multiple regression, and the resulting specifications
account for more than half of the variance.   An example of such a
specification (case chosen at random using the observed 700 mb
pattern) is shown in Figure 7.

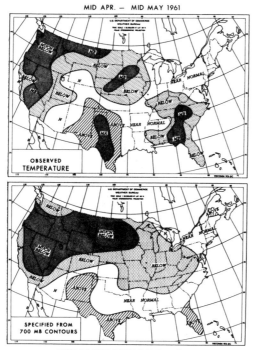

MID APR. — MID MAY 1961

OBSERVED
TEMPERATURE

SPECIFIED FROM
700 MB CONTOURS

AVERAGE SKILL SCORE (35 CASES) 31

Figure 7.   Below, an example of a specification of
monthly mean temperature made from an observed midtropospheric
circulation.   Above, the observed temperature pattern.

An experiment making seasonal predictions of temperature and
precipitation for the contiguous United States by such means has
been proceeding for about 6 years. It must be emphasized that,
while air-sea coupling is the starting point and base of the
forecast, some of the earlier discussed material is also considered—but in a secondary sense. An example of one of these
predictions, namely for the summer 1976, is shown in Figure 8.
Note that the warm dry condition of the Central and Northern
Plains and Lake Region was well forecast, as was the cool wet
spell in the Pacific Northwest. However, the extent of ·cold air
in the South was poorly forecast, as was the light rainfall in
the Southeast.

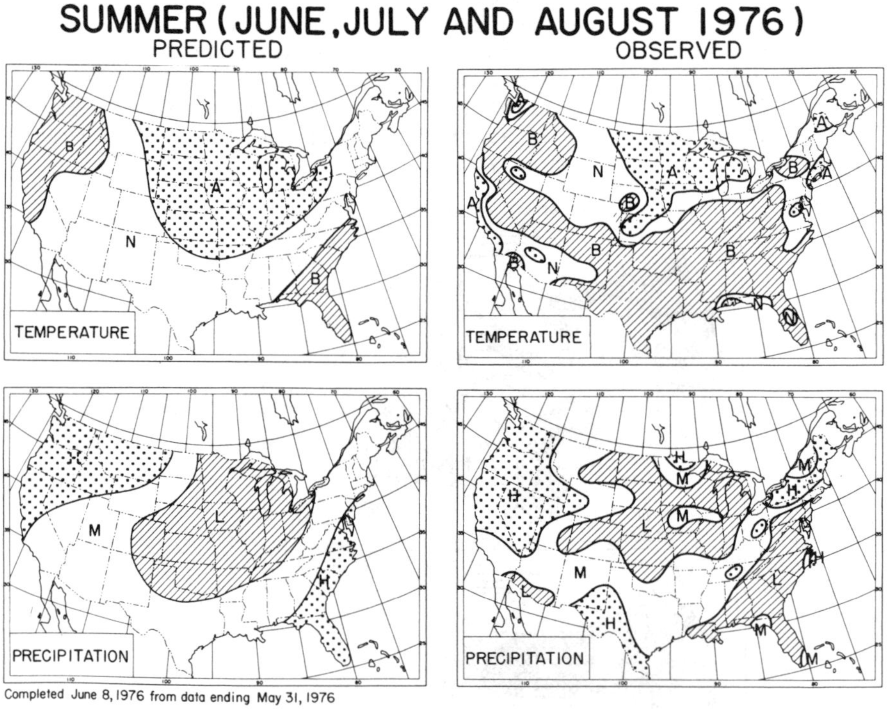

SUMMER (JUNE ,JULY AND AUGUST 1976)

Completed June 8,1976 from data ending May 31,1976

Figure 8. Predicted temperature and precipitation
anomaly classes (left) and observed patterns (right) for summer
(June, July, and August) 1976. For temperature, A stands for
above normal, N for near normal, and B for below normal. For precipitation, H stands for heavy, M for moderate, and L for light.

An idea of the general level of success of these seasonal predictions is given in Figure 9 and Table 1. Zero on both scales in Figure 9 refers to climatological probability; +1.00 would represent a perfect forecast, while—1.00 would represent a completely erroneous (out-of-phase) pattern. Table 1 gives a summary of the results in contingency form. Thus, for the predictions of temperature anomaly (at 99 equally spaced points over the United States) those numbers along the diagonal sloping from upper left to lower right are exactly in the correct category, while those in the extreme upper right and lower left boxes are 2-class errors. A similar verification is given for precipitation. From this table and Figure 9, it is clear that definite skill over and above climatological probability has been obtained, but, of course, the scores leave much to be desired. A similar

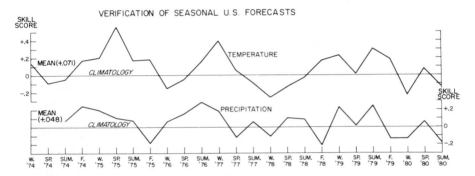

Figure 9. Verification of experimental seasonal forecasts made for 99 equally spaced points over the contiguous U.S.— temperature (above) and precipitation (below). Skill score is defined as number of points correctly forecast minus 33 (expected by chance) divided by 99 minus 33.

Table 1. Contiguous U. S. Contingency Tables for Seasonal Temperature and Precipitation Forecast Verifications (Graphed in Figure 9).

		Forecast						Forecast			
		Below	Normal	Above	Total			Light	Moderate	Heavy	Total
Observed	Below	348	211	155	714	Observed	Light	179	207	243	629
	Normal	314	293	270	877		Moderate	182	331	369	882
	Above	300	283	301	884		Heavy	160	284	322	766
	Total	962	787	726	2475		Total	521	822	934	2277

Skill = +0.071	Skill = +0.048
Temperature Forecast	Precipitation Forecast
Winter 1974 through Summer 1980	Summer 1974 through Summer 1980
(25 Seasons)	(23 Seasons)

analysis made for persistence of the anomalies from one season
to the next indicates skills only slightly above the climatology
base. Note that Figure 9 provides a gauge to evaluate the repre-
sentativeness of the forecast for the summer of 1976 that was
shown in Figure 8.

There appears to be a regional variation of forecast skills,
so that forecasts for the western third of the mainland U.S.A.
are better than elsewhere. These forecasts are shown in Figure
10 and the contingency tables are reproduced in Table 2. It will
be noted from Figure 10 that the winter precipitation forecasts
have been quite good for the last five years. Note especially
that the great drought of 1977, the following year's break in the
drought,and the next two years of above normal California precipi-

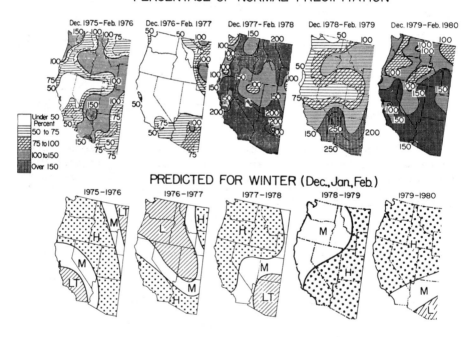

Figure 10.   Observed percentages of normal precipitation
over western third of U. S. (top) for winters of 1975-76 through
1970-80, and predicted Heavy, Moderate or Light classes (below).

tation were anticipated. Comparison of contingency Tables 2 and 1 also shows that the western forecasts were better than the total U. S. forecasts over all seasons of the past five years, for both precipitation and temperature.

Perhaps the proximity of the west to the Pacific Ocean, whose thermal structure plays an important role in the forecast, accounts for the difference.

Table 2. Western Third of U. S. Contingency Tables for Seasonal Temperature and Precipitation Forecast Verifications (Graphed in Figure 9).

		Forecast						Forecast			
		Below	Normal	Above	Total			Light	Moderate	Heavy	Total
Observed	Below	127	44	19	190	Observed	Light	55	72	60	187
	Normal	126	102	71	299		Moderate	51	95	166	312
	Above	152	80	104	336		Heavy	39	83	138	260
	Total	405	226	194	825		Total	145	250	364	759

Skill = +0.106 Temperature Forecast Winter 1974 through Summer 1980 (25 Seasons)	Skill = +0.069 Precipitation Forecast Summer 1974 through Summer 1980 (23 Seasons)

PROSPECTS FOR IMPROVEMENT AND RECOMMENDATIONS

Based on what has been written in this chapter, several avenues are suggested:

1. It is necessary to have a network of observations for both air and sea measurements over the world's oceans, or at least over the Pacific Ocean where much of the world's weather appears to be generated. This work can be a mix of ocean-weather ships, specifically equipped merchant ships, and-particularly-unmanned instrumented buoys, which have now been demonstrated to be feasible. A network of observations from points about 500 km apart would be adequate as a start; later, the data gathered could indicate whether a finer or coarser grid is necessary. Satellite measurements can supplement but cannot replace these observations, particularly the sub-surface ones that monitor the heat reservoirs of the sea and give information on the ocean currents.

2. "Air-sea interaction" should be more than a catch phrase. It is a subject that must occupy the efforts of the best young people in geophysics today. Equally important is meteorologist-oceanographer interaction. These people must not be steered into narrow specialties where they lose sight of the big problems that lie at the heart of long-range prediction. Special seminars and inclusion into academic curricula of large-scale air-sea problems on long time scales (months, seasons, and decades) are necessary despite the imprecise knowledge relative to short-period phenomena and short-range numerical weather prediction.

3. Special attempts are needed to bring meteorologists, oceanographers, and cryospheric and soil scientists together more frequently in universities and laboratories where they can analyze oceanographic and meteorological data in real time, conduct joint discussions of what went on and is going on, and try to predict what will go on in subsequent months. This will involve the use of computers and much research, but the research effort will be sparked by the satisfaction of seeing one's predictions verified. This kind of stimulus has been largely missing in the oceanographic community, where oceanographers have had to work on restricted problems mainly with data months or years old or with series of observations embracing only a small area.

These same observations and procedures, and their exploitation, will assist in most ocean-air inquiries, whether iterative or noniterative methods are employed. The ultimate long-range prediction scheme will probably be a combination of all three facets—physical, statistical, and synoptic.

Whether science will be able to achieve appreciable skill in long-range weather prediction should be known in the next 10 to 20 years, provided that enough trained people are efficiently employed and adequate data, as suggested by the World Weather Watch (WWW) and the Global Atmospheric Research Program (GARP), become available. If, however, an unbalanced program is embarked upon, making, for example, little or no use of statistics and synoptics, it is unlikely that good, practical long-range forecasts will be achieved. Considering the progress already made despite the complexity of the problem, the small number of scientists who have attacked it, and the inadequacy of data and tools in the pre-computer age, the outlook is optimistic, particularly in view of the WWW and GARP. General forecasts for periods up to a year in advance, not possible now, are quite within reach; even the general character of the coming decade's weather might be told in advance, given a fertile research environment and most importantly well trained and enthusiastic workers.

ACKNOWLEDGMENTS

   Madge Sullivan, C. K. Stidd, D. Cayan, Carolyn Heintskill, and Fred Crowe have been very helpful, respectively, in providing computational, editorial, typing services, and drafting. All are associated with the Scripps Institution of Oceanography.

   This research was sponsored by the National Science Foundation's Office for Climate Dynamics under NSF Contract No. ATM79-19237, and the National Oceanic and Atmospheric Administration under Contract No. NOAA04-8-M01-188.

   Much of this article was published in Studies In Geophysics, Geophysical Predictions, National Academy of Sciences, Washington, D. C., 1978. I thank the Geophysical Research Board for permission to reproduce this material.

REFERENCES

1. Adem, J.: 1973, "Ocean effects on weather and climate," Geofis. Intern. 13, 1.

2. Barnett, T. P. and Preisendorfer, R. W.: 1978, "Multifield Analog Prediction of Short-Term Climate Fluctuations using a Climate State Vector," J. Atmos. Sci. 35, pp. 1771-1787.

3. Baur, F.: 1948, "Einführung in die Grosswetterkunde" (Introduction to Long-Range Weather Science), Dieterich, Wiesbaden, 165 pp.

4. Bowen, D.: 1976, "Long-range weather forecasting, Water Power and Dam Construction,"July, (describes the British analogue method), pp. 31-35.

5. Gilman, D. L.: 1957, "Empirical orthogonal functions applied to thirty-day forecasting," Sci. Rep. No. 1, Contract AF19(604)-1283, 129 pp.

6. Girs, A.: 1974, "Macro-circulation Method for Long-Term Meteorological Prediction", (In Russian), Hydrometeosdat, Leningrad, 487 pp.

7. Klein, W. H.: 1965, "Application of synoptic climatology and short-range numerical prediction to five-day forecasting", U. S. Weather Bur. Res. Paper No. 46, 109 pp.

8. Namais, J.: 1953, "Thirty-day forecasting, a review of a ten-year experiment", Meteoral. Monographs 2 (6), 83 pp.

9. Namias, J.: 1964, "A five-year experiment in the preparation
    of seasonal outlooks", Mon. Wea. Rev. 92 (10), pp. 449-464.

10. Namias, J.: 1968, "Long-range weather forecasting—history,
    current status and outlook", Bull. Amer. Meteor. Soc., 49
    (5), pp. 438-470.

11. Namias, J.: 1975, "Short Period Climatic Variations, Collected
    Works of J. Namias, 1934 through 1974",U. of Calif., San
    Diego, 905 pp.

12. Nicholls, N. F.: 1980, "Long-range forecasting—values, status,
    and prospects", Unpublished manuscript.

13. Ratcliffe, R. A. S.: 1970, "New lag associations between North
    Atlantic sea temperature and European pressure applied to long-
    range weather forecasting", Quart. J. R. Meteorol. Soc., 96,
    pp. 226-246.

14. U. S. Dept. of Commerce: 1961, "Verification of the Weather
    Bureau's 30-day outlooks", Tech. Paper No. 39, 58 pp.

15. Wada, H.: 1969, "Introduction to Long-Range Forecasting", (in
    Japanese), Chijinshokan Co. Ltd., Tokyo, 234 pp.

CLIMATIC VARIABILITY AND COHERENCE IN TIME AND SPACE

Hermann Flohn

Department of Meteorology, University of Bonn
Bonn, Germany.

ABSTRACT. In a medium-sized country (area 10^4-10^5 km^2) horizontal climatic variations are smaller and less important than time variability, a fact often neglected in standard texts and tables. Extreme months, seasons and years are often responsible for scarcity of food (and thus high prices), cause high demands of water or energy supply and many economic disruptions. The frequency of such events varies with time and distance, expressed in statistical terms like variance, standard deviation or interannual variability. For selected European stations and areas running 30-year averages of long records (temperature, precipitation, runoff and water budget terms) are given, together with similar values of interannual variability. In general terms, high interannual variability occurred during the early 19th century, i.e. in the final phase of the "Little Ice Age"; in some areas (not all) low variability was observed during the early 20th century. 30-year seasonal averages of rainfall vary by 20-25 percent, in the Mediterranean by 30-50 percent. The interannual variability varies substantially with time.

Anomalies of temperature, precipitation, pressure (resp. geopotential) and wind are not only mutually correlated, they are also rather distinctly correlated in the space domain. Bjerknes and Namias investigated the possible use of sea-surface temperatures to understand the physical mechanism of extreme anomalies; in this case the autocorrelation often remains significant for periods up to 6-9 months, and large-scale teleconnections such as the "Southern Oscillation" or the Walker circulation have demonstrated unexpected and promising results. Examples of such teleconnections within the tropics and between different climatic belts are presented.

W. Bach, J. Pankrath, and S. H. Schneider (eds.), Food-Climate Interactions, 423–441.
Copyright © 1981 by D. Reidel Publishing Company.

I. Introduction

   After some early experiments to install a network of climatological stations during the 17th century (Academy of Florence
1652, Royal Society London 1662), and, on an international basis,
during the 18th century (Societas Meteorologica Palatina, Mannheim
1781), governmental station networks were founded around 1850 in
many countries, but in other countries only after 1900, in some as
late as around 1950. One of the earliest governmental networks was
founded in 1821, in the tiny grand-duché of Saxe-Weimar-Eisenach,
by Goethe, acting as minister of cultural affairs. Shortly after
his retirement, his successor terminated this 4-station network
on the pretext of high cost.

   Observations should be representative for larger areas - this
is true for atmospheric pressure, while precipitation (P) data,
temperature extremes, and surface winds are only too often locally
distorted. Some of the most important parameters - e.g. actual
evaporation (E) or evapotranspiration, net radiation - are difficult to measure with sufficient accuracy; usually they are derived
from basic data using approximative formulae. Representative
values are needed for the hydrological cycle: lake levels (especially in areas with internal drainage), runoff (which represents
P-E including storage terms). Because of the great spatial variability of convective rainfall, area-averaged series - meaningful
only if derived from significantly correlated (coherent) records -
are frequently needed.

   The value of long-term averages, carefully checked for homogeneity, is indubitable, especially for spatial comparisons.
However, from the economic point of view (e.g. agriculture, water
supply) the value of frequencies and intensities of individual
events, months, seasons or years is often greater and necessitates
a statistical treatment of variability. Since variance or standard
deviation are derived from squared deviations from an average,
individual extremes tend to exaggerate these terms. For many purposes interannual variability - taking into account e.g. the differences between adjacent values for each July - is preferable; it can
also be applied for non-Gaussian distributions. Even in well-
organized networks, random and systematic errors of measurement are
not rare and ought to be checked or smoothed by area averages.

II. Long series in European countries.

   In a recent study, requested by the European Commission,
Schuurmans and Flohn have compared a series of long records of
temperature and precipitation from European countries, especially
looking at the variability of 30-year averages, a period routinely
used in practical climatology as standard. The term "normal value"

is certainly not justified - indeed the "normal" period 1931-60 was probably the warmest of the last 500 years. 30-year time-spans such as 1931-60, 1941-70 etc., should only be recommended as "reference periods". Table 1 selects only a few temperature records: 30-year seasonal averages varied during the last 200 years by 1-1.5^{o}C during winter and by about 1°C during summer. These values are not negligible when compared with the present spatial changes of temperature: they are equivalent to horizontal displacements of 200-400 km. The interannual variability itself is variable with time, too - during this period it varied by 20-100 percent. Monthly values vary even more, but opposite trends in adjacent months indicate that this time-span is often too short to be representative. Generally speaking - but not without exceptions - interannual variability had higher values during the early 19th century; in some areas the lowest variability occurred during early or middle 20th century. A general increase of variability during the last 20 years cannot be verified (Ratcliffe et al. 1978).

Table 1 Long temperature records (^{o}C): variations of 30y averages and interannual variabilities

		Extreme variations of 30y averages				30y interannual variability	
		12-2 Wi	3-5 Sp	6-8 Su	9-11 Fa	Wi	Su
Edinburgh	1764-1960	1.5	1.5	1.0	1.2	0.8-1.9	0.6-1.1
Copenhagen	1799-1960	2.0	1.6	1.2	1.4	1.3-2.3	0.8-1.6
Bruxelles	1833-1977	1.3	1.6	1.0	1.5	1.4-2.2	0.7-1.2
München	1782-1977	1.2	1.3	1.0	1.4	1.6-2.7	0.8-1.3
Genève	1768-1972	1.5	1.5	1.0	1.3	1.1-2.1	0.8-1.3
Milano	1838-1978	1.7	1.3	1.1	1.6	1.1-1.8	0.7-1.2
Säntis	1883-1975	0.7		1.0		1.3-1.8	0.8-1.2

Table 2 shows similar statistics for precipitation. Here 30-year averages vary, in most parts of Europe, by 20-25 percent, in the Mediterranean even by 30-50 percent. This indicates clearly how unrepresentative a selected 30-year average can be. The interannual variability of these seasonal values varies by a factor of nearly two (Fig. 1); in the Mediterranean it reaches up tp 50 percent of the average. Particularly important results are derived from the long record of Milano: here the values of summer and winter rainfall revert their relative position several times, which indicates that the boundary between prevailing summer and winter rains shifts considerably. In the Netherlands rainfall of the summer tertial can also be smaller (around 1810 and 1947) than that of the winter tertial, which is characteristic for westernmost Europe. In

Table 2   Long precipitation records: variation of 30y averages and interannual variabilities (IAV, in mm)

Station	Period	Summer Tertial (5-8) Average	IAV	Winter Tertial (11-2) Average	IAV
Edinburgh	1785-1960	231-272	57-104	188-232	49- 85
London-Kew	1698-1970	188-242	38- 95	165-226	47- 85
Zwanenburg	1735-1978	223-289	54-100	205-272	37- 97
Bruxelles-Uccle	1833-1976	259-291	50-116	225-282	31- 93
Paris (Observ.Cour)[1]	1770-1972	191-241	34-102	152-208	42- 75
Münster (Westf.)	1819-1976	260-303	61-103	215-260	42-102
Stuttgart	1825-1975	282-330	53-118	140-188	50- 84
Marseille	1821-1972	99-130	44- 91	193-257	83-174
Milano	1764-1978	286-383	91-154	263-345	95-167
Roma	1783-1978	108-162	43- 86	298-409	105-183
Area Averages(ca.30-40 000km^2)					
Southeast England	1840-1969	216-241	52- 81	215-271	56- 86
Lower Rhine (F.R.G.)	1806-1975	248-306	47- 84	207-258	34- 85
Upper Rhine	1864-1970	325-372	52-113	189-252	59- 85
For comparison:Runoff (51 000 km^2) in 10 m^3/s					
Rhine, Karlsruhe	1821-1974	136-150	17- 34	99-114	17- 39

1) 1770-1810 Observ. Terrace (reduced); IAV only since 1811

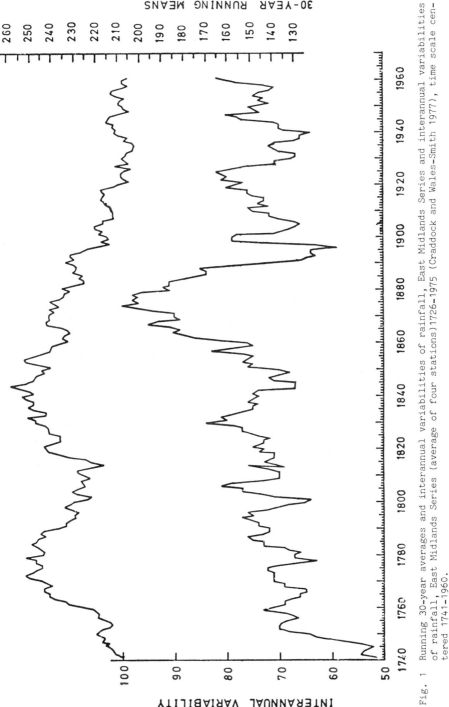

Fig. 1 Running 30-year averages and interannual variabilities of rainfall, East Midlands Series and interannual variabilities of rainfall, East Midlands Series (average of four stations)1726-1975 (Craddock and Wales-Smith 1977), time scale centered 1741-1960.

many areas of western and central Europe monthly extremes can
occur, in individual years, in every month of the year, and a
reasonably stable annual trend can only be derived after averaging over several decades. Fig. 2 gives monthly area-averages of
precipitation in southern Germany (about 100,000 km^2) from 1981-
1970, together with running 12-monthly averages. While in individual months rainfall can approach zero even in such an area,
aperiodic 12-month averages fluctuate more than annual values.

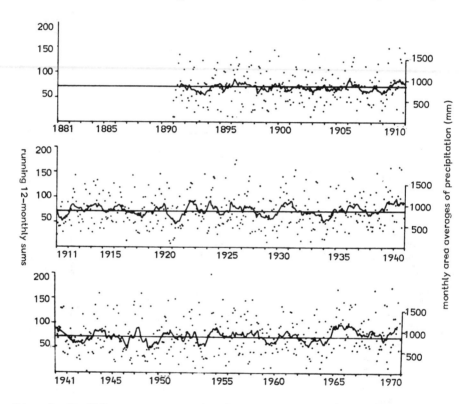

Fig. 2   Monthly area-averages of precipitation (S-Germany, 23 sta-
         tions 1891-1970) and running 12-monthly sums (right scale)
         (Flohn 1973). Note the near-zero value of October 1943 and
         the large 12-month anomalies 1920/1, 1939-1941 and 1965/6.

Runoff of the Rhine river is controlled, at Basel and Karlsruhe (Table 2), mainly by summer melting of mountain snows, with
only small variations in the 30-year averages, while further downstream cool-season precipitation gains equal weight. The interannual variability increases with increasing catchment area, thus
indicating a comparatively high spatial coherency of P. Its time
variation increases by a factor 2.2 - 2.7, mainly because of the
negative correlation between P and E.

Table 3 gives, for comparison, several other terms involved in the water balance, using the coefficient of variability for comparison. The largest variability is that of soil moisture of the summer tertial (Wigley-Atkinson 1977), an important parameter for agriculture due to the negative E/P correlation. On the other hand annual potential evaporation (assuming unlimited availability of water) and annual actual evaporation are both rather constant in time. Their annual values are mainly determined by the available radiation during the warm season.

Table 3 Long record parameters of atmospheric budgets and their variations

	Av.	Max.	Min.	CV[1] %	Period
London-Kew					
ann. precipitation(mm)	613	970	308	17.1	1876-1970
pot. evaporation(mm)**	547	658	439	5.6	1876-1976
soil moisture deficit(5-8,mm)**	82.3	126	0	32.8	1871-1976
Karlsruhe					
ann. precipitation(mm)	764	1115	457	19	1834-1975
act. evaporation(mm)**	454	487	389	5.5	1931-1970
sunshine duration(5-8,h)	945	1142	773	10.0	1921-1960
global radiation(5-8, W/m^2)**	224	256	192	5.7	1895-1977

* coefficient of variation = σ/M (standard deviation σ in percent of average M)

** computed from empirical formulae (5-8 = May - August)

This term is represented by the global radiation at Karlsruhe, with exactly the same coefficient of variation. Monthly values of sunshine duration are highly correlated with temperature during the summer months (June-September 0.67-0.84) while the correlation with rainfall is mainly negative (February-May -0.50 to -0.55, July-September -0.63 to -0.71); during winter both correlations are weakly negative (Wacker 1979).

Very long-term series of global or especially of net radiation are not available as direct measurements. However, strong correlations between global radiation and sunshine duration, between terrestrial radiation and cloudiness or other observed quantities allow a rational estimate of long series with errors mostly

in the order of 5 percent (Wacker, 1979). Such data can be used,
with only slight increase in error, to estimate evaporation or
soil moisture.

Time series of oceanic evaporation, to be estimated with the bulk
aerodynamic formula, are hitherto available only from coastal
light ships or from stationary weather ships. In many cases they
are strongly influenced by local factors such as the wind distribution. Even in the central North Atlantic, 6-month averages of
evaporation during the winter half-year with highest winds fluctuate quite strongly (between 77 and 125 percent of the average)
from year to year (Flohn and Rodewald 1975). At least at a regional scale, ocean evaporation cannot be considered as time-invariant;
this is even more true for the (much smaller) flux of sensible
heat which depends strongly on thermal stability. At regions with
coastal (or equatorial) upwelling, such as along the coast of northwest Africa, interannual variations of the fluxes of latent and
sensible heat are rather high and obviously play a great role in
the mechanism of climate. Convective activity, clouds and precipitation are decreasing with increasing thermal stability, which
reduces also drastically marine evaporation (Hastenrath and Lamb
1978, Henning and Flohn 1980).

III. Climatic fluctuations during the 17th and 19th centuries.

    Some of the long quasi-homogenous records (Manley 1973, Baur
1975) reveal quite serious climatic fluctuations, especially when
evaluated from the point of view of climatic impact on agriculture.
The coldest (1683-92) and warmest (1943-52) decade of the famous
Central England records show temperature differences (Fig. 3)
equivalent to a horizontal shift of several hundred km. In the
warmest decade, January has about the same temperature as March
in the coldest decade; the same is true for March/April in the
warmest decade compared with April/May in the coldest. During
autumn the time difference between equal temperatures in both series
is only two weeks.

    Late spring is a characteristic property of the climate during
the so-called "Little Ice Age"; in the centuries between 1660 and
1850, the frequency of very cold spring months (March, April, May
with a negative deviation of $-3^{\circ}C$ and more) was 3-4 times larger
than between 1850 and 1910; after 1910 only one event of this kind
was observed. This difference is coherent with the observed strong
drop of sea surface temperatures around the Faroes and with the
repeated advance of Arctic sea-ice towards Scotland and Norway during the Little Ice Age between 1550 and 1850 (Lamb 1979).

    During the peaks of the "Little Ice Age" the vegetation per-
iod was drastically reduced. If it is defined as a period with a
temperature average above $5^{\circ}C$ ($6^{\circ}C$), the vegetation period lasted

only 200 (209) days during the coldest decade, compared with 277 (247) days during the warmest decade. In many marginal agricultural areas in high latitudes, such shortening by 5-6 weeks had a prohibitive effect on agriculture production.

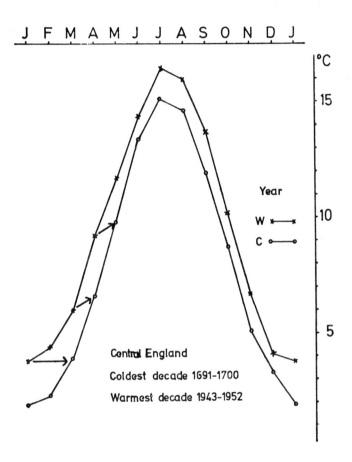

Fig. 3 Central England, temperatures (Manley 1973) averaged for the warmest (1943-52) and coldest (1691-1700) decade of record.

IV. Past cereal price fluctuations.

With the generous support of the Rockefeller Foundation in the 1930's many long series of cereal prices in European countries have been evaluated and published (Abel 1966, 1974). Additional series have been published, e.g. for Great Britain (Titow 1960, 1970; Hoskins 1964, 1968). Together with many other "proxy data", such as weather diaries, vine harvest dates and quantities, density

measurements of wood at the alpine tree limit, glacier fluctuations - as documented e.g. in Oeschger et al. 1980 - they promise
to be highly useful for a detailed study of the climatic history
of Europe since about 1500 (partly since 1200) and of the economic
impact of extreme weather events. Such a comprehensive study is
now in progress (Pfister in Oeschger et al. 1980) for Switzerland,
and it could be extended, with lesser details, to other large
European areas.

From data collected by the author (Flohn 1981) only two examples are given here. Fig. 4 shows, for comparison, three selected
cereal price index records (1400-1450) for Winchester in southern
England (Titow 1970), for Brugge in Belgium (Verlinden 1965) and
Frankfurt/Main in Germany (Elsas 1936, 1940). In spite of the
distance and of different governments, several peaks coincide quite
well; the available proxy data suggest a sequence of unusually
cold and wet spring/summers. The peak of 1437/8 coincides well
with a severe famine in Kyoto, Japan (T. Yamamoto 1971), thus indicating - as in other cases - quite distant teleconnections.

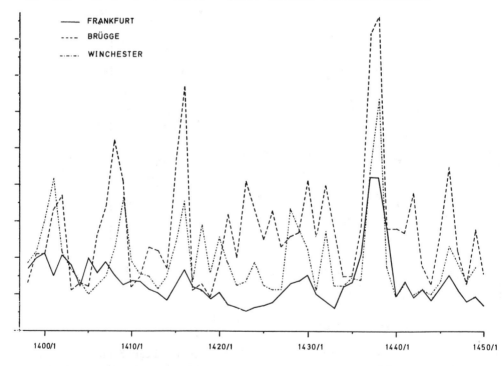

Fig. 4   Yearly rye price index at Winchester (S.-England), Brugge
         (Belgium) and Frankfurt (Germany), harvest years 1400-50

Fig. 5 gives rye price index data averaged for five German
towns. Except for some data during the catastrophic 30-years war
in Germany (1622 currency devaluation, 1634 peak of the war in

central/southern Germany), a number of excessive peaks (rising to 200-400 percent of the "normal" price shortly before and after the peaks) can be distinguished. Each of these coincides with a year or a group of years with marked climatic anomalies, mostly cold springs and cold/wet summers. The famous peak 1816/17 - after the eruption of Tambora 1815 - brought bad harvest simultaneously in the eastern USA (1816 = year without summer), in Europe and Japan; the economic consequences have been described by Post (1977). 1739/40 was the coldest year during the instrumental period and very similar to 1962/63. The years around 1770, with well documented quasi-permanent blocking highs over the North Sea and cold lows over the Alps, resulted not only in summer snowfall in the mountains at altitudes as low as 800 m but also in serious food shortages in Switzerland (Pfister 1975), Austria and southern and central Germany. A population loss of 6 percent (out of some 100,000 people) in Saxony alone (Abel 1974) and severe famines for the southern Sahara have also been reported (Nicholson 1980a).

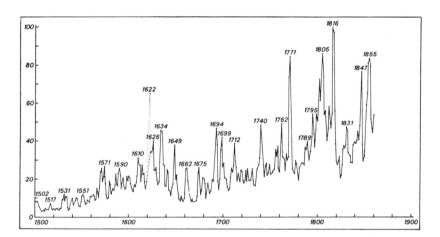

Fig. 5 Preliminary rye price index (not corrected for inflation) averaged for five German cities (Leipzig, Frankfurt/Main, Würzburg, Augsburg, Munich) 1500-1860(data from an unpublished manuscript by H. Philippi, cf. Elsas 1936, 1940).

V. Climatic anomalies, their spatial coherency and teleconnections.

From the economic point of view - especially when looking at food production for a still expanding world population - it is necessary to know how the productivity of different regions is interrelated. Many organizations and authorities tacitly assume a random behaviour in such a sense that negative and positive deviations of yields are smoothing themselves out in a global comparison. However, the example of 1972 has demonstrated that climatically caused deficiencies in different productive areas may accumulate. A failure of the global cereal production by only a few percent was accompanied by a price increase in the order of a factor 3.

Indeed climatic anomalies are not randomly distributed. They are controlled - in spite of the undeniably high number of degrees of freedom of the climatic system - by some spatial coherencies known as teleconnections. One of the best examples is the see-saw of temperature anomalies at both sides of the Atlantic, between Europe and the Labrador/Greenland area (van Loon-Rogers 1978), but there are many others. Apart from some investigations by Namias (1975) and Hastenrath and Heller (1977), Eickermann and Flohn (1962) Doberitz (1968/9) and Fleer (1981), this topic has found little interest. Most effective are large-scale anomalies of the sea surface temperature, which effectively displace the patterns of differential heating, i.e. of the varying fluxes of sensible and latent heat.

From the point of view of climatic impacts, our main interest must concentrate on the largest anomalies with important, in some cases even catastrophic, consequences. Climatic records reflect only insufficiently real climatic catastrophes, such as droughts and floods, harvest failures and famines. The cause of the latter can be pests - e.g. phytophtora and the famous potato famine of 1845-47 (Bourke 1980) - but also in these cases climatic effects were responsible. A recent example of such catastrophes was the famous flood of the river Arno, which devastated Florence in 1967. This was the result of a widespread rainfall of over 700 mm in 1-2 days. A similar catastrophe inundated large areas of Algeria and Tunisia in September/October 1969. Four consecutive cyclones crossed the Sahara pouring down up to 1200 mm of rainfall, thereby exceeding the long-term average by a factor of 30-40 (Flohn 1975). The statistical recurrence of such an event is 3-5000 years.

Many climatic anomalies show a tendency to occur almost simultaneously in distant parts of the globe, sometimes accompanied by such a catastrophes. To mention only a few examples for 1972: an extreme cooling of the American section of the Arctic, drought conditions in India, northeast Brazil and in the Sahel belt of Africa, unusually moist conditions in Australia and an El Nino along the west coast of South America and in the equatorial Pacific. In that year the climatic system demonstrated very well the role of teleconnections, i.e. of significant spatial coherences. Such teleconnections (on a global scale) have been investigated by Hildebrandsson, Exner, Eliot and Walker (cf. Namias 1968). Nowadays, with a more complete (but still deficient) data base, with a multitude of statistical tools and the access to powerful computers we should be able, in principle, to investigate these teleconnection on a global scale. Examples for Southeast Asia are given in Takahashi-Yoshino (1978).

Coherency in space has to be complemented by coherency in time. The variability of climatic anomalies is not purely random, but there is a tendency for extreme anomalies to cluster and to

occur in groups of similar properties. Recent experience can
demonstrate this clustering. The sequence of three very cold
winters (1976-79) in the eastern and central United States, or
the three greatest iceberg seasons of this century (1971-73)
around Newfoundland. The prolonged series of droughts in the
Sahel (Nicholson 1979, 1980) has hardly been interrupted since
1969 (only by a few years with near-normal rainfall). The unusual series of 7 mild winters in western and central Europe
(1971/2-1977/8) should also be mentioned. Occasionally even anomalies of an opposite sign follow one another immediately. The
most famous example is the flood year of 1917 and the drought
year of 1918 in India. As described by the Walker circulation
(Bjerknes 1969), these anomalies occurred almost simultaneously
with the driest and the rainiest 12 month period at Nauru (Equatorial Pacific, 167°E), with absolute extremes of 95 mm and more
than 5200 mm of rainfall. Short gaps in the 81-year record (1892-
1972) have been filled with (weighted) observations from Ocean
Island and Fanning Island, giving high spatial correlations above
0.7. At the northern hemisphere extreme deviations from the average, both negative and positive, tend to occur in periods with
a dominant meridional circulation, i.e. with recurrent blocking
highs and ridges. Extreme anomalies of temperature and rainfall
are more frequent in periods with these meridional patterns (Lamb
1977), and their recurrence is by no means a rare phenomenon.

Regarding food-climate relations, droughts represent the
greatest risk in semi-humid and semi-arid countries with rainfed
agriculture, when during the growing season actual evapotranspiration exceeds rainfall and when soil moisture is depleted. Typical areas are the central plains of the American Midwest (including Canada), the cereal regions of the USSR from the Ukraine to
Kasachstan, the Sahel-Sudan belt of Africa, NE-Brazil, large portions of India/Pakistan, central China and Australia. In humid
western and central Europe the greater risk is due to cold springs
and persistent rainfall during summer and harvest time, especially
when related to weather-induced pests (Bourke 1981). In marginal
lands near the northern boundary of food production the risks are
due to late snowcover, killing frosts in late spring and to the
shortness of the growing season (see Fig. 3).

Regional studies of teleconnections of rainfall anomalies
have been carried out, for example the drought conditions in den-
sly populated NE-Brazil. Two long rainfall records are available:
Fortaleza (1849) and Quixeramobim (1896), with a remarkable rainy
season during autumn (peak in March/April) and high interannual
variability. Markham and McLain (1977) have indicated that a correlation exists between rainfall during the first part of the
rainy season and the sea surface temperature (SST) 1-2 months
prior to that in the equatorial upwelling regions. This confirms
earlier results obtained by Eickermann and Flohn (1962) who found

a highly significant correlation of +0.77 (35 pairs) between rainfall at Luanda/Angola during summer (Nov.-February) and that at
Fernando de Noronha during the following four months. Similar
correlations were found for other stations.

The mechanism of this relationship is complex and is not
discussed here. Other teleconnections have been investigated by
Namias (1972) with cyclonic activity in the Newfoundland region.
A see-saw (negative) correlation between the drought in NE-Brazil
and the El Nino phenomenon (ceased upwelling, warm water and rainfall) at the usually arid Peru-Ecuador coast was found by Doberitz
(1969) and Caviedes (1973). Global pressure anomalies described
by Berlage (1966) as Southern Oscillation - cf. also Bjerknes
(1969) and Flohn (1975) - show a similar reversal in the sign of
the correlation with Djakarta as reference station. Fleer (1981)
has analysed these world-wide correlations with the powerful technique of cross-spectrum analysis based on 290 station records in
the Tropics and Subtropics. He showed that in this data a quasi-
periodicity near 5 years is dominant.

Such teleconnections are particularly impressive in the Tropics.
Time-lags of several months are not infrequent, due to the seasonal
displacement of rainfall belts and to the rather high persistence
of sea-surface temperature anomalies of up to 6-8 months (Namias
1970, 1974, Flohn and Rodewald 1975). Of particular interest are
years with quite large, sometimes almost incredible rainfall anomalies in distant areas of the globe. Some of the excessive years
are: 1911-12, 1916-18. 1925, 1940-42, 1957/8, 1965. They merit
a series of global case studies to analyse both the geophysical
mechanism of climatic anomalies and their impact on food production in different areas. Even more excessive anomalies have been
observed during the 18th and 19th centuries, namely in 1877/8
(severe famines in NE-Brazil and India) and in 1899 (famine in
India, marked drop in rainfall over wide areas of Africa, see
Flohn 1974). Historical data are available from the Nile (Riehl
1979) and other African areas (Nicholson 1980a), and especially
from China some 500 years ago (Wang and Zhao 1981) and Japan
(Yamamoto 1971).

The role of severe climatic anomalies for famine, social un-
rest and revolutions of the past has frequently been underrated.
Lamb (1977, 1979) as well as Neumann (1977) have given impressive
examples from European history. Weather induced harvest failures
have not only contributed to the Ethiopian revolution (after 1972),
but also to the French revolution of 1789 (Neumann 1977), and the
European revolution of 1848.

In recent decades, since about 1920, such extreme anomalies
have been comparatively rare - the Little Ice Age with its several
peaks between 1560 and 1860 and its predecessors around 1315, 1430
and 1480 is indeed characterized by a high frequency of quite unusual events.

Teleconnection studies of my collaborators at Bonn have shown spatial interrelations of precipitation in Germany and Turkey, within the Indio-Pakistan subcontinent including Sri Lanka, within northern Africa (Nicholson 1979, 1980) and both flanks of the Sahara (Flohn and Nicholson 1980). In contrast to former papers e.g. by Winstanley (1973) who combined a few distant and incoherent rainfall series to inhomogenous area averages, the correlations between India and the Sahel are not very convincing. Also of interest are those between the south coast of Anatolia and central India (June-September -0.37, December-March +0.32, 45 pairs) and between the Maghreb coast and central Europe (November-April -0.42, 119 pairs). A negative correlation between the equatorial Pacific and India (Walker circulation) is confirmed for the monsoon months (June-October). Along the southern flank of the Sahara a significant positive correlation exists in zonal direction (Nicholson 1979) over more than 5000 km, in contrast to the Mediterranean coast, where longitudinal differences lead to negative space correlations. This zonal coherency (rainfall Sahel-Rep. Sudan, June-September +0.51, 69 pairs) is especially marked in the correlation of annual runoff: Nile (at Aswan) versus Chari-Logone (to Lake Chad) +0.74 (41 pairs), Nile-Senegal +0.44 (70 pairs), Senegal-Chari +0.57 (43 pairs).

Teleconnection studies of pressure and/or geopotential 700 mb or 500 mb are more rewarding (many examples for 700 mb have been published by Namias 1975), if quasi-homogenous grid-point series are available. They should include also investigations on the stability of such spatial coherency in time, since earlier investigations show many examples of instability. Correlations between different climatological parameters - temperature and precipitation, global radiation and sunshine as important parameters for impact studies, pressure and geopotential as indicators for large-scale atmospheric processes - indicate that in some seasons (especially during the growing season) these parameters are mutually coherent. Only the largest (and thus most effective) climatic anomalies show convincing spatial coherency. In many other years the great number of degrees of freedom within the climatic system seems to obscur such simple relations.

A comparative investigation of droughts in the 6 grain-producing regions of the USSR (Table 4a) indicate that in 8 years (out of 57) drought occurred simultaneously in the 4 regions of the European territory; in 5 years it expanded also to Western Siberia and Kazachstan. Only 7 years remained drought-free in all the 6 regions (Rauner 1979).

During a 107-year period only 7 effective drought periods (with slightly different threshold values) occurred simultaneously in the grain region of the USSR and the USA (Table 4b). The same was observed only in 4 years between Western Europe and the USA, as well as between Western Europe and the USSR. In contrast to this, the year 1976 was a good example of a coincidence between dry Western Europe and moist USSR.

Table 4   Drought years in the USSR and other grain regions
(Rauner 1979).

---

a) Simultaneous drought in 4 regions of the USSR, European
territory (1920-76)
1920 - 1924 - 1939 - 1946 - 1950 - 1963 - 1968 - 1975
do. in 6 regions, including Western Siberia and Kazachstan
1920 - 1924 - 1963 - 1968 - 1975

b) Simultaneous drought in USSR and USA (1870-1976)
1890 - 1901 - 1924 - 1931 - 1939 - 1955 - 1967
do. in Western Europe and USA
1874 - 1887 - 1904 - 1934
do. in Western Europe and USSR
1906 - 1911 - 1921 - 1959

In three typical regions of the USSR (Ukraine, Volga, Kazachstan, with records between 92 and 107 years) the relative frequency of drought was 42 percent, from which the majority (25 percent of all years) occurred in groups lasting 2-5 years. Also in other areas (perhaps with the exception of humid climates) clustering is a quite typical pattern of climatic anomalies. In the Great Plains of the USA the years 1894-95, 1917-19, 1933-36 and 1954-56 are good examples.

REFERENCES

Abel, W.: 1966, Agrarkrisen and Agrarkonjunktur. 2. Auflage. Parey, Hamburg, Berlin 290 pp.

Abel, W.: 1974, Massenarmut und Hungerkrisen im vorindustriellen Europa. Parey, Hamburg, Berlin, 427 pp.

Baur, F.: 1975, Beil. Berliner Wetterkarte 24.6.1975.

Berlage, H.P.: 1966, Med. Verh. Kon. Nederl. Meteor. Inst. 88.

Bjerknes, J.: 1969, Monthly Weather Review 97, pp. 163-172.

Bourke, A.: 1981, Chapter 6 in "Whither Climate Now?" (ed. H. Flohn, R. Fantechi, in press).

Caviedes, C.: 1973, Proc. Assoc. Amer. Geographers 5, pp.44-49.

Craddock, J.M., Wales-Smith, B.G.: 1977, Meteor. Mag. 106, pp.97-111.

Doberitz, R.: 1968, 1969, Bonner Meteor. Abhandl. 8 and 11.

Eickermann, W., Flohn, H.: 1962, Bonner Meteor. Abhandl. 1.

Elsas, M.: 1936, 1940, Umriß einer Geschichte der Preise and Löhne in Deutschland. 2 Bände, Leiden.

Fleer, H.: 1981, Bonner Meteor. Abhandl.26.

Flohn, H.: 1973, Naturwissenschaften 60, pp. 340-348.

Flohn, H.: 1974, Ann. Meteor. N.F. 9, pp. 25-31.

Flohn, H.: 1975, Bonner Meteor. Abhandl. 21.

Flohn, H.: 1981, in: "Climate and History", Symposium Norwich (in print).

Flohn, H., Nicholson, Sh.: 1980, Palaeoecology of Africa 12, pp. 3-21.

Flohn, H., Rodewald, M.: 1975, Beil. Berliner Wetterkarte 17. 17.7.1975.

Hastenrath, St., Heller, L.: 1977, Quart. Journ. Roy. Meteor. Soc. 103, pp. 77-92.

Hastenrath, St., Lamb, P.M.: 1978, Heat budget Atlas of the Tropical Atlantic and Eastern Pacific Oceans. Madison, Wisc.

Henning, D., Flohn, H.: 1980, Contrib. Atmos. Phys.53, pp. 430-441.

Hoskins, W.G.: 1964, 1968, Agricult. Hist. Review 12, pp. 28-46; 16, pp. 15-31.

Lamb, H.H.: 1977, Climate, Present, Past and Future. Methuen, London Vol. II.

Lamb, H.H.: 1979, Quatern. Res. 11, pp. 1-20.

Manley, G.: 1974, Quart. Journ. Roy. Meteor. Soc. 100, pp. 389-405.

Markham, C.G., McLain, D.R.: 1977, Nature 265, pp. 320-323.

Namias, J.: 1968, Bull. Amer. Meteor. Soc. 49, pp. 438-470.

Namias, J.: 1970, 1974, Journ. Geophys. Res. 75, pp. 565-582, 5952-5955, 79, pp. 797-798.

Namias, J.: 1972, Tellus 24, pp. 336-343.

Namias, J.: 1975, Short Period Climatic Variations. Collected Works 1934-1974, 2 Vol. San Diego.

Neumann, J.: 1977, Bull. Amer. Meteor. Soc. 58, pp.163-168.

Nicholson, Sh.: 1979, 1980, Monthly Weather Review 107, pp.620-623; 108, pp. 473-487.

Nicholson, Sh.: 1980a, in M.A.J. Williams, H. Faure: The Sahara and the Nile. Balkema, Rotterdam, pp. 173- 200.

Oeschger, H., Messerli, B., Svilar, M.: 1980, Das Klima. Analysen und Modelle, Geschichte und Zukunft. Springer Verlag Berlin, Heidelberg, 296 pp.

Pfister, Chr.: 1975, Beiheft 2, Jahrb. Geogr. Ges. Bern.

Post, J.D.: 1977, The Last Great Subsistence Crisis in the Western World. Baltimore.

Ramaswamy, C., Pareek, R.S.: 1978, Tellus 30, pp. 126-135.

Ratcliffe, R.A.S. et al.: 1978, Quart. Journ. Roy. Meteor. Soc. 104, pp. 243-255.

Rauner, Yu. L.: 1979, Paper USSR-USA Symposium on Climate Modelling, Climate Change and Statistics, Tbilisi (USSR).

Riehl, H. et al.: 1979, Monthly Weather Review 107, pp. 1546-1553.

Schuurmans, C., Flohn, H.: 1981, Chapter 3 in: "Whither Climate Now?" (eds. H. Flohn, R. Fantechi, in press.)

Takahashi, K., Yoshino, M.M. (Eds.): 1978, Climatic Change and Food Production. Univ. of Tokyo Press, XI + 433 pp.

Titow, J.: 1960, Econ. Hist. Reviews 12, pp. 360-407.

Titow, J.: 1970, Annales Econ. Soc. Civil. 25, pp. 312-350.

van Loon, H., Rogers, J.C.: 1978, Monthly Weather Review 106, pp. 296-310.

Verlinden, C.: 1965, Dokumenten voor de Geschiedenis von Prijzen en Lonen in Vlaanderen en Brabant, Deel II.

Wacker, U.: 1979, Dipl. Thesis. Bonn.

Wang Shao, Zhao Zengci: 1981, "Climate and History", Symposium Norwich (in print).

Wigley, T.M.L., Atkinson, T.C.: 1977, Nature 265, pp. 431-434.

Winstanley, D.: 1973, Nature 243, pp. 464-465; 245, pp. 110-114.

Yamamoto, T.: 1971, Geophys. Mag. 35, pp. 187-206.

DISCUSSION

Dr. Sakamoto was asked if his rather simple model could be used by a planning agency to prepare for the possibility that food aid would be needed. Responding he pointed out that communication is often poor in developing countries and that meteorological input is helpful. Even these simple methods have been useful in planning. Asked whether his model has been used to modify agricultural practices during a growing season, Dr. Sakamoto said that it has not been used operationally, but only in "what-if" exercises.

A participant said that the University of Nebraska has a computer terminal system that a county extension agent, an individual farmer or someone in agribusiness can use by putting his own farm data into a model and using it to follow the progress of the crop, recommend irrigation timing and make other decisions as the season progresses. The "what-if" game can be used to consider options by asking the computer what if the rest of the growing season is wet or dry. A number of these centralized models are being developed with access to groups and individuals. This kind of use of models is just beginning, but he predicts that it will become more and more common.

A participant played the devil's advocate by asking how the misuse of such models can be prevented. Another participant replied that there is no greater risk of misuse of these models than there is with any other information source. He believes that the only hope for a rational system of grain distribution is for everyone to have access to all the information that is available. He said that the information is not prohibitively costly or complicated and suggested that countries that have not learned to use it will have to learn. The great secret of success of American agriculture is the application of knowledge through the extension service, universities and corporate channels.

One participant suggested that because it is so important to have the right information at the right time in the commodities market, the free flow of information might be restricted. The Soviet grain deal in the early 1970s was mentioned as an example of suppression of information that gave some parties an advantage over others. It was stated emphatically that a situation like that is unlikely to happen again because so much information is widely available now. For example, information about the size of the Soviet crop has been released publicly ever since then, starting with estimates very early in the season. This information is now available without restriction.

W. Bach, J. Pankrath, and S. H. Schneider (eds.), Food-Climate Interactions, 443–445.

More data are needed to derive and verify these models. What kinds of international agreements are needed to make such data available? Dr. Sakamoto replied that there simply are no data for many countries that they work with. Another participant said that the quality of the historical meteorological data in most countries is at least an order of magnitude better than the quality and length of record of agronomic data. Most developing countries do not have much information about their own agriculture. It was suggested that the utilization of data by developing countries might be increased by bringing them into the process of gathering the data.

Responding to Dr. Namias' paper, a participant said that Australian weather forecasters will not go out on a limb. He said that even uncertain predictions of certain weather and climate events would be better than nothing, for example climatic events that have a probability of 75% or more. Dr. Namias replied that this kind of forecasting had been considered until recently, but that the subject of long-range prediction is becoming more dignified. He expects more support for this field to come from the oceanographic community, especially for the southern hemisphere.

Asked if there are biological precursors associated with any weather phenomena that are useful in predicting weather, Dr. Namias said that most of these ideas seem to be old wives' tales. They have tried to verify them without much success. But perhaps they have not been examined seriously enough.

Dr. Namias was asked if climate is becoming more variable and severe. He replied that he does not see any upward trend in variability. There were some extreme events in the 1970s, but there have been similar episodes in the past. He suggested that a climatic regime characterized by high variability may last several years, then shift to a low-variability regime. In response to the question of whether there is any evidence that human activities have any effect on climate Dr. Namias said that we cannot detect any CO_2 signal now as far as large-scale climatic effects are concerned. He feels, however, that human activities might soon be large enough to produce a signal.

In response to Dr. Flohn's paper, a participant said that the central problem in long-range forecasting is the question of what causes persistence and particularly persistence recurrence - what can provide a memory for the atmosphere? The reservoir of heat in the ocean is important. The top 200 meters of the ocean contains as much heat as the entire atmosphere and the time scale of thermal anomalies in the ocean is an order of magnitude slower than that of the atmosphere. Once a thermal regime is set up, it does not change from day to day or week to week, and it may not change from season to season. Snow and ice change albedo and air-mass refrigeration and influence the storm boundary, and the sun may play a role in climatic fluctuations. The character of the soil has a strong influence. The quest for long-range prediction research is for things with long memories that interact with the atmosphere.

ECONOMIC CONSEQUENCES OF FOOD/CLIMATE VARIABILITY

Pierre Spitz

Food Systems and Society Project
United Nations Research Institute for Social Development
Geneva, Switzerland

ABSTRACT. The analysis of economic consequences of food/climate variability is restricted in this paper to the year-to-year variability of cereal yields. This requires a definition of the level of analysis of both the temporal and spatial limits of the economic and social entity. It is argued that the tools and methods have not been adequately developed as social and natural sciences tend to respond to the effective social demand of the most powerful countries and social groups. The issue of food/climate variability was of prime concern to many French "economists" from the late 17th century to the French Revolution, when agriculture was the backbone of the economy. Periodic famines and food scarcities forced administrators, statesmen and scholars to make an attempt to understand these crises, which threatened to impoverish the nation and to lead to social disruption

With a few exceptions, there was no interest in these issues between the French Revolution and the 1940s crisis, not only because industrialization was becoming the major issue but also because advocates of "laissez-faire" policies thought their policies would take care "naturally" of the problems while their opponents oversimplified the relationship between nature and society.

The recent food crises in poor countries have produced empirical studies that still feed from the 19th century Nature/Society controversy. By treating climate as a resource in a food sytems approach it is hoped that new theoretical developments and alternative policies may result.

On the social sciences side, it is felt that self-provisio-

W. Bach, J. Pankrath, and S. H. Schneider (eds.), Food-Climate Interactions, 447–463.

ning in poor countries has a theoretical and practical importance
not yet fully comprehended. For this reason, the concept of forces of retention (food as a basic need) struggling against forces
of extraction (food as a commodity) is introduced. This dual
nature of food, with low average yields, accounts for the extreme
sensitivity to climate variability of the poor farmers in poor
countries.

On the natural sciences side, the most urgent task is to
develop a better scientific knowledge of inter cropping and relay cropping systems which make full use of climatic resources.
This requires a drastic change in agricultural research priorities
and funding towards a more democratic agricultural research geared to the needs of a majority of poor farmers in rainfed agriculture who are particularly vulnerable to climate variability
rather than to a minority of rich farmers in irrigated areas.

I.   INTRODUCTION

The United Nations Research Institute for Social Development
(UNRISD,Geneva), to which I belong, is conducting a research
programme on "Food Systems and Society" with special emphasis
on Eastern India and Mexico. Several scores of researchers drawn
from many disciplines are studying the functional relationships
between the components of these systems. This research is so
complex that it will take several years before conclusions can
be drawn. UNRISD has also conducted studies on the impact of
climatic variability on the food systems of Eastern India and
Mexico. This research is to be carried out within the framework
of the World Climatic Programme of the World Meteorological Organization in asscociation with the United Nations Development
Programme.

Much effort has been devoted to this research programme
because a piecemeal approach to the food problems of less industrialized countries seems incapable of leading to the formulation of development policies that would really ensure a guaranteed food supply to all of mankind from season to season and
from year to year. On the other hand, the reason why a specific
programme on the impact of climate has been undertaken is because
this field of research has been particularly neglected until
recently. As these projects are still in progress, I can do no
more than make a few comments by way of definition and discussion.

II.   CONSEQUENCES FOR WHOM?

We need a definition of the term "food/climate variablity".
I define this variability as a short-term variation from one
year to the next. The food production with which I shall deal
here is plant production, excluding animal production. Furthermore, I shall concern myself primarily with grain production,
and concentrate my analysis on yield variations because of its

importance. In such a brief analysis it is essential that no
more than one factor should be varied at any one time, while
holding the others constant (an increase or decrease in cultivated areas in response to established climatic variations, technical changes, etc.). This proviso "ceteris paribus" is clearly
one of the limitations which systems analysis attempts to overcome.

The systems analysis that we cannot attempt here does, however, remind us of a requirement which should apply to all analysis in economic and social sciences, namely the need to define
the level on analysis and the temporal and spatial limits of
the system under consideration. This is rarely done. We often
read of cost estimates on climatic disaster (e.g. drought in
the Sahel) which ignore the reference period. This is probably
because society tends to attribute agricultural success to man's
ingenuity and its failure to nature. Bad weather makes headlines.
Good weather is taken for granted. When lean years come after
a sequence of fat years, normality tends to be attributed to
the former, not to the latter. Climatic variability is perceived
more often than not by the negative consequences, the costs to
the economy rather than the benefits, the shortfalls in agricultural production, farmers incomes, exports, rises in consumer
prices and increased imports. How is this variation measured?
Against which norm?

The only way to study the economic consequences of climate/
food variablity is through the analysis of a specific situation
with time, space and society specified. The consequences will
clearly be very different according to the economic and social
entity selected and the period chosen. The consequences are also
different within the selected economic and social entity according to whether one tries to pinpoint the overall repercussion
on that entity or on one of its sub-groups, for example on consumers, agricultural producers or another social group. This
delimination in time, space and society is most often implicit.
The few nineteenth and twentieth century economists who concerned
themselves with the impact of climate on the economy usually
neglected to be explicit concerning their choice of "scenario"
in time and space. Thus, Keynes, having discussed the writings
of Jevons wrote: "The agricultural causes of fluctuation are
much less important in the modern world for two reasons. In the
first place agricultural output is a much smaller proportion
of total output. And in the second place the development of a
world market for most agricultural products, drawing upon both
hemispheres, leads to an averaging out of the effects of good
and bad seasons, the percentage fluctuations in the amount of
the world harvest being far less than the percentage fluctuations
in the harvests of individual countries"(1).

On the one hand, the choice of historical time scale, i.e.
in the "modern" and the "ancient" world, is fairly vague. On
the other hand, although Keynes' point is of general nature, he
is referring implicitly to the United Kingdom, a country possessing sufficient purchasing power to take advantage of world market development for obtaining the lacking agricultural resources
as a result of a "bad" year.

There also has to be the political will to prevent regions
or social groups within the entity under consideration from suffering famine. Such was not the case of the United Kingdom in
1846 when the Irish famine occurred although the world market
was already well developed. Neither the supplies nor the means
of transport were lacking. And yet a million Irish perished,
and hundreds of thousands had to got into exile, while ships
continued to export grain produced by tenant farmers from Ireland to England. The Commissariat Officer wrote from Waterford
on April 24, 1846: "The barges leave Clonmel once a week for
this place, with the export supplies under convoy which, last
Tuesday, consisted of 2 guns, 50 cavalry and 80 infantry escor-
ting them on the banks of the Suir as far as Carrick". Cecil
Woodham-Smith, who quotes this remark, comments: "It was a sight
which the Irish people could not understand and would not forget"(2).

Consequently, the scope afforded by the development of the
world market for making up bad harvests in some areas by good
harvests in others is essentially a theoretical one. In fact,
the extension of the balance of forces on which the operation
of the world market depends can produce opposite effects. That
was the case when, following the opening of the Suez Canal in
1870-71, British India began to export cereals. The Suez route
had reduced the cost of transporting Indian wheat so that it
became less expensive than American wheat. Cereals were the main
export crops of India between 1883 and 1914. Their destination
was the United Kingdom, which was India's principal supplier for
several years. This abrupt increase was not matched by an increase in Indian grain production, but was brought about by the
extension of the world market and the unequal relations between
a metropolitan country and its colony. From what source did these
cereals come? "From surplus stocks that formerly we stored
against times of scarcity and famines"(3).

During the three years from 1876-77 to 1879-79, which were
years of poor harvests, India exported 3.75 million tons of cereals. In these years some of the most severe famines in the history of India occurred resulting in six million deaths.

The few examples given above show that the economic consequences of food/climate variability are greatly dependant on institutional and political conditions, and that they will differ

depending on the entity or economic and social group in which
one is interested, and on the level of analysis.

In order to illustrate this latter point more precisely,
let us first consider tow groups: producers of foodstuffs and
consumers. The aim of consumers is to keep the price as low as
possible. A poor harvest tends to raise prices, and, consequently, adversely affects their purchasing power. The position in
which food producers find themselves is entirely different. What
they are most interested in is not the price, but the income
derived from the sale of the food that they produce, i.e. for
cereals, to which we are confining ourselves, the product of
price times yield. Should the price increase more than the yield
declines, income will rise when yield declines. In other words,
a poor harvest will in this case enable the producer to increase
his income. Under the opposite conditions, a good harvest will
result in a decrease in income.

Food self-provisioning still plays an important role in
less industrialized countries. Here we have to distinguish between:
(a) those producers of food, who, after having set aside
sufficient food to feed their families throughout the year, have
an excess, which they sell on the market, and who are interested
in the gross income (price times quantity), and
(b) those producers who do not have sufficient food. For
a part of the year their situation is the same as that of urban
consumers, since they have to buy foodstuffs to balance a bad
harvest. In a bad year for agriculture they are in a worse position than urban consumers because not only must they buy more
food than in a "normal" year, but additionally their income from
agriculture also declines. (NB: the income of urban consumers
is, in general, less sensitive to variations in agricultural
output). The vast majority of the poor peasants of the third
world are category(b) producers. A decline in ouput, following
a drought, for example, forces them into debt to category(a)
producers, to whom they sell their labour power at a lower wage
rate, and to whom they sell what they still possess, including
their means of production, at knock down prices; they are also
forced to mortgage their land and, eventually, to sell it. The
shocks repeatedly inflicted by climatic variability thus accentuate the long-term tendency to polarize these two groups of
producers and to eliminate the weakest, who are driven into the
cities(4). It should be noted that this continuous movement does
not necessarily imply that the pattern of land distribution al-
ters. The individuals are different within the established pattern of land distribution, and the changes that take place are
in accordance with the demographic evolution of each group, the
laws governing the inheritance of land, movements of the children of rich farmers towards employment outside agriculture,
and the children of poor farmers towards unemployment in the towns.

Contemporary analyses of the food problems of less industrialized countries tend to forget these special effects of
climatic variability on agricultural incomes. And yet these
effects were the subject of many analyses during the formative
period of political economy, namely in eighteenth-century Europe,
and more particularly in France. A few examples from that period may illustrate how diverse the possible consequences of
the variability of the harvest (essentially for producers in
category (a)) are, and they bear witness to an initial theoretical preoccupation that might be succeeded by profit in the
less industrialized countries today.

III. EFFECTS OF FOOD/CLIMATE VARIABILITY ON FARMERS' INCOMES

Setting a price for grains is very different from industrial price setting. The variable volume of grain output coupled
with relatively constant demand has the consequence that a
"good" year, with a higher than normal output tends to bring
down prices and may cause a negative effect on farmers income;
a "bad" year may have the opposite effect. When is the producer's income highest: in a "poor" year, in a "normal" year, in
a "good" year?

This question attracted the attention of early economists,
especially the eighteenth century French economists. According
to Quesnay, leader of the Physiocrat school, "when the wheat
harvest is such that it provides a surplus of three or four
months of consumption over the needs of a year, the wheat price
falls so low that this super-abundance ruins the farmer"(5).

Quesnay gives ghe following table of variations in wheat
yields, prices and income befcre tax:

Table I: Actual food price variations according to Quesnay(1757)

Crop year	Yield[1]	Price[2]	Gross income	Expenditures	Income before tax
very good	7	10	70	60	10
good	6	12	72	60	12
average	5	15	75	60	15
poor	4	20	80	60	30
very poor	3	30	90	60	30
total for 5 years	25	87	387	300	87
average	5	17.4	77.4	60	17.4

1) Net yields (after deducting the seeds). In "setiers" (measure
    of capacity) per "arpent" (measure of area).
2) In "livres" (pounds).

According to Quesnay, during famines large farmers become rich in an "ocean of misery" while, when the yields increase, the gross income and the net income decrease. For him, "the farmer makes a small profit only in bad years(...)", in "very good" years, the income before tax is one third of what it is in "very poor" years: he calls this the "misery of abundance".

It is striking that, according to Quesnay, an increase in yields brings about a decrease in gross income and net income. He used this table as an argument for the new agricultural policy he was promoting. I read Quesnay's writings as a plea to feudal absentee landlords to stop wasting their time and money in Versailles, return to their estates and strengthen their deteriorating economic base against the encroachment of a new and enterprising bourgeoisie whose increasing power was a growing threat to the foundations of the political regime. We would say today that for him only an agricultural revolution of a capitalist type could avoid a political revolution. He therefore needed to demonstrate that the prevailing prices were not stimulating grain production and that a policy of high agricultural prices (as a result of free trade), coupled with technical improvements and investments in agriculture would benefit the landlords. This is reminiscent of the arguments of "green revolutionists" in the late 1960s and early 1970s in India and Mexico.

To make his demonstration more convincing Quesnay proposed an "ideal" table of yields, prices and profits:

Table II: Ideal food price variation according to Quesnay(1757)

Crop year	Yield	Price	Gross income	Expenditures	Income before tax
very good	8	16	128	66	62
good	7	17	119	66	53
average	6	18	108	66	42
poor	5	19	95	66	29
very poor	4	20	80	66	14
total for 5 years	30	90	530	330	200
average	6	18	106	66	40

Quesnay's calculation that an increase of 10% in expenditures (from 60 to 66) results in a flat increase of one "setier" per "arpent" for any year is somewhat difficult to appreciate.

By shifting the spectrum of yield variations from 3-7 to
4-8, he avoids the highest price (30 pounds) corresponding to
the yield of 3. The price of 20 for the year yielding 4 is
kept while the prices for the years yielding 5, 6 and 7 are considerably raised (respectively by 27%, 50% and 70%). The variation of prices is reduced from 3 to 1 to 1.25 to 1. According
to Quesnay this was the effect of free trade, particularly between France and England. In this country, he writes, "agriculture has made very great progress" and price variations are
very small "from 18 to 22 pounds". Free trade should bring
French prices nearer this level.

It is through these manipulations of yields and prices that
Quesnay tries to demonstrate that good harvests, which benefit
consumers, would also benefit landlords who, under this new
agricultural policy, would obtain their highest profits with
an abundant harvest. This should not only encourage agricultural innovation but also result in the highest possible yields.

Turgot, who was Louis XVI's Minister of Finance in 1774,
defined a normal year as a year with a yield of 5 "setier" for
one "arpent". The price of the "setier" in a normal year is
25 pounds, the normal income therefore 25 x 5 = 125 pounds. For
Turgot, 125 pounds is also the maximum income, as it decreases
symmetrically in bad years and in good years as shown in
Table III:

Table III: The Turgot Table

Crop year	Yield	Price	Gross income	Expenditures	Income before tax
very good	7	15	105	96	9
good	6	20	120	96	24
average	5	25	125	96	29
poor	4	30	120	96	24
very poor	3	35	105	96	9
average	5	25	125	96	29

The "normal" year for Turgot is therefore the "average"
year, and at the same time the year of highest profit(6).

Quesnay and Turgot had a famous predecessor in Gregory King
(1648-1712) whose work was apparently unknown to them. King
states that when wheat production declines by 10%, 20%, 30%,
40% or 50%, wheat prices increase by 30%, 80%, 160%, 280%, or
450%. There is no mention in King's work(7) of the effect of a

production increase on prices. This reflects the preoccupation of his time with scarcity and famine situations. We cannot therefore compare directly Turgot's and King's observations. But if we assume that King takes as a starting point a normal year, a 20% decrease in yield produces an 80% increase in price, while the same decrease in yield for Turgot (from 5 to 4 "setiers") produces an increase from 25 to 30 pounds, which represents only 20%.

Similarly, for King a decrease in yield of 40% produces an increase in price of 280%, but only 40% for Turgot.

Why is there a much more dramatic increase in prices for King than for Turgot? If both observations are correct, many factors contribute to explain this difference in late 17th century England and in the 1760s in France. Quesnay points out that at his time of writing English price fluctuations were small. We are not, however, concerned here with economic history but with economic theory and the reasons which might explain King's law. I submit that this law refers to a situation in which the amount of marketed food is a very small percentage of total food production.

I have shown elsewhere[8] the importance of the ratio between self-provisioning and marketed surplus. Let us summarize the argument: Food has a dual nature, as a basic need and as a merchandise. In a society in which self-provisioning still plays an important role, we find both efforts to extract food from the producers and efforts by the producers to retain it. The forces of extraction are of a composite nature – from cash requirements to land tenure relations, of which sharecropping is the most extreme example. The mechanics of the extraction of surplus food cannot be understood without taking into account the countervailing resistance of the self-provisioning peasant. As long as peasants cannot exercise some control through political organizations over the prices of their produce they sell on the market and the prices of their necessary purchases (agricultural inputs and food), their attempt to retain some food for self-provisioning is an attempt to stay alive in an environment of fluctuating yields and prices.

To illustrate this conflict between the forces of retention and the forces of extraction, let us assume that one hundred units of food are produced in year 1, and a reduction to ninety units occurs the following year due to drought. If in year 2 the level of self-provisioning remains the same, the forces retaining food for self-provisioning are stronger than the market forces of extraction (Table IV). If in year 2 the level of "surplus"[9] remains the same (Table V), the forces of extraction are stronger. The balance between extraction and retention is

therefore shown by the way in which the decrease in production
is traded off between self-provisioning and marketing.

Table IV Case in which the forces of retention are stronger
          than the forces of extraction (units of food).

	Production	Self-provisioning	Surplus
Family X			
year 1	100	50	50 [1)
year 2	90	50	40
Family Y			
year 1	100	80	20 [2)
year 2	90	80	10

1) A reduction of 10% in production corresponds to a reduction
    of 20% in the marketed surplus.

2) In the case of Family Y the same reduction of 10% in pro-
    duction entails a reduction of 50% in the marketed surplus.

Table V: Case in which the forces of extraction are stronger
          than the forces of retention (units of food)

	Production	Self-provisioning	Surplus
Family X			
year 1	100	50 [1)	50
year 2	90	40	50
Family Y			
year 1	100	80 [2)	20
year 2	90	70	20

1) In this case the impact of drought, reducing production by
    10%, is to reduce the level of self-provisioning by 20%.

2) Under the same conditions, the level of self-provisioning
    is only reduced by 12.5%.

     The fact is well known, and has often been studied at the
level of international trade that a decline in production (with
the amount of retention remaining constant) has greater reper-

cussions on the "marketed surplus" and therefore on price when
the "marketed surplus" represents only a small proportion of
production. In contrast, this phenomenon seems to be less well
understood at the national and, above all, at the microeconomic
level in an agricultural situation where self-provisioning still
plays a major role.

Commentators of King's law, including Alfred Marshall(10),
Henri Guitton(11), or J.Milhau(12) failed to see that the strong
non-proportionality effect between output and prices implied
by King's observations relates to a situation of a high percentage of self-provisioning to total production (coupled with an
absence of government price controls).

Schumpeter, commenting on King, writes:
"More important for economics proper (as a) performance
that illustrates the curious obtuseness ... of economists:
Gregory King's law of demand for wheat ... The remarkable
thing is that King, though he did not attempt any further
refinements, evidently understood the problem perfectly;
that he worked with deviations from a normal is a particularly interesting touch. Still more remarkable is it
that, in spite of the general notoriety that 'King's law'
was to gain, it did not occur to economists either to improve upon it - though all that was required was to proceed further on a line unmistakably chalked out - or to
apply the same method to other commodities until the work
of H.L.Moore (1914)"(13).

What are the reasons for this "obtuseness" of modern economists? Economists of the seventeenth and eighteenth centuries
wrote about fluctuations in food production and prices because in
largely agricultural economies, where food scarcities and famines were frequent, the main problem was the food/climate variability. What should the government do in order to avoid famines and the resulting economic disruptions and social upheavals?
Given wide fluctuations in grain output, should they regulate
the international and internal trade of cereals or not? Each
writer had his own solution - from total free trade to rigid
controls(14). Economic thinking was thus responding to the major
problems of the times in Europe. But as industrialisation developed, economic thinking was less and less concerned with agricultural and food problems except during a food "surplus" crisis such as the one connected with the Great Depression. The
approaches followed today have been shaped by the history of
the now dominant countries. They do not correspond to the problems of developing countries in which serious food crises and
famines still exist and in which self-provisioning, insignificant in developed countries, still plays an important role.

This is why we are so ill-equipped at a theoretical level
to deal with the economic consequences of food/climate variability in developing countries and why, at the same time, we find
more relevance in eighteenth century writings than in many contemporary analyses. The lines "chalked out" by early economists
still have to be elaborated upon.

IV. POLICIES AIMED AT REDUCING THE ADVERSE EFFECTS OF YEAR-TO-YEAR CLIMATIC VARIABILITY

Very early in human history the variability of the harvest
led different social groups and societies to set aside reserves
in years of plenty in order to cope with difficult years. In
the essentially agricultural and cereal-growing ancient societies, the carrying over of grain stocks from one year to another
was closely connected to the creation of the State and the development of a taxation system. The subject is so vast and complex that it cannot be dealt with in this brief contribution.
The preoccupation with grain policy by which eighteenth century
Europe was characterized and which contributed so powerfully
to the development of political economy, ended in the 19th century, and neither the Irish nor the Indian famines were conducive
to a better analysis of the economic consequences of the food/
climate variability. Only Marx outlined new directions for such
an analysis in his writings on the Irish famine, and these were
rapidly forgotten, even by those who still use his thoughts today. We have noted that the crisis which struck the industrial
countries in the thirties spawned new analyses. However, the
Third World famines of the 1970s did neither lead to any profound theoretical studies nor to new policies. Here we are unable to examine price policies at the national and international
levels, foodaid, and policies to build foodstocks. These require detailed case studies with a precise specification of time,
society and place. On the other hand, two major policies will
be referred to here because they are of a general nature and
of fundamental importance although too often overlooked: the
reduction of inequality through agrarian reform and the development of democratic agronomic research.

Man does not live on averages. Necker, the Genovese banker
who succeeded Turgot as the Minister of Finance to Louis XVIth,
expressed this in a striking manner. For him, variations from
one year to another or within the same year
      "are a source of worry for those who live on their work.
      Landlords or their stewards, can reach a general balance
      in their accounts by off-setting the income of one year
      by that of another; but ordinary people cannot regulate
      their way of life the same way: a man who is haunted by
      the fear of losing his means of subsistence cannot be ex-
      pected to think of the present year in terms of the next
      and of today in terms of tomorrow"(15).

The reduction of inequality, in particular agrarian reform aimed at a more equitable distribution of land, is an important part of the redistribution of economic and political power. It is evident that in a more egalitarian society the variability of the harvest would diminish polarization between the category (a) and (b) producers referred to above and, more generally, reduce exploitation of those lacking reserves by those possessing these (large landowners, grain merchants, moneylenders). Agrarian reforms aimed at providing more security for farmers and share-croppers also eliminated a poor harvest as a cause for eviction. The redistribution of political power that is implied by agrarian reform in the interests of the people (as opposed to bureaucratic agrarian reform) makes it equally possible to obstruct the acquisition of land by the strongest farmers assisted by poor harvests, food crises and famines. To prevent this economic and social polarization, state aid must be applied at a very precise moment in the agricultural calendar. It is too late when land has been mortgaged and tools sold in the period preceding the next harvest: the situation cannot be rectified by soup kitchens, the distribution of food, relief work, or even by loans. The time for intervention by the State to protect the means of production of the weakest is immediately after harvest. More often than not, however, politicians act only when they feel themselves threatened, for example when a large-scale protest march by peasants on the way to town.

Agrarian reforms have given rise to many analyses and much research. On the other hand, as regards policies aimed at reducing the adverse effects of variability of the harvest, another sphere has received far less attention, namely, agronomic research and the systems of agriculture proposed in the name of "modernization". There are, admittedly, many and frequently contradictory anlyses of the "Green Revolution", but there are very few alternatives offered. What is essential is that agronomic research should be reorganized in a completely different way. Like the history of economic analysis, the history of agronomic research was shaped by the highly specific needs first of Europe, and then of North America. The key problem of European agriculture in the middle of the nineteenth century was compounded by the scarcity of land and the abundant supply of labour, the opposite being the case in North America. These specific conditions provided a different orientation to agronomic research: mechanization in North America, the use of fertilizers in Europe; increased productivity of labour, as formulated by McCormick, on the one hand, increased productivity from the soil, as formulated by Liebig, on the other. It was not until after the Second World War that these two approaches came together, with European agriculture adopting mechanization, and American agriculture making increasing use of fertilizers. Mechanization, especially mechanized harvesting, requires not only that a field should contain a single crop species, but also that there should be the greatest possible homogeneity, so that all plants should mature at the same time. Equally, therefore, one needs a single variety that should be as pure as possible. Mechanization has pushed towards monovarietal cultivation. The development has been the other way round in the Third World. The introduction of new and highly homogeneous varieties of cereals in the Third World is making it necessary to harvest all fields at the same time. Should there be, as is often the case, overriding climatic reasons for the harvest to be gathered as rapidly as possible, then the demand for labour' becomes very great during the harvesting period. The hired agricultural labourers are then in a position to exert pressure on the owners for increased wages. Faced with this social "threat", some landowners decide to mechanize their harvest operations, even if such an investment is not economically very attractive. This is what happened in India, in the Punjab, during the 1970s: the homogenization of high-yield varieties led to mechanization, increased use of tractors, combine harvesters and the simultaneous reduction in the employment of agricultural labour during the harvest period. When such a development took place in the third world, it skipped the historical phase of the reaper, passing directly to that of the combine harvester. This had two effects: a reduction in the labour requirement for threshing, and the immediate release of grain for the market. During the reaper phase which was still the rule in Europe in the years following the second world war, threshing, farm by farm, was a process that took several months. Consequently, grain came on the market gradually and the pressure on prices was very different.

The use of fertilizers is only one of several possible ways of increasing soil productivity and yield. Their use in Europe was developed in combination with more careful cultivation, more sensible crop rotation and the use of organic manure. And yet the use of fertilizers has angled agronomic research towards the selection of varieties giving maximum yields in monovarietal cultivation. Thus, although North American and European agriculture have developed in very different ways, they have both favoured cultivation that is mono-specific (one plant species per field) and monovarietal (a single variety).

Agriculture in the Third World is very different: like the former agriculture of Europe it is in general plurivarietal and, most often, except in rice farming, plurispecific.

Heterogeneity is a way of spreading the risks arising from climatic factors and the hazards of plant diseases and pests, which are themselves often affected by climatic factors. The Irish famine is the most dramatic example of the vulnerability of a mono-specific crop. The potato harvest was devastated by

Phytophthora infestans, the development of which was favoured
by the weather conditions of 1846.

The line of development pursued thus far has been inspired
by the Euro-American model of homogenization, in conjunction
with the pursuit of maximum yield, without having sufficient
regard to the variability of yield. When the harvest is very
poor in Europe or North America, the farmers are vocal in de-
manding monetary compensation, which these industrialized societies are sufficiently rich to give. Insurance against poor
harvests is a part of the economic and social fabric. In the
Third World, on the other hand, the outcome of a bad harvest
for a poor farmer may be the loss of his land, emigration or
death. Consequently, stability of yield, i.e. the reduction of
variation is an essential problem in the Third World. This stability should not be pursued for an isolated variety, but for
the whole system of cultivation. The traditional peasant system
of mixed cropping involving the use of many varieties guarantees
resistance to climatic variations, but at the cost of a yield
from unit area that is too low. Absolute priority must be given
to the operational study of these systems so as to raise the
over-all productivity level. Such research calls for a dialogue
between scientists and farmers, in which each group learns from
the other. Only endogenous research of this type, in which social and cultural values are respected, can give free rein to
the creativity of the farmers and of the scientists. Every agricultural research institute in the Third World now has a research
programme on "inferior" cereals, dry farming, and cultivation
systems, and is also attempting to carry out research jointly
with farmers. The amounts allocated for these activities are,
however, a small proportion of the total research budget, the
bulk of which goes to the "dominant" agriculture, that is practised by a minority of rich farmers. From the nineteenth century onwards demand from the industrialized countries directed
research towards cash crops. It is the effective social demand
of the rich farmers of the Third World which directs research
towards obtaining the maximum yield from the so-called "superior"
cereals (wheat,rice) in irrigated agriculture. The hundreds of
millions of poor peasants who grow "secondary" cereals on non-
irrigated land, i.e. under conditions in which climatic variability has the most dramatic consequences, have to be content with
a few high-sounding declarations and poor budgetary allocations
for the study of cultivation systems, mixed cropping, relay cropping, etc. The study of these systems calls for an enormous
scientific effort if consideration is to be given to all the
interactions between different plant varieties and species and
their environment at different times during the year. To carry
out such research the greater part of the budget rather than
the lesser part would have to be devoted to it. Were this done
the proportionate allocation would thus be a reflection of so-

ciety (the greater part for the majority, the lesser part for the minority) rather than, as at present, the reverse. Admittedly, a minority of farmers can produce sufficient food for a whole population, at a very high cost in terms of energy. However, agriculture remains the main source of employment in the countries of the Third World, and will do so for a long time to come. This majority, which depends on agriculture for its livelihood, needs a different pattern of development, one less costly in terms of energy, less at the mercy of climate variations, and one which will provide sufficient food and stable income from season to season and from year to year in conformity with the social and cultural standards of each group. As I have already stressed, the reasons why so little attention has been paid to problems of food/climate variability both in the economic and social sciences and in applied research is because these problems only dramatically affect the poor peasants of poor countries and not the rich farmers or the rich countries. If some attention has been paid lately to the economic consequences of food/climate variability, it is in connection with climate change which is of great interest to the industrialized countries.

The most extraordinary advances have been made in electronics and space research. However, while research stations are placed in orbit and even while factories are run in space, men, women and children will continue to die in some seasons and in some years if agricultural research priorities remain the same, if development patterns continue to increase inequality, and if the creativity that is latent in all social groups is not freed from patterns of thought and action that took shape under the highly specific conditions of a few countries during a very brief period in human history.

REFERENCES

1. Keynes, J.M., The General Theory of Employment Interest and Money, in: The Collected Writings of John Maynard Keynes, MacMillan, St. Martin's Press for the Royal Economic Society, Vol. VII, p. 329-331, London, 1973.

2. Woodham-Smith, C., The Great Hunger - Ireland 1845-1849, New England Library, Times-Mirror 1962, p. 72

3. Watt, G., The Commercial Products of India, p. 1088, London, 1908.

4. For further development see Spitz, P., Drought and Self-provisioning, UNRISD Working Paper, Geneva, 1980 and in: Ausubel, J., and Biswas, A. (ed.), Climatic Constraints and Human activities, IIASA Proceedings Series, Pergamon Press, Oxford, 1980.

5. Quesnay, Grains, in François Quesnay et la Physiocratie, INED, Paris, 1958, Tome II.

6. Turgot, Quatrième Lettre (1770), in: Oeuvres de Monsieur Turgot, Ministre d'Etat, Paris, 1808, Tome VI, p. 143.

7. King. G., Natural and Political Observations and Conclusions upon the State of England of 1696, in: Chalmers, An Estimate of the Comparative Strength of Great Britain, 1804.
On Quesnay, Turgot and King, see also Faure, E., Les bases économiques et doctrinales de la politique économique de Turgot, Revue Historique du Droit Français, no. 3, 1961.

8. Spitz, P., Drought and Self-provisioning, op. cit.

9. The term "surplus" is used for convenience. For a discussion of this concept see Spitz, P., Livelihood and the Food Squeeze", CERES, FAO Review on Agriculture and Development, May-June 1981, Rome.

10. Marshall, A., Principles of Economics, 8th edition, MacMillan and Co., London, 1927.

11. Guitton, Henri, Essai sur la Loi de King, Nancy, 1938.

12. Milhau, J., Etude économétrique du prix du vin en France, Montpellier, Causse, Graille et Castelnau, 1935.

13. Schumpeter, J., History of Economic Analysis, London, George Allen and Unwin, 1967.

14. See Spitz, P., The Public Granary, CERES, FAO Review on Agriculture and Development, November-December 1979, Rome.

15. Necker, J., Sur la législation et le commerce des grains, Paris 1775, in: Oeuvres complètes, (compiled in 1820-21, reprinted in 1971), New York International Publications Services.

STRATEGIES TO DEAL WITH CLIMATE/FOOD INTERACTIONS
IN DEVELOPED COUNTRIES

Wilfried P. Thalwitz

World Bank
1818 H. Street, N.W., Washington, D.C. 20433

"The problem is that while everyone talks about the weather no
one is prepared to do anything about it."
--Mark Twain

ABSTRACT. This paper explores the linkage between climate and
food with special emphasis on the problems of developing
countries. The following sections will:

...Review the global trade implications of weather induced
   production variability on food production.

...Discuss responses to different weather regimes in
   developing countries.

...Outline actions which developed countries could take
   to ameliorate prospective developing country food
   problems.

The paper does not address the potential impact of long term
changes in weather at either the worldwide or regional level
and assumes that weather over the next twenty years will be
roughly equivalent to that of the previous two decades. Such
an assumption is clearly problematic but recent projections
undertaken by OECD for its Interfutures Project lend support
to this hypothesis. In addition, the technical biological relationship between climate and plant physiology will not be
discussed. This is a complex issue in itself and for our purposes it is sufficient to suggest, following well known studies,
that yield variability is largely determined by weather conditions.

465

W. Bach, J. Pankrath, and S. H. Schneider (eds.), Food-Climate Interactions, 465–475.

Global Climate/Food Link

World food production has grown substantially over the past
quarter century. The most rapid increases have been in the
developed country exporters (including Australia, Canada and
the US) where grain production has increased by about 4% a year
over the last two decades compared to growth rates of 2.8% worldwide. Developing countries as a whole have increased food output
faster than their rate of population growth. On a per capita
basis, the developing world produces about 5% more food today
than it did 20 years ago. However, aggregate statistics on
food production trends can be misleading and mask two types of
problems.

> ...First, is the fact that while aggregate trends are
> encouraging, the food situation of many individual
> developing countries has deteriorated. In some areas
> political disturbances have prevented effective utiliza-
> tion of potentially productive natural resources. Else-
> where political instability, ineffective domestic
> policies and climatic conditions have contributed to
> serious food situations.

> ...Secondly, long term trends can hide the role of variations
> above or below trend increases with temporary shortages
> and price instability being the result.

To understand the importance of these variations one must
focus on the demand side of the world food economy. In the early
1960s many analysts suggested that population growth would be
the critical determinant of increased demand for food exports.
In fact, rising incomes in the more affluent countries of Europe,
East Asia and Latin America, not population growth, proved most
important. As incomes rose so did consumption of higher value
foods including tropical products, vegetables and, most important
of all, livestock products from grain fed animals. The resulting
increase in livestock production had a profound effect on international trade. In the early 1950s only about 5% of total grain
production entered world trade and very little of this was used
for feed. Since then international trade in grain has grown by
about 6% a year. The proportion of total grain consumption going
to livestock has grown from less than 20% in the 1950s to more
than 40%. Today more grain is fed to animals than is consumed by
the 1.4 billion people living in low income countries.

This shifting consumption pattern has been accompanied by
rises in aggregate production instability. For the world as a
whole average deviation from trend food output in the 1970s was
twice that experienced in the 1950s. Weather related yield
variations were roughly 70% higher in the 1970s than in the 1960s.
The reason for this rising instability is not climate changes but
the expansion of cropped areas into regions where rainfall and
temperature makes food production more uncertain.

A graphic illustration is presented in the USSR. The average
variation from trend wheat yields in traditional producer regions
of the Ukraine is a little over 10% versus average yield variations
of over 30%, in the "new lands" of Kazakhstan. In the US production/
climate linkages encourage stable output but farmers have been
bringing relatively more marginal lands into production thus
raising average yield variations. Production uncertainty in
exporters other than the US is smaller than that found in
Siberia but still relatively high. At a global level·poor weather
in either the US or USSR is the critical factor determinant of
worldwide production shortfalls. For the rest of the world aggregate variation is relatively small; poor weather in one area
being compensated for by good weather elsewhere. In particular,
developing countries are a stabilizing influence on global
production trends.

Concurrent production shortfalls as the result of weather in
the US and USSR first had significant trade repercussions in
the early 1970s. In the 1950-1970 period the international food
system was relatively stable. While production fluctuated
moderately, overall export levels remained steady and international
prices tended to decline in real terms. Exporting countries maintained farmer support programs partly by accumulating large
stocks of grain. In 1972 the USSR decided for the first
time to keep feedgrain consumption stable in the face of a large
net importer. At the same time, exporters reduced their level
of stockpiles while EEC importers kept internal prices artificially
low and feed consumption high. Abnormally poor weather in the US
corn belt, where yields fell by over a quarter, resulted in a
doubling of export prices and a widespread sense that the world
had experienced a "food crisis".

Developing country importers payed much of the cost for
the instabilities created by the policies of the USSR and EEC
combined with the vagaries of the weather in other countries.
Since the middle 1970s developing countries have had difficulty mounting effective programs to feed their people. The
most critical bottlenecks related to weather induced disruptions
of transport. Inadequate physical and administrative infrastructure
have often precluded movement of food supplies into rural areas.
Shipments which did arrive often disrupted local markets, reducing farmer incentives to the detriment of subsequent crops.
Imports for urban demand have been stepped up to roughly the
maximum physical and administrative carrying capacity of existing public distribution systems.

Weather and Developing Country Food Systems

Almost all developing countries have taken steps to provide
for food security that assures all individuals (including
children) access throughout the year to minimally required nutri-

tion at prices which are affordable. Food insecurity can result
from: (i) poor weather affecting either production or transport;
(ii) political disturbances; and (iii) reduced access to imports as
a result of exporter embargoes or prohibitive price levels.
Effective measures to reduce food insecurity require consideration
of:

Village Food Security. Many village farming systems based
on traditional food crops are essentially self-sufficient with
regard to food. In these circumstances, security against poor
harvests is largely a local matter. Crops like cassava, which
need not be harvested annually, provide a backstop when weather
is poor. Storage is an integral aspect of such agricultural
systems but varies widely in terms of type and efficiency
according to the particular ecological zone, and the level of
technology including the extent of irrigation. While the trend
in crop production is up, variability in output is frequently increased by the use of new technology. When weather turns poor,
as has recently happened in South Asia, the decline in output
can be at least as large as in traditional farming systems.
Where output is variable, more storage and better transport is
needed. This has been the case in areas like Northeastern Brazil
and the Sahel where food cannot be moved quickly and efficiently
into rural areas during times of shortage. The usual response
in these areas to poor harvests and inadequate stockpiles is
for hungry people to migrate into urban centers where food can
be distributed.

Micro famines and shortages. Understanding is limited of
price formation in small-scale, modernizing farming systems. Field
work in various countries supports the view that interseasonal
variations in the output of a particular production/marketing
unit (generally an isolated village) can cause serious hardship
to small producers and increase malnutrition. ICRISAT has documented the disincentive effect of variable weather conditions and
the inability of many local marketing systems to move surpluses
to deficit areas.

National Security Reserves. Emergency stockpiles have had
important benefits in times of tight supply resulting from poor
weather. India's recent experience confirms this. However,
the relatively high cost of maintaining such reserves ($45-80
per ton per year) has prompted a serious reexamination of their
economic efficiency. Even more important low income countries
have found that inadequate internal distribution systems frequently
prevented the timely use of existing stockpiles outside of
urban centers. Increasing imports has proven equally effective
and far less costly a mechanism for maintaining per capita food
consumption than using emergency stockpiles. A poor crop will
provide adequate food for rural population for some months
immediately after the harvest, and most importing countries

have adequate stocks to cover urban demand for the two months
it takes to arrange imports. With certain exceptions, present
buffer capacity in developing countries is adequate. Incremental managerial and financial resources could better be used
to improve the efficiency of the food distribution pipeline.

Global Food Security and Market Stabilization Reserves.
In assessing the level of total global interseasonal stocks
needed to guard against production shortfalls or price swings a
number of critical conditions must be considered: (i) overall
production remains relatively stable in the developing countries
as a whole; (ii) substantial global production variations and
associated export price fluctuations result primarily from low
yields in the USSR or North America; (iii) support policies in
the USSR, EEC and Japan which keep domestic feedgrain consumption
steady have the effect of transferring domestic production instabilities into the world market; and (iv) a cataclysmic fall
in world output that could only result from an unprecedented level
of uniformly poor weather could be compensated for by diverting
part of the 500 million tons of grain used to feed livestock. The
indications are that buffers for stabilizing commercial export
prices or mitigating the effects of world production shortfalls
will have only a marginal impact on food security in individual
developing countries, particularly where those most prejudiced
by shortage are found largely in rural areas.

Short term food security measures should not divert attention from the overwhelming importance of increasing agricultural
output in developing countries. Successful projects to this end
have reflected the agro-climatic situation of particular farming
areas as well as social, economic and political conditions.
Through the middle 1960s most growth in global food output was
associated with expanding the area being cultivated. The single
largest such expansion took place in Siberia where roughly 20
million ha of land subject to extreme weather variability were
brought into production. Although less concentrated the expansion of agricultural areas in developing countries was even
more spectacular. From 1955 to 1975 over 150 million ha of new
farm land was brought into production, an area larger than total
acreage devoted to cereals in the US, Canada, the EEC and Japan
combined. This expansion was related to demographic pressures
and the application of technologies which permitted settlement
in formerly inaccessible and unhealthy tropical areas. Yet,
the sustainability of agriculture in all of these new lands had
not been demonstrated.

The climatic variability inherent in any rainfed farming
system has made raising productivity of the rural poor difficult.
Agricultural research has previously generated technologies to
maximize potential production through the introduction of high

yielding varieties of seeds and the application of pesticides
and fertilizers. However, field surveys indicate that small
farmers in rainfed, largely subsistence, areas act to reduce
risks associated with adverse weather and are not concerned about
maximizing potential output. Periodic droughts effectively dis-
courage small farmer expenditures for production augmenting land
improvements or inputs.

In the middle 1960s the advent of the Green Revolution
technologies, and the synergism between water and fertilizer,
fundamentally altered the structure of agricultural production
in many developing countries where production became less vulnerable to weather. Area expansion became relatively less important as a source of growth. In South Asia about 75% of total
incremental output was the result of higher yields or double
cropping. In the high growth regions of Subsaharan Africa more
than half of incremental production was the result of higher
yields; elsewhere (including the Sahel) the figure was about a
third. The rate of area expansion continued to slow in the 1960s
and the 1970s. Most of the expansion of cultivated area occurred
in Subsaharan Africa.

FAO has estimated that almost 80% of total cereal yield
increase since the middle 1960s in developing areas is due to
incremental fertilizer use and better water management. Nutrient
consumption of chemical fertilizers has increased by about 15%
each year. High growth developing countries use twice as much
fertilizer per hectare as lower growth countries and use water
more effectively at the farm level. India today uses seven
times as much fertilizer per hectare of farmland as it did 15
years ago and the area irrigated has increased by over a third.
The low income developing countries as a group consume three
times as much fertilizer as they did in the mid-1960s and probably
twice as much water from irrigation systems. Despite these growth
rates, fertilizer application and water usage remains much below
optimum levels.

Improved water management and expansion of irrigation has
proven the most important element in successful strategies to
reduce food insecurity and raise agricultural productivity. India
is a classic illustration. Government decisions after disastrous
droughts in the middle 1960s led to changes in producer prices
which encouraged necessary on-farm investments as well as a
massive public program to expand irrigation. This has led to
a doubling in the size of the irrigated area, from 30 million ha
to about 60 million ha today. The effects of this immense investment in infrastructure which reduces the weather's influence have
been great. A country which many analysts once thought faced
hopeless food problems and inevitably increasing production
shortfalls is today self-sufficient in food production and less
vulnerable to droughts. Rainfall needed for cereal production
was 50% below average in 1979/80. Similarly poor monsoons in the

past have led to harvests that were 20% below trend but this
year production was only marginally below trend and stocks built
up in earlier years completely covered the shortfall.

It is difficult to generalize about the constraints to
increased and more stable production by large numbers of small
producers in ecologically difficult circumstances. However,
several general points have emerged from recent experience:

...There is no substitute for suitable price policies.
Farmers require a credible assurance of adequate
returns before undertaking the effort required to increase
productivity.

...Domestic resource mobilization is important. In most
countries the scale of public investment in agriculture
has not kept pace with requirements and in some areas
has not yet matched physical depreciation rates.
Typically, the investment rate in agriculture, in proportion to value added, is about half that for the economy
as a whole despite evidence that the returns to agricultural investments are no less, and frequently higher, than
those in other sectors.

...The weak administrative capacity of authorities in
implementing agriculture projects has proven to be a
critical bottleneck. Government priorities in the
allocation of scarce managerial resources are frequently
as important to project success as the availability of
financial resources.

...Low cost investments can have a large impact on agricultural productivity. The two most important examples
are extension and research. Well-designed, low-cost
extension programs can raise small farmer yields by a
third. Returns from adaptive agricultural research are
similarly large.

...Private sector investments in agriculture are important
and depend critically on a favorable economic environment in the sector. Experience with irrigation projects
has shown that on-farm private investments which account
for a small proportion of total expenditure, are crucial.
Private investments in marketing and distribution systems
for production inputs have proven equally important.

...All high growth regions within the low income countries
have had the advantage of better developed distribution
infrastructure and markets. Experience has shown that
these are prerequisites for subsistence farmers to
begin producing and selling surpluses.

Strategies for Developed Countries

The above analysis indicates various linkages between climate and food systems in developing countries. It provides a backdrop for exploring the specific actions which developed countries could take to alleviate the impact of climate variability on food supplies in developing countries. This section outlines actions which developed countries have been taking to this end and discusses future priorities. It includes brief discussions of:

The importance of resource transfers and technical assistance for projects to increase the output and improve the distribution of food in developing countries

The need to ensure that projects do not increase production instabilities or food insecurity

Developed country trade practices and farm support policies that adversely affect developing country food security

Resource Transfers

The need for increasing the flow of international and domestic resources into the food section is obvious. Judging from the incomplete and tentative data available, investments in this sector have increased in recent years and the developed world has substantially stepped up aid for agriculture. In constant prices, aid commitments more than doubled in the 1973-78 period. The lion's share of this increase was due to an acceleration in World Bank Group lending which, in terms of disbursements, accounts for well over half of the total aid for agriculture in developing countries (disbursements of $1.5 billion in FY1980, up from less than $300 million in FY 1973). In fact, it appears that aid to agriculture has risen more rapidly than has the public sector investment made by developing countries themselves.

Capital flows are not the only or most important form of development assistance in agriculture. Technical assistance is vital in two ways. First, it creates the indigenous capacity for preparing and implementing investment projects and builds absorptive capacity for incremental capital transfers. Secondly, technical assistance can help countries increase output by raising the efficiency with which existing productive assets are used. This is often more important than financing new investments. Despite this, bilateral donor technical assistance for agriculture has been declining since the early 1970s. While this trend has been mitigated by larger contributions to the international agricultural research system and increased activity by international development agencies such as the

World Bank, region banks and UNDP the numbers indicate that serious problems could emerge in the middle part of this decade.

Environmental Management and Protection

Agricultural production must be organized to ensure that natural resources are utilized on a sustainable basis. Where the ecological consequences of new technologies have no,t been adequately considered serious repercussions often result. In the Sahel, for example, the introduction of new more productive livestock technologies increased herd size with the result that water and grazing resources were overtaxed. The resulting desertification has had an adverse effect on long-term climatic conditions.

The lack of yield augmenting technologies appropriate to the specific agro-climatic situation of many low income areas indicates the need for special research and development efforts. The most pressing need is for work on tropical rainfed agriculture and on developing improved technology for traditional, non-cereal, food crops such as cassava. Priority areas where developed country assistance could prove most helpful include:

Integrated Water Management: Irrigation systems must be designed or rehabilitated to increase the efficiency with which water is used. The "mining" of ground water which involves pumping water resources more rapidly than the acquifer is naturally recharged, has already affected millions of hectares in developing (and developed) countries. Another problem is inadequate drainage in gravity irrigation systems which results in lower yields due to salinity and water logging in many areas of the Middle East, South Asia and the Far East.

Soil Erosion: The problem of land degradation and soil erosion is extremely serious and affects 45 million sq. km of drylands which support approximately 700 million people.

Preserving Forests: Deforestation exacerbates problems of soil erosion and flooding and demand for higher cost non-renewable energy. In order to balance new growth of wood with the annual cutting rate in developing countries, 10-15 million ha of forest should be planted each year whereas only 5 million ha is actually being planted.

Agriculture Protection

Developed countries have had a large influence on developing country food security through the indirect consequences of their farm support policies. The favored status of agriculture in the

EEC and Japan has distorted the patterns of global production
of food which would have otherwise applied. Developed countries
especially the EEC and Japan, spend large amounts on farm
support programs of their own which result in uneconomic
allocation of resources needed to produce food. This year
alone the cost of the Common Agriculture Policy is roughly
twice the total incremental public investment for developing
countries food production. These policies have had a depressing
impact on output in developing countries. Higher energy and
other factor costs may force many developed countries to make
long overdue adjustments in their farm sectors. By facilitating
a removal of the market distortions an international framework
to repeal such policies would benefit developing exporters with
a potential for efficient food production. For example, lower price
support prices in Japan would probably increase export markets
for Thailand and other efficient rice producers.

Conclusion

     In conclusion, the developing countries can increase the
productivity of their rural areas and meet the food needs of
their poor only through intensifying agricultural output. The
key will be improved cultivation practices that preserve the
resource base, and developing new technologies suited to changing
economic situations and specific agro-climatic situations. Success can only come at the level of individual farmers. Government policies which have discriminated against the rural sector
and small farmers will have to be abandoned. The institutional
capacity for implementing rural development and food production
projects must be built up. The commercialization of food production which this implies will strain existing food distribution capacity in the process of development. Dislocations of
substantial numbers of people are possible and the well being
of large numbers could be jeopardized. All of this could
create extreme political and social problems unless specific
actions are taken including improved food distribution infrastructure -- especially administrative capacity -- within
developing countries. Such programs would represent a major
step forward in providing food security.

     The founders of the post war economic order established
institutions to prevent the problems which spawned the inflation
of the 1920s and the depression of the 1930s. The International
Monetary Fund and GATT were founded to assure that no country
was forced to resort to beggar the neighbor policies in order
to achieve balance of payments equilibrium. The World Bank
was created to assure that long term capital flows were available
to assist countries to develop. In recent years, financing rural
development and food production has become the fulcrum of its
program. Judging from the Bank's experience, there is reason

to believe that developing countries can meet most of their
food production and consumption requirements. India's increase
in total production in the past 10 years offers a very heartening
example in this regard. However, improved food security in the
developing world cannot be achieved without a concerted effort
from the international community.

STRATEGIES TO INCREASE FOOD PRODUCTION IN DEVELOPING COUNTRIES

Werner Treitz

Agriculture, Forestry, Fishery and Rural Development,
Ministry of Economic Co-operation, Bonn, Germany

ABSTRACT. The present world population of approximately 4.4 billion will increase to more than 6 billion by the year 2000. The increase over the next 20 years will be approximately equivalent to the total world population of 1950. Population growth will occur mainly in developing countries where many people are already suffering from hunger and malnutrition.

Various estimates suggest that during the last decade the number of people not receiving sufficient food increased from 300 million to 1 billion. Most of the undernourished people live in rural areas of the developing countries. Approximately 250 million women in developing countries are suffering from anaemia, resulting in lower survival chances for newborn babies and other health hazards.

Food supplies have become especially precarious in Subsaharan countries of the African continent where 193 million people, or 60 % of the total population, were suffering from serious calorie deficits in 1975.

Until World War II the developing countries were net exporters of cereals. Due to rapid population growth, food imports have increased steadily over the last two decades. Projections indicate that the quantity of grain imported by the developing countries, excluding Argentina and Thailand, will increase from 42.6 million t (1973 - 75) to 79.8 million t in the year 2000.*

---

* The Global 2000 Report to the President, Vol. 1, p. 19.

W. Bach, J. Pankrath, and S. H. Schneider (eds.), Food-Climate Interactions, 477–480.

In order to avoid serious food shortages and disasters worldwide agricultural production must increase some 60% by the end of
the century.*

In spite of great difficulties during the last two decades,
food production has increased considerably in many developing countries, especially in Asia and South America.

This increase can be explained by the following factors:

. unused land or land used for other purposes was reclaimed
  for agricultural production;
. the increase in the use of fossil energy for fertilizers,
  pesticides and irrigation resulted in higher yields;
. the commercial initiatives of millions of small farmers compensated for poor agricultural policies in many countries;
. although agricultural research into food production has
  been limited, its results have become significant since 1970,
  especially as far as wheat and rice are concerned.

The required growth of food production will have to be brought
about mainly by more intensive use of existing farmlands in developing countries, since the area available for expanding agricultural production is very limited. Only 30% of the necessary production can be expected from an additional 200 million hectares which
is presently unused. Most of the available arable area is already
under cultivation and the cost of developing new farmland is extremely high. What is more, there is reason to believe that part
of the increase in food production during the last two decades
has been achieved at the expense of the ecological balance. Taking into consideration the estimated rate of desertification,
soil losses, deforestation, salination, water logging and deterioration of natural vegetation in many developing countries, it is
clearly important to base the intensification of agricultural
production on ecologically sound criteria, systems and methods.

The requirements for intensifying agricultural production
include:

. higher inputs of fertilizer, pesticides and other resources
  on the more productive land, in addition to the development
  of energy-saving methods;
. improved irrigation including rehabilitation of existing
  systems and drainage;
. agricultural research.

---

* Agricultural Production, Research and Development Strategies
  for the 1980s. Conclusions and Recommendations of the Bonn Conference, October 8 - 12, 1972, DSE - GTZ - BMZ - Rockefeller
  Foundation, 1980, p. 7.

On a global basis consumption of chemical nutrients increased from approximately 1 million t in 1906 to more than 80 million t by the end of the 1970s*

Most of the fertilizer is used in agricultural production by the industrial countries of Europe, North America and Japan where consumption per hectare is approximately five times higher than in the developing countries. In 1978 developing countries applied approximately 28 kg per hectare of crop area which corresponds to a fourfold increase over 1964.

Pinstrup-Andersen** estimates that 54% of the increased grain yields in developing countries during the period from 1948 to 1973 can be credited to rising fertilizer usage and may be even higher in the years thereafter.

Of the nitrogen used in agriculture only 50% is actually utilized by the plant. Consequently, technologies designed to improve the efficiency of nitrogen application will not only increase food production but will also protect natural resources.

Tropical and subtropical agriculture is more affected by insects, plant diseases and pests than agriculture in temperate zones. As a result, plant protection, including the utilization of chemical pesticides, is of even greater significance in developing countries than in Europe or North America. In order to increase food production in these countries on a sustained basis, research schemes should be established which will improve resistance of plants to pests and diseases leading to an integrated pest management.

In general irrigation systems are very attractive investment schemes in many developing countries where rainfall for crop production is insufficient and where water is a limiting factor. However, many irrigation systems are not effectively used. Taking into consideration the high capital costs of new investments and the low efficiency of existing projects, the rehabilitation of existing facilities, the construction of small irrigation and drainage schemes have first priority. These measures could accelerate increased food production in many deficit countries.

Better technologies to increase yields per unit of water, including drip irrigation, are also important in the intensification of food production in developing countries. Furthermore, irrigation development based on groundwater, possibly using so-

---

* Haldore Hanson, Biological Resources, Report prepared for the Bonn Conference, ibid., p. 15.

** Pinstrup-Andersen, The role of fertilizer in meeting developing countries food needs, Tulsa, Oklahoma, 1976.

lar energy, is important to a great number of countries facing
high population growth and low agricultural productivity.

The agricultural research required to ensure food supplies
as well as an appropriate standard of nutrition and the protection of natural resources must cover plant breeding, the development of improved farming systems and the identification of new
crops, which are presently underused, and which have a high potential for calorie and protein production. As far as plant breeding and the development of new crops are concerned, the following goals are important:

- higher yields per unit of land,
- reduction of the vegetation period,
- greater resistance to insects and diseases,
- better utilization of water, nutrients and other inputs, and
- higher tolerance towards salination, water shortage and
  other climatic stress.

Research on biological nitrogen fixation appears to be of
special significance. Nitrogen available in the atmosphere can
be converted into an organic form by certain micro-organisms.

Nitrogen fixing bacteria (rhizobia) colonize in the roots
of legumes and provide nutrients for many important food crops.
Important research is being conducted into this subject, including genetic engineering, with the intent of using biological fixation for non-leguminous foods (cereals) and for mixed farming
systems.

Plant breeding for improving food production depends on the
preservation of existing genetic materials of principal food crops.
This requires the establishment and operation of genetic resource
centres and seed collections, and the preservation of the natural
areas. In industrial countries a number of such gene banks have
been established. For tropical food crops some international research centres and other institutions maintain collections, but
more of these gene banks are required in order to ensure world
food security on a long-term basis.

DISCUSSION

Responding to Dr. Spitz's paper, a participant said that one definition of a drought, suggested by an economist working in India, is an unusual change in the price of agricultural products. That is a very functional definition, and it does not always work. If you examine the price variations of three grains- rice, wheat and sorghum - in Maharashta during the very bad drought period of the early 1970s, you get three totally different price curves. Rice was very important in both urban and rural areas, so the price really reflected the problems that were occurring. Sorghum was important only in the rural areas - people in urban India do not eat sorghum. The wheat price was entirely dependent on food aid, which supplied wheat from abroad, so the price did not reflect the drought. However, the economic consequences of serious droughts are disastrous. One of the most important con- sequences is that a drought causes farmers to underinvest in their own activity. If ways can be found to reduce the effects of climatic variability, the farmers will invest more labor or money, producing more food and helping themselves and the ur- ban people as well.

Dr. Spitz agreed that this variation in the price of diffe- rent crops is very important. The distinction between urban and rural consumption is useful, and another factor is the ratio of the portion of the crop that is sold to the amount kept for self-provisioning. When the market surplus is a small fraction of the total production, there is an amplification of the effects of drought. The question of underinvestment is also important, because it is necessary to invest in ways to reduce the impact of future climate fluctuations, such as irrigation systems and grain storage facilities. We are really talking about two kinds of investment, namely short-term by the farmer and long-term by the state.

Another participant said that, while it is a wise exercise to look at history, some things have changed radically in re- cent decades in ways that make the lessons of history inappli- cable. International grain trade used to be at a low level; needs were usually met internally or not at all. Now the eco- nomic structure has changed so that what happens in Brazil has a significant impact on the price of grain everywhere. We need information and policies that allow grain to be distributed in- ternationally in a way that was not possible in past years. It is a big mistake to think of the food and climate problem only in terms of famine years or great shortages. The impact of cli-

W. Bach, J. Pankrath, and S. H. Schneider (eds.), Food-Climate Interactions, 481–485.
Copyright © 1981 by D. Reidel Publishing Company.

mate on the grain situation is equally great in good crop years
when it appears to the policy makers that we have too much grain
and they have to worry about getting rid of it. There has been
a remarkable change in the perception of the grain situation
since the big crop of 1979. Everybody thought the grain problem
was solved, but now, with grain crop failures in several parts
of the world in 1980, we are suddenly talking about shortages
and price increases. He predicted that if the 1981 crops are
good, there will be a return to the other attitude. Our fundamental problem, he said, is an inability to settle down to a
rational policy that will last more than one year at a time.

Again Dr. Spitz agreed that it is important to note that
good years cause problems, too. Historians have done some intensive research on the problem of long-term grain reserves. Underground grain storage, with grain being kept for 20, 50 or 60
years, has been an important way of maintaining reserves. Another
important point is that we have not been studying history for
the sake of history. We are using economics as it has been shaped
by problems over the centuries. It is true that some problems
that were very important in eighteenth-century France are not
relevant now because an industrial society requires another kind
of analysis. Because the international market in grain is such
a small percent of the total world production, a drop of only
1 or 2 % in production can be amplified into a wild fluctuation
on the market. But some of the policy options that were there
historically are still there.

A participant pointed out that in 1971, just before the
drought, the Australian government imposed a reduction on wheat
acreage to one-third of its previous quota, because they were
overloaded with grain. By late 1972, Australia was buying grain
from Canada to fulfill its export contracts. The question of
how to handle a grain surplus was solved rather elegantly in
Britain in 1931. They fixed a figure at a certain level for an
average return on the English wheat crop, and a subsidy was provided to meet that level. If the crop went above that level,
the subsidy was reduced proportionately. That approach, based
on the need that Britain has always had to import wheat on the
world market, provided a stabilizing factor. The policy that
has been substituted since then of trying to set a price level
that will encourage constant production seems to reverse the
idea of getting some sort of balance into the production.

Responding to Dr. Thalwitz's paper, a participant said
that the decline in technical assistance is partly caused by the
refusal of developing countries to get technical assistance.
Another participant said that Dr. Thalwitz appears to think
that the developing countries should be brought into the western
market system as soon as possible by technology transfer, capi-

tal assistance and so on and that western way of thinking should
be imposed on them. He suggested to contemplate the possibility
of adapting our market system to third-world needs. Dr. Thalwitz
replied that when climatic variation causes a sudden food gap,
there are two ways of dealing with it. One is to forget about
the market completely and operate through charity. The other
possibility is using the market.

A participant said that he was surprised by comments on
the decline of technical assistance because he feels that it
is really increasing in money terms. The World Bank is planning
to spend $250 to $300 million a year for capital investment and
research. Agricultural development is an important part of the
German government's development policy. He suggested that technical assistance may be declining in terms of effective delivery
rather than effort. He also responded to Dr. Thalwitz's criti-
cism of irrigation projects. There are problems with large projects, and it is often easier to establish the structure than
to get the water to the farmers' fields. But small irrigation
projects can be valuable and should not be neglected. Another
participant cautioned that irrigation is not a panacea. In many
arid areas irrigation is done with fossil water that is not replenished as it is used. Even where there are large untapped
rivers, shifts in climate patterns could remove the precipitation
that feeds them. The climate can be unstable on a scale of 50
to 100 years, which is the scale of capital investment in large
irrigation projects. Dr. Thalwitz replied that it is important
not to have an up-front investment that is so big that it can
not be utilized. A good approach is to use existing irrigation
structures and work to increase their utilization through better
on-farm water management. This is often better than building
new structures.

A participant remarked that it is easy for discussions of
this sort to drift into technicalities of production, but that
it is important to concentrate on the economics of scarcity and
the maldistribution of scarce commodities between continents.

Another participant said that the European Community imports
a lot of cassava for animal feed from Thailand. There is pressure
to stop this because the cassava could feed so many Asian people.
But if the trade was stopped, the European Community would probably be blamed for not being open to world trade and for keeping
Thai farmers from taking advantage of European market opportunities and thereby raising incomes in a developing country. Cassava has been the only cheap fodder that was available for farmers
in Indonesia and the Philippines. Now that it is being exported,
the price has gone out of sight. This is a serious dilemma that
should be analyzed carefully. It was pointed out that food aid
is a disincentive to small farmers. He recommended that when

food aid is given, at least an equal amount should be provided in the form of fertilizer and other inputs that will promote local agriculture.

A participant said that the most important thing that must be done is to set priorities and that the highest priority should go to studies of mixed cropping.* Agricultural research in the industrialized countries has been geared to monoculture, but traditional mixed farming has a built-in resistance to climatic variability, pests and diseases. We should learn how mixed cropping really works by focusing scientific scrutiny on the body of knowledge that has been built by peasants over many centuries. Slash-and-burn agriculture is another system that is very complex and that has not been studied adequately. He recommended that a great amount of money and resources be devoted to an effort to understand these traditional farming systems and to assess their usefulness in responding to current problems.

Another participant said that he sees a dilemma in the use of biomass for producing liquid fuel from cassava and sugar cane grown on several hundred thousand acres of farmland because this use of the land may compete with food production. The system by which countries interact seems to put pressures on countries like the Philippines, to produce liquid fuel instead of mixed crops because of rising oil prices.

One participant from a developing country made the important point that there is a great need in developing countries to produce leaders who are honest, sincere and dedicated. Presently, still much of the food aid and monetary assistance that goes to developing countries is not used properly, but goes into the hands of only a few people.

Finally, the point was raised what question to ask the models. If we ask the model to give us a scenario of the future that everyone can believe, we are asking a very dangerous question. But if the question is where are the vulnerable sectors and where are the sectors that are not vulnerable, then the model may be able to give some useful answers. But the detailed evolution of the future will depend on so many random factors that it is impossible to have any confidence in long-term prediction. Even when you ask the right question of the model, we have to go on and ask how do we know that the sensitivity of the model is correct. This can be tested by using the model variations that have actually occurred in the past, such as in 1972. See what the model says would happen, then look at what really hap-

---

*Editor's note: The following is the summarization of some of the important points that were raised in general discussion toward the end of the Workshop.

pened. If they are the same, then we can have higher confidence in the model's ability to predict the future. That kind of verification has not been done sufficiently with these models, so even if they tell us something that seems very obvious, skeptics have good reason to remain skeptical.

One thing that links the modeling business with the real world is the sensitivity analysis. One of the purposes is to throw out variables that really do not make a difference.

The model is only as good as the equations, especially the empirical coefficients in those equations. Since the modelers have a fairly good idea how the equations behave, they can be accused of building ideological bias into the structure of the model. If we want to persuade people to listen to us, we must work very hard to douplicate real cases. We must try to bring expert judgment into the process.

INDEX